The History of Statistics

Stephen M. Stigler

THE HISTORY OF STATISTICS

The Measurement of Uncertainty before 1900

THE BELKNAP PRESS OF HARVARD UNIVERSITY PRESS

Cambridge, Massachusetts, and London, England

Printed in the United States of America
Seventh printing, 1998

Set in APS-5 Baskerville and designed by Marianne Perlak

Library of Congress Cataloging-in-Publication Data

Stigler, Stephen M.
The history of statistics.

Bibliography: p.
Includes index.
1. Statistics—History. I. Title.
QA276.15.S75 1986 519.5′09 85-30499
ISBN 0-674-40340-1 (cloth)
ISBN 0-674-40341-x (paper)

To Virginia

Acknowledgments

IN THE DECADE in which I have worked on this book I have encountered many remarkable people to whom I owe thanks for intellectual stimulation, encouragement, and moral and financial support. Some of the greatest pleasures associated with this project have come from encountering the likes of De Moivre, Laplace, Quetelet, Galton, Pearson, and Yule. They have sometimes taken me far afield, as to the Paris Observatory to analyze the original barometric readings from the 1820s for comparison with Laplace's figures, or to London to read Francis Galton's mail, trips made doubly enjoyable by the company of those first-class minds of the past. These same individuals have also been the sources of occasional frustration, as when Quetelet's evidently hasty and often erroneous calculations serve to mask his intentions. But even an initial frustration could have a happy result, as when a confusing derivation by Laplace could be seen to be a revealing error, or when an inconsistency between Galton's statements about his procedures and the results of his own calculations cast new light upon the way he understood a problem. I have attempted to acknowledge those intellectual debts, and those to more recent writers, in the text and bibliographical notes, and it remains to express my gratitude to a variety of unpublished but no less important sources of inspiration and encouragement.

I have benefited in ways that are difficult to measure from conversations and correspondence with George Barnard, Churchill Eisenhart, William Kruskal, and E. S. Pearson. Without the additional encouragement of Robert K. Merton, Fred Mosteller, George Stigler, and my colleagues at the universities of Wisconsin and Chicago, it is doubtful I could have completed the book in finite time. Lorraine Daston, Arthur Dempster, William Kruskal, Ted Porter, and George Stigler read portions of the manuscript, and I hope the final version reflects the high quality of their suggestions.

I am indebted to many libraries and librarians for assistance beyond the call of duty, not only to those currently on duty who helped locate periodicals and manuscripts from even the barest of descriptions, but also to the sometimes anonymous cataloguers and collectors of the past. I put particular demands upon the staffs of the University of Chicago libraries, the

University of Wisconsin libraries, and the Manuscript Library at University College London, and their response was more than equal to those demands, in terms both of locating hard-to-find items and of indulging my extravagant requests for copying. I am also grateful to a host of other hosts to visiting scholars, including the New York Public Library, the Archives of the Paris Academy of Sciences, the Manuscript Library of the University of Basel, the Library of the Royal Society of London, the John Crerar Library (in two locations), the Newberry Library, the Calderdale Central Library, and the libraries of Harvard University, Columbia University, Cornell University, Stanford University, the University of California at Berkeley, Iowa State University, the University of Illinois, Yale University, Rutgers University, Princeton University, the University of Minnesota, and the Paris Observatory. I am grateful to the Librarian of University College London for permission to reproduce selections from the correspondence of Francis Galton, Francis Edgeworth, Karl Pearson, and Udny Yule, to the Librarian of the Royal Society of London for permission to reprint parts of three letters from the Royal Society Archives, to Martin Wittek of the Bibliothèque royale Albert 1er for permission to reproduce their picture of Quetelet, to June Rathbone of the Galton Laboratories for showing me Galton's quincunx, to Elaine Harrington of the Cornell University Libraries for information from the *Cornell Register,* to Sarah Pearson for permission to reproduce the picture of her grandfather, and to the Warden and Fellows of All Souls College, Oxford, for permission to publish their photograph of Edgeworth. In several cases friends provided bibliothecarial assistance by helping to retrieve items I could not otherwise have seen; I thank David Butler, Claire Friedland, Charles C. Gillispie, Ivor Grattan-Guinness, David Heilbron, Agnes Herzberg, Bernard Norton, Earl Rolph, Glenn Shafer, Oscar B. Sheynin, David Stigler, and Harriet Zuckerman. My gratitude to numerous antiquarian booksellers has already been given written expression in the form of bank drafts.

During the course of this project I have received support from many sources. The research was begun with the help of a grant from the National Science Foundation, and this support continued in various forms throughout the decade. During 1976–77, I was able to make major headway by virtue of a fellowship from the John Simon Guggenheim Foundation, supplemented by funds from the University of Wisconsin Research Committee. The actual writing was begun during my 1978–79 fellowship year at the Center for Advanced Study in the Behavioral Sciences, where both the setting and the administrative support are ideal for undertaking such projects. The support and encouragement of my family have been no less important; the carefully rationed but apparently inexhaustible patience of my wife, Virginia, has made possible the completion of a project that might otherwise have been overwhelmed by the exuberant companionship of Andrew, Geoffrey, Margaret, and Elizabeth.

I am indebted to many for their skillful and patient secretarial assistance, including in particular Anne Sutton, Virginia Stigler, Michelle Schaaf, and Barbara Boyer, although to none more than to the incomparable Mitzi Nakatsuka. I thank Richard Askey, R. R. Bahadur, N. J. Cox, R. W. Farebrother, Oscar B. Sheynin, and David Wallace for calling to my attention a few errors that crept into the first printing. Finally, I am grateful to Harvard University Press; Michael Aronson's interest in this project has helped shape it from the beginning, Jodi Simpson's editing helped smooth the edges, and the final form exhibits the craft of many hands.

Contents

Introduction 1

PART ONE
The Development of Mathematical Statistics in Astronomy and Geodesy before 1827 9

1. Least Squares and the Combination of Observations 11
   - Legendre in 1805 *12*
   - Cotes's Rule *16*
   - Tobias Mayer and the Libration of the Moon *16*
   - Saturn, Jupiter, and Euler *25*
   - Laplace's Rescue of the Solar System *31*
   - Roger Boscovich and the Figure of the Earth *39*
   - Laplace and the Method of Situation *50*
   - Legendre and the Invention of Least Squares *55*

2. Probabilists and the Measurement of Uncertainty 62
   - Jacob Bernoulli *63*
   - De Moivre and the Expanded Binomial *70*
   - Bernoulli's Failure *77*
   - De Moivre's Approximation *78*
   - De Moivre's Deficiency *85*
   - Simpson and Bayes *88*
   - Simpson's Crucial Step toward Error *88*
   - A Bayesian Critique *94*

3. Inverse Probability 99
   - Laplace and Inverse Probability *100*
   - The Choice of Means *105*
   - The Deduction of a Curve of Errors in 1772–1774 *109*
   - The Genesis of Inverse Probability *113*
   - Laplace's Memoirs of 1777–1781 *117*
   - The Error Curve of 1777 *120*
   - Bayes and the Binomial *122*
   - Laplace the Analyst *131*
   - Nonuniform Prior Distributions *135*
   - The Central Limit Theorem *136*

4. The Gauss–Laplace Synthesis *139*
Gauss in 1809 *140*
Reenter Laplace *143*
A Relative Maturity: Laplace and the Tides of the Atmosphere *148*
The Situation in 1827 *157*

PART TWO
The Struggle to Extend a Calculus of Probabilities to the Social Sciences *159*

5. Quetelet's Two Attempts *161*
The de Keverberg Dilemma *163*
The Average Man *169*
The Analysis of Conviction Rates *174*
Poisson and the Law of Large Numbers *182*
Poisson and Juries *186*
Comte and Poinsot *194*
Cournot's Critique *195*
The Hypothesis of Elementary Errors *201*
The Fitting of Distributions: Quetelismus *203*

6. Attempts to Revive the Binomial *221*
Lexis and Binomial Dispersion *222*
Arbuthnot and the Sex Ratio at Birth *225*
Buckle and Campbell *226*
The Dispersion of Series *229*
Lexis's Analysis and Interpretation *233*
Why Lexis Failed *234*
Lexian Dispersion after Lexis *237*

7. Psychophysics as a Counterpoint *239*
The Personal Equation *240*
Fechner and the Method of Right and Wrong Cases *242*
Ebbinghaus and Memory *254*

PART THREE
A Breakthrough in Studies of Heredity *263*

8. The English Breakthrough: Galton *265*
Galton, Edgeworth, Pearson *266*
Galton's Hereditary Genius and the Statistical Scale *267*
Conditions for Normality *272*
The Quincunx and a Breakthrough *275*
Reversion *281*
Symmetric Studies of Stature *283*

Data on Brothers *290*
Estimating Variance Components *293*
Galton's Use of Regression *294*
Correlation *297*

9. The Next Generation: Edgeworth *300*
The Critics' Reactions to Galton's Work *301*
Pearson's Initial Response *302*
Francis Ysidro Edgeworth *305*
Edgeworth's Early Work in Statistics *307*
The Link with Galton *311*
Edgeworth, Regression, and Correlation *315*
Estimating Correlation Coefficients *319*
Edgeworth's Theorem *322*

10. Pearson and Yule *326*
Pearson the Statistician *327*
Skew Curves *329*
The Pearson Family of Curves *333*
Pearson versus Edgeworth *338*
Pearson and Correlation *342*
Yule, the Poor Law, and Least Squares: The Second Synthesis *345*
The Situation in 1900 *358*

Appendix A. Syllabus for Edgeworth's 1885 Lectures *363*

Appendix B. Syllabus for Edgeworth's 1892 Newmarch Lectures *367*

Suggested Readings *370*

Bibliography *374*

Index *399*

Illustrations

Adrien Marie Legendre; from an old lithograph, reproduced in The Biometric Laboratory's *Tracts for Computers,* no. 4, 1921 *11*
Abraham De Moivre; from *Biometrika,* vol. 27, 1925 (frontispiece) *62*
Jacob Bernoulli; reproduced with the permission of the Museum für Völkerkunde und Schweizerisches Museum für Volkskunde, Basel *62*
Pierre Simon Laplace; from an old lithograph in the author's possession *99*
Carl Friedrich Gauss; from Werckmeister, 1898–1899, vol. 1, plate 116 *139*
Adolphe Quetelet; from an old lithograph, in the Bibliothèque royale Albert 1er, Brussels, Cabinet des Estampes, Odevaere, 4°, SII 53642; copyright Bibliothèque royale Albert 1er, Brussels *161*
Wilhelm Lexis; from *Illustrirte Zeitung* (Leipzig), vol. 143, 1914, p. 394 *221*
John Arbuthnot; from Aitken, 1892 *221*
Hermann Ebbinghaus; from *Zeitschrift für Psychologie,* vol. 51, 1909 *239*
Gustav Theodor Fechner; from Werckmeister, 1898–1899, vol. 3, plate 250 *239*
Francis Galton; from Pearson, 1914–1930, vol. 1, p. 242, plate LXI *265*
Francis Ysidro Edgeworth; reproduced with the permission of the Warden and Fellows of All Souls College, Oxford *300*
Karl Pearson; from Pearson, 1938, plate III *326*
George Udny Yule; from Stuart and Kendall, 1971, reproduced with the permission of the Royal Statistical Society *326*

Figures

1.1 Tobias Mayer's original drawing of the moon *19*
1.2 The moon *20*
1.3 An oblate earth *41*
1.4 Boscovich's algorithm *48*
1.5 Legendre's 1805 appendix *58*
2.1 Two pages from J. Bernoulli's *Ars Conjectandi* *70*
2.2 De Moivre's 1738 translation of his 1733 paper *74–75*
2.3 Thomas Simpson's 1757 graph of a triangular error density and the density of the mean error *96*
3.1 Diagrams accompanying Laplace's 1774 memoir on inverse probability *107*
3.2 Laplace's 1777 error curve *121*
3.3 Portion of a posterior density based on Laplace's 1777 error curve *122*
3.4 Bayes's "Billiard Table" *125*
4.1 Gauss's 1809 derivation of the normal density *142*

4.2 Laplace's 1810 statement of the central limit theorem *144*
4.3 Laplace's 1812 derivation of the bivariate normal limiting distribution of two linear functions of observational errors *149*
5.1 Quetelet's 1827 diagram showing the variations over different regions in Belgium in rates of birth, death, and marriage *167*
5.2 Quetelet's 1827 diagram showing the variations over different months in temperature and the rates of birth and death *168*
5.3 Quetelet's analysis fitting a normal distribution to data on the chest circumferences of Scottish soldiers *207*
5.4 Quetelet's 1846 rendition of a symmetric binomial distribution *209*
5.5 Portion of Quetelet's table of a symmetric binomial distribution *210*
5.6 Bertillon's depiction of a bimodal distribution of heights *217*
6.1 Lexis's drawing of a mortality curve *224*
7.1 An illustration of Fechner's 1860 argument for a normal psychometric function *247*
8.1 Galton's illustration of the "law of deviation from an average" *269*
8.2 Two of Galton's renditions of the ogive *270*
8.3 Galton's illustration of the relationship of eminent men within families *273*
8.4 Galton's original quincunx (1873) *277*
8.5a Galton's letter to George Darwin (12 January 1877), explaining the two-stage quincunx *278*
8.5b Pearson's drawings of the two-stage quincunx *279*
8.6 Galton's 1877 published drawing, showing both the two-stage quincunx and the law of reversion *280*
8.7 Galton's smoothed rendition of his height data *287*
8.8 Galton's regression line *295*
9.1 Edgeworth's 1885 illustration of Galton's 1875 insight *312*
9.2 Galton's graphical determination of a correlation coefficient *320*
10.1 Yule's schematic drawing based on a correlation table *350*

Tables

I.1 Cassini's 1740 comparison of data on the obliquity of the ecliptic with values found by linear interpolation *6*
1.1 Mayer's twenty-seven equations of condition, derived from observations of the moon's crater Manilius *22*
1.2 Mayer's three equations, derived from his twenty-seven equations by summation in groups *23*
1.3 Laplace's Saturn data *34*
1.4 Boscovich's data on meridian arcs *43*
1.5 Boscovich's pairwise solutions for polar excess and ellipticity *45*
1.6 Intermediate calculations for Boscovich's 1760 solution *49*
1.7 Measurements of the French meridian arc, made in 1795 between Montjouy and Dunkirk *59*

2.1 Values of the normal integral, as computed by De Moivre in 1733 and as computed exactly by modern methods *82*
2.2 Simpson's calculated values of the probability distribution of the mean of six errors, compared with the probability distribution of a single error *93*
3.1 Roots of Laplace's equation *117*
4.1 Mean diurnal variation in barometric pressure at Paris, 1816–1826 *155*
4.2 Mean change in barometric pressure, 9:00 A.M. to 3:00 P.M., 1816–1826 *156*
5.1 Quetelet's data on the conviction rate in the French courts of assize *175*
5.2 Quetelet's conviction rates, broken down by year and state of accused *176*
5.3 Quetelet's analysis of the relative degree of influence of the state of the accused upon the conviction rate *177*
5.4 Poisson's data on French conviction rates for all of France, for the department of the Seine, for crimes against persons, and for crimes against property *189*
5.5 Bessel's comparisons of the distributions of the absolute values of three groups of residual errors *204*
5.6 Distribution of heights and chest circumferences of 5,732 Scottish militia men *208*
5.7 Quetelet's analysis of 1817 data on the heights of French conscripts *216*
5.8 Relative frequency distribution of the heights of 9,002 French conscripts from the department of Doubs, 1851–1860 *218*
7.1 Fechner's notation for the numbers of right cases for his two-hand lifted weight experiment *249*
7.2 Fechner's data, giving the values of r for the four main conditions of the two-hand lifted weight experiment *250*
7.3 Values of t derived from Table 7.2 for the two-handed series *251*
7.4 Fechner's estimated effects of lifting time, according to order lifted and location of heavier weight *252*
7.5 Ebbinghaus's table of the theoretical law of error *258*
7.6 An example of Ebbinghaus's test of fit to a normal distribution *259*
8.1 Galton's 1885 cross-tabulation of adult children, by their height and their midparent's height *286*
8.2 Galton's "Special Data" on the heights of brothers *292*
9.1 The data from which Galton in 1888 found the first published correlation coefficient *319*
10.1 Yule's "correlation table" showing the relationship between pauperism and prevalence of out-relief *347*

The History of Statistics

Introduction

STATISTICIANS have an understandable penchant for viewing the whole of the history of science as revolving around measurement and statistical reasoning. This view, which stops far short of insisting that science is *only* measurement, is not entirely parochial; many distinguished scientists and philosophers have shared it. The Victorian physicist Lord Kelvin went so far as to characterize nonnumerical knowledge as "meagre and unsatisfactory," echoing the phrase "small and confused" that the English scholar and wit John Arbuthnot had used for the same purpose nearly two centuries earlier.[1] Similar sentiments can be found in the works of writers as diverse as Immanuel Kant, Leonardo da Vinci, and Roger and Francis Bacon. Yet in the twentieth century we have become increasingly sensitive to the fact that even in the sciences that depend crucially upon measurement (whether these include all sciences or only a majority of them), measurement alone is not enough. To serve the purposes of science the measurements must be susceptible to comparison. And comparability of measurements requires some common understanding of their accuracy, some way of measuring and expressing the uncertainty in their values and the inferential statements derived from them.

Modern statistics provides a quantitative technology for empirical science; it is a logic and methodology for the measurement of uncertainty and for an examination of the consequences of that uncertainty in the planning and interpretation of experimentation and observation. Statistics, as we now understand the term, has come to be recognized as a separate field only in the twentieth century. But this book is a history of statistics before 1900. Thus my subject is not the entire development of a single discipline but rather the story of how that discipline was formed, of how a logic common to all empirical science emerged from the interplay of mathematical concepts and the needs of several applied sciences. In ex-

1. For an excursive history of the Kelvin Dictum ("when you can measure what you are speaking about, and express it in numbers, you know something about it; but when you cannot measure it, when you cannot express it in numbers, your knowledge is of a meagre and unsatisfactory kind"), its antecedents (including John Arbuthnot's 1692 statement, "There are very few things which we know; which are not capable of being reduc'd to a Mathematical Reasoning; and when they cannot, it's a sign our Knowledge of them is very small and confus'd"), its uses, and rejoinders to it, see Merton, Sills, and Stigler (1984).

ploring the details of that emergence we come to a realization of what constitutes the essence of this twentieth-century discipline and of how the empirical methods of all the sciences have achieved an intellectual identity that goes far beyond the sum of the constituent parts, each science borrowing strength from another in a self-exemplification of what better statistical procedures do to measurements.

Modern statistics is much more than a toolbox, a bag of tricks, or a miscellany of isolated techniques useful in individual sciences. There is a unity to the methods of statistics. The same computer program that analyzes the data of a geophysical scientist today might be used by an economist, a chemist, a sociologist, a psychologist, or a political scientist tomorrow. And this unity is not superficial: Even if the interpretations given these analyses differ subtly with the field, the concepts employed in those interpretations and their logical consequences and limitations are much the same. Yet the beautiful unity we see in modern statistics can pose vexing problems and seeming paradoxes for a historian of statistics. This unified logic has not arisen from any single source. Rather it is a unity bred from diversity, and it is this diversity that first strikes and can confuse the eye.

For example, elementary statistics texts tell us that the method of least squares was first discovered about 1805. Whether it had one or two or more discoverers can be argued; still the method dates from no later than 1805. We also read that Sir Francis Galton discovered regression about 1885, in studies of heredity. Already we have a puzzle—a modern course in regression analysis is concerned almost entirely with the method of least squares and its variations. How could the core of such a course date from both 1805 and 1885? Is there more than one way a sum of squared deviations can be made small?

Consider a second example. By the 1830s statistical methods were widely used in astronomy, and we can find reasonably accessible texts from that period that bear at least a cousinly resemblance to modern elementary texts. Yet it is only in the twentieth century that we find these same methods making substantial inroads into the social sciences. Were nineteenth-century social scientists unable to read? And if they were neither illiterate nor too dense to see the need for the quantification of uncertainty in their data, why did they ignore what is so obvious a century later? The resolutions of these apparent paradoxes (why regression followed least squares by about eighty years, why social science was so slow to come to a statistical methodology) are subtle—and surprisingly closely linked. Indeed, one principal goal of this narrative is to explore and develop that link.

Any history must restrict its scope in some way. If all sciences require measurement—and statistics is the logic of measurement—it follows that the history of statistics can encompass the history of all of science. As attractive as such an imperialistic proposition is to a statistician, some limits must be imposed. I have chosen to limit my inquiry by emphasizing the role of statistics in the assessment and description of uncertainty, accuracy, and variability, by focusing on the introduction and development of explicitly probability-based statistical methods in the predisciplinary period before 1900. In some respects the emphasis upon explicit probability narrows the focus unduly. An explicit use of probability is not a necessary condition to some understanding of accuracy, either historically or in the present day. Many measurements carry with them an implicitly understood assessment of their own accuracy. Indeed, it could be argued that any measurement that is accepted as worth quoting by more than one person may have some commonly understood accuracy, even if that accuracy is not expressed in numerical terms. But this argument can be an extremely difficult one to apply in historical terms: It is often simpler and arguably more objective to attempt to recognize an explicit probability statement than to determine when a measurement is widely accepted and a common worth attached to it.

One dramatic early instance of a numerical assessment of accuracy that was not given in terms of explicit probabilities was the Trial of the Pyx. From shortly after the Norman Conquest up to the present, the London (later Royal) Mint maintained the integrity of its coinage through a routinized inspection scheme in which a selection of each day's coins was reserved in a box ("the Pyx") for a later trial. Even in the earliest indentures between the mint and the king the contract stated that the trial would allow a tolerance in the weight of a single coin and, by linear extrapolation, in the aggregate weight of the entire contents of the Pyx. Thus as early as 1100 an economic necessity had led to an institutionalized numerical allowance for uncertainty, uncertainty in how the value of an entire coinage could be judged by that of a sample, in the presence of unavoidable variability in the production process.[2]

Although such early examples are fascinating, they are isolated instances of human ingenuity and contribute little to our understanding of the development of the field of statistics. The procedures of the Pyx were not abstracted from their single application; they were not developed and integrated into a formal scientific discipline. On the other hand, the procedures that form the core of the present study were all that and more: They laid the foundation for a logic of science that, in addition to being a discipline itself, has helped frame and answer questions in virtually every modern science.

2. The Trial of the Pyx was not without its flaws. The use of linear extrapolation was a major one: if 1 coin was allowed a tolerance of 5 grains, an aggregate of 100 coins would be allowed a tolerance of 500 grains, rather than the $\sqrt{100} \times 5 = 50$ grains modern theory might suggest. The story of the Pyx, including Isaac Newton's connection to it, is told in Stigler (1977b).

The development of statistics chronicled in this book begins about 1700. There are earlier threads we shall not pick up. John Graunt's 1662 *Observations upon Bills of Mortality* contained many wise inferences based on his data, but its primary contemporary influence was more in its demonstration of the value of data gathering than on the development of modes of analysis. In another direction, the development of mathematical probability dates from correspondence between Fermat and Pascal in the 1650s and a short tract Christian Huygens printed in 1657. But those early works do not attempt to come to grips with inferential problems and do not go beyond the games of chance that were their immediate interest. Rather we shall follow two lines of development: first, the combination of observations in astronomy and geodesy; and second, the use of probability for inferential purposes, from their separate beginnings up to their remarkable synthesis about 1810. We shall then examine the development of these procedures as they spread to biology and the social sciences by the end of the nineteenth century.

Over the two centuries from 1700 to 1900, statistics underwent what might be described as simultaneous horizontal and vertical development: horizontal in that the methods spread among disciplines, from astronomy and geodesy, to psychology, to biology, and to the social sciences, being transformed in the process; vertical in that the understanding of the role of probability advanced as the analogy of games of chance gave way to probability models for measurements, leading finally to the introduction of inverse probability and the beginnings of statistical inference. This development culminated with the incestuous use of probability models to validate inferences derived from them and then, eventually, to extend the domain of application for statistical inference.

Our two main themes, the combination of observations and the uses of probability models in inference, cut across both the horizontal and vertical dimensions. Before the middle of the eighteenth century there is little indication in extant literature of a willingness of astronomers to combine observations; indeed, as we shall see in the case of Euler in 1748, there was sometimes an outright refusal to combine them. Astronomers took simple averages of nearly perfectly replicated determinations of the same quantity; but the idea that accuracy could be increased by combining measurements made under different conditions was slow to come. They feared that errors in one observation would contaminate others, that errors would multiply, not compensate.

It was an essential precondition for probability-based statistical inference that in order to be combined, measurements should be made under conditions that could be viewed as identical, or as differing only in ways that could be allowed for in the analysis. During the course of the eighteenth century astronomers progressed from simple means to linear models and, with the aid of Newtonian theories, were able to reach a mature statistical theory in the first two decades of the nineteenth century. Psychologists were able to substitute control over their experimental material — experimental design — for the discipline of Newtonian theory. Social scientists, however, had neither route readily available, and for much of the century they groped toward some acceptable way of overcoming the inherent diversity of their material. They tried to overcome the conceptual barrier to combination of observations by brute force, with massive data bases that they hoped would inform them of all important causal groupings. But the result was frustration, even among those pioneers who made some progress. For example, the economist William Stanley Jevons defended himself in 1869 against charges that he was wrong to combine prices of several very different commodities into one index in a study of the variation in the value of gold with these words:

> Not a few able writers . . . are accustomed to throw doubt upon all such conclusions, by remarking that until we have allowed for all the particular causes which may have elevated or depressed the price of each commodity we cannot be sure that gold is affected. Were a complete explanation of each fluctuation thus necessary, not only would all inquiry into this subject be hopeless, but the whole of the statistical and social sciences, so far as they depend upon numerical facts, would have to be abandoned. (Jevons, 1869; 1884, p. 155)

But this melodramatic defense was not accompanied by a conceptual structure that could disarm his critics. Twenty years passed before such a structure appeared, and even more time elapsed before it was widely adopted.

Now, contemporaneously with the earliest developments in the combination of observations in astronomy, there were major advances in the use of probability in inference, starting with its use in the assessment of accuracy of the values determined by simple means. Broadly speaking, accuracy can be assessed in two ways: by using information external to the data at hand and through analyses internal to the data. Both approaches can be found in early astronomical work, sometimes within the same investigation. For example, in 1740 Jacques Cassini compared a list of observations, made over nearly 2,000 years, of the inclination of the earth's equator to its orbit about the sun (the "obliquity of the ecliptic") with a list created by a simple linear interpolation from the earliest observation to the latest (Table I.1). He presented a two-stage argument as to why this comparison showed that the change in that angle had not been uniform in time. First, Cassini said, the discrepancies between the two lists were too large to be admitted as errors. In particular, Ptolemy's value differed from the interpolated value by more than 4′, Pappus's by more than 14′. This assessment

Table I.1. Cassini's 1740 comparison of nearly 2000 years' data on the obliquity of the ecliptic with values found by linear interpolation.

Source	Date	From observation	From linear interpolation
Eratosthenes	230 B.C.	23°51′20″	23°51′20″
Hipparchus	140 B.C.	23°51′20″	23°50′17″
Ptolemy	140 A.D.	23°51′10″	23°47′0″
Pappus	390	23°30′0″	23°44′7″
Albategnius	880	23°35′0″	23°38′21″
Arzachel	1070	23°34′0″	23°36′8″
Prophatius	1300	23°32′0″	23°33′27″
Regiomontanus	1460	23°30′0″	23°31′35″
Copernicus	1500	23°28′24″	23°31′7″
Waltherus	1500	23°29′16″	23°31′7″
Danti	1570	23°29′55″	23°30′18″
Tycho	1570	23°31′30″	23°30′18″
Gassendi	1600	23°31′0″	23°29′57″
Cassini	1656	23°29′2″	23°29′19″
Richer	1672	23°28′54″	23°29′6″
Paris Observatory	1738	23°28′20″	23°28′20″

Source: Cassini (1740, p. 113).

is an external assessment of accuracy, based on Cassini's general experience with determinations of this type. Second, if it were claimed to the contrary that the more ancient observations could in fact err by that much, that claim would be inconsistent with the close agreement among the three oldest, which differed by only 10″. Such an assessment is an internal assessment, implicitly assuming that the determinations were independent[3] (Cassini, 1740, p. 113). Similar informal internal assessments were not unusual at the time. For example, Isaac Newton, in a discussion of data on the comet of 1680 wrote: "From all this it is plain that these observations agree with theory, so far as they agree with one another" (Newton, 1726, p. 518).

External assessments of accuracy have always been important in science. They can provide the only access to the measurement of systematic errors or biases. With the development of modern Bayesian inference they can be incorporated with internal assessments, thus considerably sharpening the result. But it is with internal assessments that we find the conceptual developments that are of most interest to this historical investigation—

3. Recent scholarship suggests that Ptolemy's value was in error by about 10′, and the close agreement with Eratosthenes and Hipparchus was due to Ptolemy's use of their values as a basis for his (Goldstein, 1983, p. 3).

the first inferential uses of probability. The mathematical simplicity of the binomial probability model that had proved so useful in the analysis of games of chance was not matched by a corresponding conceptual simplicity when inference problems were addressed. Instead, the first important breakthroughs came in the mathematically more complex but conceptually simpler realm of the theory of errors, an application of the calculus of probability to the internal assessment of the accuracy of observations. By the early 1800s, according to Joseph Fourier, the French scientists in Egypt were able both to estimate the height of the great pyramid of Cheops by adding together the measured heights of the 203 steps to the top and to judge the accuracy of this figure by multiplying the estimated error in measuring a single step by the number found from probability theory, $14 = \sqrt{203}$ (Fourier, 1829, pp. xxvii–xxviii; 1890, pp. 569–570).

The role of probability theory in the historical development of statistics was far more extensive than simply that of a refinement to the already developed combination of observations in astronomy, or even that of a tool for deriving combination schemes better than simple means—though it certainly succeeded as both of these. More important, it provided the framework for the growth and spread of those early statistical methods, first by adding a firm theoretical foundation for the existing linear procedures and thus permitting their growth and then by furnishing a conceptual structure for both the combination of observations and the making of inferences with social science data.

These last steps did not come easily. As I have mentioned, the social sciences lacked not only the discipline that a universal law of gravitation had given astronomy but also the compensating control over experiments that psychology had enjoyed. Even though many "laws" of society were known by the mid-nineteenth century, and some of these (particularly in economics) were quantitative, the trip from the *Mécanique céleste* to the *Mécanique sociale* was not to be a short one. The laws of social science, unlike those of astronomy, did not capture all major operating causes and influences; instead they usually related pairs of variables, other forces being assumed constant. But what could be held constant in theory could not be done so easily in practice; and if observations were to be sufficiently homogeneous in practice to permit combination and analysis, something else was needed. Quetelet and Lexis made valiant attempts to use probability models to validate simple grouping, trying to "reason backward" and infer homogeneity from some of its consequences. For very different reasons they failed to accomplish the larger goal; Quetelet found homogeneity everywhere, Lexis found it nowhere.

The success that came later with Galton, Edgeworth, Pearson, and Yule was remarkable, however; for beyond providing social science with a statistical methodology, they created a discipline of statistics. Galton's regression (as finally developed by Yule) was not simply an adaptation of least squares to a different set of problems; it was a new way of thinking about multivariate data, a new set of concepts that by sheer force of intellect created for biology and the social sciences a surrogate for the missing universal law. The new laws of regression and correlation were laws of a kind different from those astronomers knew; and the methods were a transformation of the earlier methods, not a simple adaptation of them. These new laws provided a new way to look at the problem, a way that potentially allowed an internal analysis to control for effects of measured variables even when no external theory provided guidance.

These laws were far from completely developed by 1900. The exploration of their potential and limitations has given twentieth-century statistics a research program that shows no sign of approaching exhaustion. Even so, the promise evident by 1900 has been fulfilled. In time, in the works of Eddington and Chandrasekhar these statistical laws would return to and transform astronomy, just as they would transform economics, genetics, psychology, and sociology. As a modern discipline this cross-disciplinary logic of empirical science has been a magnificent success. As a study in the history of science it exhibits its own beautiful symmetry, with the successful combination of observations by astronomers in the mid-1700s mirrored by a similar triumph by social scientists a century and a half later and with advances in probability, which were inspired by astronomical concepts, in return inspiring a revolutionary conceptual framework in the social sciences. Galton's own words from the introduction to *Natural Inheritance* seem particularly apt:

> The road to be travelled over . . . is full of interest of its own. It familiarizes us with the measurement of variability, and with curious laws of chance that apply to a vast diversity of social subjects. This part of the inquiry may be said to run along a road on a high level, that affords wide views in unexpected directions, and from which easy descents may be made to totally different goals to those we have now to reach. I have a great subject to write upon. (Galton, 1889, p. 3)

PART ONE

The Development of Mathematical Statistics in Astronomy and Geodesy before 1827

1. *Least Squares and the Combination of Observations*

Adrien Marie Legendre (1752–1833)

THE METHOD of least squares was the dominant theme—the leitmotif—of nineteenth-century mathematical statistics. In several respects it was to statistics what the calculus had been to mathematics a century earlier. "Proofs" of the method gave direction to the development of statistical theory, handbooks explaining its use guided the application of the higher methods, and disputes on the priority of its discovery signaled the intellectual community's recognition of the method's value. Like the calculus of mathematics, this "calculus of observations" did not spring into existence without antecedents, and the exploration of its subtleties and potential took over a century. Throughout much of this time statistical methods were commonly referred to as "the combination of observations." This phrase captures a key ingredient of the method of least squares and describes a concept whose evolution paced the method's development. The method itself first appeared in print in 1805.

Legendre in 1805

In March of 1805 political Europe focused its attention uneasily on France. The 1801 Peace of Amiens was crumbling, and preparations were under way for a new round of war, one that would begin that autumn with the battle of Trafalgar and the opening of the Napoleonic campaigns with victories at Ulm and Austerlitz. Scientific Europe also looked to France; there the discipline was intellectual, not martial, and was subsequently more sure and longer lived than that of the new emperor of the French.

In March of 1805 Laplace celebrated his fifty-sixth birthday, prepared the fourth volume of his *Traité de mécanique céleste* for press, and, perhaps, began to return his thoughts to the completion of a book on probability he had first contemplated more than twenty years before. Also in March of 1805 another French mathematical scientist, Adrien Marie Legendre, sent to press the final pages of a lengthy memoir that contained the first publication of what is even today the most widely used nontrivial technique of mathematical statistics, the method of least squares.

Legendre (born 18 September 1752, died 10 January 1833) was a mathematician of great breadth and originality. He was three years Laplace's junior and succeeded Laplace successively as professor of mathematics at the Ecole Militaire and the Ecole Normale. Legendre's best-known mathematical work was on elliptic integrals (he pioneered this area forty years before Abel and Jacobi), number theory (he discovered the law of quadratic reciprocity), and geometry (his *Eléments de géométrie* was among the most successful of such texts of the nineteenth century). In addition, he wrote important memoirs on the theory of gravitational attraction. He was a member of two French commissions, one that in 1787 geodetically joined the observatories at Paris and Greenwich and one that in 1795 measured the meridian arc from Barcelona to Dunkirk, the arc upon which the length of the meter was based. It is at the nexus of these latter works in theoretical and practical astronomy and geodesy that the method of least squares appeared.

In 1805 (the appendix we shall discuss is dated 6 March 1805) Legendre published the work by which he is chiefly known in the history of statistics, *Nouvelles méthodes pour la détermination des orbites des comètes.* At eighty pages this work made a slim book, but it gained a fifty-five-page supplement (and a reprinted title page) in January of 1806, and a second eighty-page supplement in August of 1820. The appendix presenting the method of least squares occupies nine of the first eighty pages; it is entitled "Sur la méthode des moindres quarrés." For stark clarity of exposition the presentation is unsurpassed; it must be counted as one of the clearest and most elegant introductions of a new statistical method in the history of statistics. In fact, statisticians in the succeeding century and three-quarters have found so little to improve upon that, but for his use of $\int$ instead of Σ to signify summation, the explanation of the method could almost be from an elementary text of the present day. Legendre began with a clear statement of his objective:

On the Method of Least Squares

> In most investigations where the object is to deduce the most accurate possible results from observational measurements, we are led to a system of equations of the form
>
> $$E = a + bx + cy + fz + \&c.,$$
>
> in which, a, b, c, f, &c. are known coefficients, varying from one equation to the other, and x, y, z, &c. are unknown quantities, to be determined by the condition that each value of E is reduced either to zero, or to a very small quantity. (Legendre, 1805, p. 72)

We might write this today as

$$E_i = a_i + b_i x + c_i y + f_i z + \ . \ . \ .$$

or

$$a_i = -b_i x - c_i y - f_i z - \ . \ . \ . + E_i,$$

but we would probably join Legendre in calling the E's "errors." When the number of equations equaled the number of unknowns, Legendre saw no problem. But when there were more equations than unknowns, it became impossible to choose values for the unknowns that would eliminate all the errors. He noted that there was an element of arbitrariness in any way of, as he put it, "distributing the errors among the equations," but that did not stop him from dramatically proposing a single best solution:

> Of all the principles that can be proposed for this purpose, I think there is none more general, more exact, or easier to apply, than that which we have used in this work; it consists of making the sum of the squares of the errors a *minimum*. By this method, a kind of equilibrium is established among the errors which, since it prevents the extremes from dominating, is appropriate for revealing the state of the system which most nearly approaches the truth. (Legendre, 1805, pp. 72–73)

Minimize the sum of the squares of the errors! How simple—but was it practical? Legendre wasted no time in writing down the equations he derived by differentiating the sum of squared errors $(a + bx + cy + fz +$

&c.)^2 + $(a' + b'x + c'y + f'z + \&c.)^2 + (a'' + b''x + c''y + f''z + \&c.)^2$ + &c. with respect to $x, y, \ldots$, namely,

$$0 = \int ab + x \int b^2 + y \int bc + z \int bf + \&c.$$

$$0 = \int ac + x \int bc + y \int c^2 + z \int fc + \&c.$$

$$0 = \int af + x \int bf + y \int cf + z \int f^2 + \&c.$$

> where by $\int ab$ we understand the sum of the similar products $ab + a'b' + a''b''$ + &c.; and by $\int b^2$ the sum of the squares of the coefficients of x, that is, $b^2 + b'^2 + b''^2$ + &c., and so on. (Legendre, 1805, p. 73)

Lest there be any doubt as to how to form these, the "normal equations" (a name introduced later by Gauss), Legendre restated the rule in italicized words:

> In general, *to form the equation of the* minimum *with respect to one of the unknowns, it is necessary to multiply all the terms of each equation by the coefficient of the unknown in that equation, taken with its proper sign, and then to find the sum of all these products.* (Legendre, 1805, p. 73)

The equations, as many as there were unknowns, could then be solved by "ordinary methods."

The practicality of Legendre's principle was thus self-evident; all it required was a few simple multiplications and additions and the willingness to solve a set of linear equations. Because the latter step would be required in any case, even if the number of observations equaled the number of unknowns, the formation of the "equations of the minimum" was all that Legendre added to the calculations. It was a small computational price to pay for such an appealingly evenhanded resolution to a difficult problem. In case more convincing was needed, Legendre added these points:

(1) If a perfect fit were possible, his method would find it.

(2) If it were subsequently decided to discard an equation (say, if its "error" were too large), it would be a simple matter to revise the equations by subtracting the appropriate terms.

(3) The arithmetic mean was a special case of the method, found when there is a single unknown with constant coefficient $b = b' = \ldots = 1$.

(4) Likewise, finding the center of gravity of several equal masses in space was a special case. By analogy with this last case Legendre closed his introduction of the method with these words, "We see, therefore, that the method of least squares reveals, in a manner of speaking, the center around which the results of observations arrange themselves, so that the

deviations from that center are as small as possible" (Legendre, 1805, p. 75).

Legendre followed this explanation with a worked example using data from the 1795 survey of the French meridian arc, an example involving three unknowns in five equations. The clarity of the exposition no doubt contributed to the fact that the method met with almost immediate success. Before the year 1805 was over it had appeared in another book, Puissant's *Traité de géodésie* (Puissant, 1805, pp. 137–141; Puissant did use Σ for summation); and in August of the following year it was presented to a German audience by von Lindenau in von Zach's astronomical journal, *Monatliche Correspondenz* (Lindenau, 1806, p. 138–139). An unintended consequence of Legendre's publication was a protracted priority dispute with Carl Friedrich Gauss, who claimed in 1809 that he had been using the method since 1795 (Chapter 4).

Ten years after Legendre's 1805 appendix, the method of least squares was a standard tool in astronomy and geodesy in France, Italy, and Prussia. By 1825 the same was true in England.[1] The rapid geographic diffusion of the method and its quick acceptance in these two fields, almost to the exclusion of other methods, is a success story that has few parallels in the history of scientific method. It does, however, raise a number of important questions, including these: Did the introduction of the method create a historical discontinuity in the development of statistics, or was it a natural, if inspired, outgrowth of previous approaches to similar problems? What were the characteristics of the problems faced by eighteenth-century astronomers and geodesists that led to the method's introduction and easy acceptance? In what follows I shall attempt to answer these questions by examining several key works in these fields. In particular, I shall argue that least squares was but the last link in a chain of development that began about 1748, and that by the late 1780s methods were widely known and used that were, for practical purposes, adequate for the problems faced. I shall show how the deceptively simple concept that there was a potential gain to be achieved through the combination of observational data gathered under differing circumstances proved to be a major stumbling block in early work and how the development of these methods required the combination of extensive empirical experience and mathematical or mechanical insight.

1. The earliest English translation was by George Harvey (1822), "On the Method of Minimum Squares." Before the mid 1820s, it was common for Legendre's name for his method, "Moindres quarrés," to be translated into English as "minimum squares" or "small squares" rather than as "least squares."

Cotes's Rule

By the middle of the eighteenth century at least one statistical technique was in frequent use in astronomy and navigation: the taking of a simple arithmetic mean among a small collection of measurements made under essentially the same conditions and usually by the same observer (Plackett, 1958). But it is worth emphasizing that, widespread as the practice was, it was followed only in a narrowly conceived set of problems. Astronomers averaged measurements they considered to be equivalent, observations they felt were of equal intrinsic accuracy because the measurements had been made by the same observer, at the same time, in the same place, with the same instrument, and so forth. Exceptions, instances in which measurements not considered to be of equivalent accuracy were combined, were rare before 1750.

One possible exception is a rule found in a work of Roger Cotes published in 1722 (published posthumously, for Cotes died in 1716).

> Let p be the place of some object defined by observation, q, r, s the places of the same object from subsequent observations. Let there also be weights P, Q, R, S reciprocally proportional to the displacements which may arise from the errors in the single observations, and which are given from the given limits of error; and the weights P, Q, R, S are conceived as being placed at p, q, r, s, and their centre of gravity Z is found: I say the point Z is the most probable place of the object, and may be most safely had for its true place. (Cotes, 1722, p. 22; based on the translation by Gowing, 1983, p. 107)

Cotes's rule can be (and has been) read as recommending a weighted mean, or even as an early appearance of the method of least squares (De Morgan, 1833–1844, "Least Squares"). However, it has about it a vagueness that could only be cleared up by one or more accompanying examples, examples Cotes did not provide. To understand the genesis of the method of least squares, we must look not just at what investigators say they are doing (and how the statement might be most charitably interpreted in the light of later developments) but also at what was actually done. Cotes's rule had little or no influence on Cotes's immediate posterity. In the literature of the theory of errors its earliest citation seems to be that by Laplace (1812, p. 346; 1814, p. 188).

Tobias Mayer and the Libration of the Moon

The development of the method of least squares was closely associated with three of the major scientific problems of the eighteenth century: (1) to determine and represent mathematically the motions of the moon; (2) to account for an apparently secular (that is, nonperiodic) inequality that had been observed in the motions of the planets Jupiter and Saturn; and (3) to determine the shape or figure of the earth. These problems all involved astronomical observations and the theory of gravitational attraction, and they all presented intellectual challenges that engaged the attention of many of the ablest mathematical scientists of the period.

It seems most appropriate to begin with two works: "Recherches sur la question des inégalités du mouvement de Saturne et de Jupiter," written by Leonhard Euler and published in 1749; and "Abhandlung über die Umwalzung des Monds um seine Axe und die scheinbare Bewegung der Mondsflecten," written by Tobias Mayer and published in 1750. Although these works were not the first to consider their respective subjects (the inequalities of the motions of Jupiter and Saturn and the libration of the moon), they were among the best of the early treatments of these subjects. Because they were widely read, they greatly influenced later workers and, from a statistical point of view, form a unique and dramatic contrast in the handling of observational evidence. Together they tell a story of statistical success (by a major astronomer, Mayer) and statistical failure (by a leading mathematician, Euler). They show why the discovery of the method of least squares was not possible in the intellectual climate of 1750, and they highlight the conceptual barriers that had to be crossed before this climate became sufficiently tropical to support the later advances of Legendre, Gauss, and Laplace.

We shall consider Mayer's statistical success first. Notwithstanding the principal "monthly" regularity in the motion of the moon about the earth, its detailed motion is extraordinarily complex. In the eighteenth century the problem of accurately accounting for these minor perturbations in the moon's movements, either by a mathematical formula or by an empirically determined table describing future lunar positions, was of great scientific, commercial, and even military significance. Its scientific importance lay in the general desire to show that Newtonian gravitational theory can account for the movements of our nearest celestial neighbor (within the always decreasing limits of observational error) if allowance is made for the attraction of other bodies (such as the sun), for periodic changes in the earth's and the moon's orbits, and for the departures from sphericity of the shapes of the earth and moon.[2] But it was the potential commercial and military usefulness of a successful accounting of the moon (as an aid to navigation) that was primarily responsible for the widespread attention the problem received. Over the previous nineteen centuries, from Hipparchus and Ptolemy to Newton and Flamsteed, the linked development of theoretical and practical astronomy had played the key role in freeing ship's navigators from a dependence upon land sightings as a way of deter-

2. As eloquent testimony to the difficulty of the problem, we have Newton reportedly telling Halley that lunar theory "made his head ache and kept him awake so often that he would think of it no more" (Berry, 1898, p. 240).

mining the ship's position. The developments of better nautical instruments—including the sextant in 1731—and a more accurate understanding of astronomical theory, increasingly enabled navigators to map their ships' courses across previously trackless seas. By 1700 it had become possible to determine a ship's latitude at sea with relative precision by the fixed stars—simply by measuring the angular elevation of the celestial pole above the horizon. The determination of longitude, however, was not so simple. Indeed, in 1714 England established the "commissioners for the discovery of longitude at sea," a group that by 1815 had disbursed £101,000 in prizes and grants to achieve its goal. The two most promising methods of ascertaining longitude at sea were the development of an accurate clock (so that Greenwich time could be maintained on shipboard and longitude determined by the comparison of the fixed stars' positions and Greenwich time) and the creation of lunar tables that permitted the determination of Greenwich time (and thus of longitude) by comparison of the moon's position and the fixed stars.

Johann Tobias Mayer (1723–1762) had already made a name for himself as a cartographer and practical astronomer by the time he undertook a study of the moon in 1747 (a study that eventually led to the preparation of lunar tables that were to earn his widow £3,000 from the British commissioners in 1765). The specific work of Mayer that most influenced statistical practice was his study, published in 1750, of the librations of the moon.

The popular notion that the moon always presents the same face to the earth is not literally true. The moon in fact is subject to "libration": The face viewed from earth varies, so that over an extended period of time about 60 percent of the moon's surface is visible from earth. Two sources of this libration were known to Galileo: the apparent diurnal libration due to the earth's rotation and a libration in latitude due principally to the fact that the moon's axis of rotation is not perpendicular to the earth's orbital plane about the sun. By the time of Mayer's work it was known that the earth's location was at a focus, not the center, of the moon's elliptical orbit. Thus the moon's rotation at a uniform speed produced a third type of libration, one of longitude.

Over the period from April 1748 to March 1749, Mayer made numerous observations of the positions of several prominent lunar features; and in his 1750 memoir he showed how these data could be used to determine various characteristics of the moon's orbit (Figure 1.1). His method of handling the data was novel, and it is well worth considering this method in detail, both for the light it sheds on his pioneering, if limited, understanding of the problem and because his approach was widely circulated in the major contemporary treatise on astronomy, having signal influence upon later work.

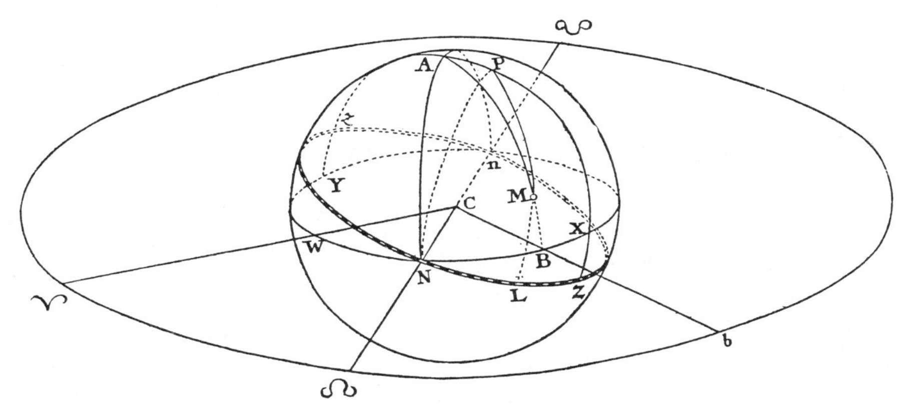

Figure 1.1. Tobias Mayer's original drawing of the moon. (From Mayer, 1750, table VI.)

Mayer's method for the resolution of inconsistent observational equations can be discerned in his discussion of the position of the crater Manilius. Figure 1.2 represents the moon, which Mayer considered as a sphere. The great circle QNL represents the moon's true equator, and P is the moon's pole with respect to this equator, one end of its axis of revolution. The great circle DNB is that circumference (or apparent equator) of the moon that is seen from earth as parallel to the plane of the ecliptic, the plane of the earth's orbit about the sun, and A is the pole of the moon with respect to DNB, its apparent pole as viewed by an earthbound astronomer oriented by the ecliptic. The point γ (the point on the circle DNB in the direction from the moon's center C toward the equinox) was taken as a reference point. The circle DNB and the pole A will vary with time, as a result of the libration of the moon, but they form the natural system of coordinates at a given time. The equator QNL and the pole P are fixed but not observable from earth. Mayer's aim was to determine the relationship between these coordinate systems and thus accurately determine QNL and P. He accomplished this by making repeated observations of the crater Manilius. Now, in Figure 1.2, M is the position of Manilius, and PL and AB are meridian quadrants through M with respect to the two polar coordinate systems. Mayer was able to observe the position of M on several occasions with respect to the constantly changing coordinate system determined by DNB and A; that is, subject to observational error he could at a given time measure the arcs $AM \equiv h$ and $\gamma B \equiv g$.

To determine the relationship between the coordinate systems, Mayer sought to find the fixed, but unknown, arc length $AP \equiv \alpha$, the true

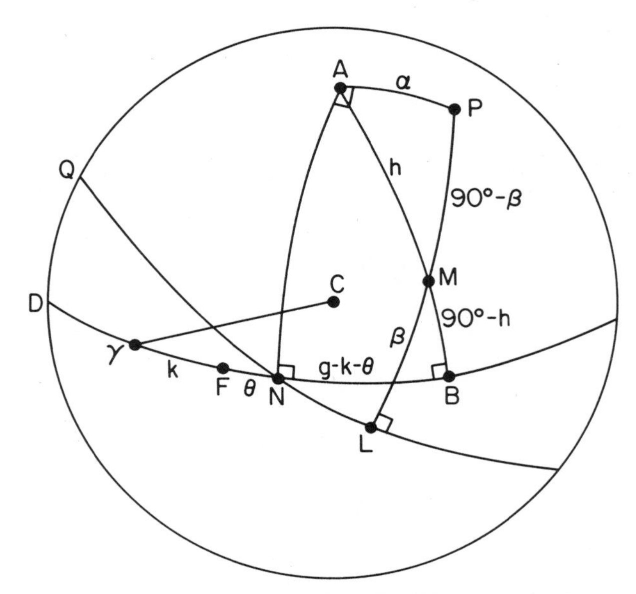

Figure 1.2. The moon. M, The crater Manilius; NL, the moon's equator; P, the moon's equatorial pole; NB, the circumference parallel to the plane of the ecliptic; A, the pole of NB; Cγ, the direction of the equinox from the moon's center C; F, the node of the moon's orbit and the plane of the ecliptic. See text for more details.

latitude of Manilius $\beta = \mathrm{ML}$, and the distance θ between the unknown node or point of intersection of the two circles (N) and the known point of intersection F of the plane of the orbit of the moon and the circle DNB. He let $k = \gamma\mathrm{F}$ be the observed longitude of F. Then g, h, and k were observable and varied from observation to observation as a result of the motion of the moon (and observational error); and α, θ, and β were fixed and unknown, to be determined from the observations. Because NAP forms a right angle, a basic identity of spherical trigonometry implies that these quantities are related nonlinearly by the equation.

(A) $\sin \beta = \cos \alpha \cos h + \sin \alpha \sin h \sin(g - k - \theta)$.

Now, Mayer knew that both α and θ were small (in the neighborhood of 2 or 3 degrees), and he proceeded, via several trigonometric identities, to derive an almost linear approximation to this equation under the supposition that $\cos \alpha$, $\cos \theta$, and $\cos(\beta - 90° + h)$ were approximately 1.0, and

$\sin(\beta - 90° + h) \cong \beta - 90° + h$. He was led[3] to the equation

$$\text{(C)} \qquad \beta - (90° - h) = \alpha \sin(g - k) - \alpha \sin \theta \cos(g - k),$$

which provided at least an approximate relationship between the observations and the unknowns.

At this point Mayer was in sight of his goal. He only needed to take observations for three different days, solve the resulting three linear equations for β, α, and $\alpha \sin \theta$ (and then solve for θ), and he would be done. His problem, however, was that he suffered an embarrassment of riches—he had twenty-seven days' observations of Manilius. The resulting twenty-seven equations are given in Table 1.1.

The form of Mayer's problem is almost the same as that of Legendre; Legendre might have written E for the discrepancy (due to the linear approximation and observational error) between the two sides; he would then have had

$$\mathrm{E} = (90° - h) - \beta + \alpha \sin(g - k) - \alpha \sin \theta \cos(g - k).$$

In Mayer's form the equations came to be called the *equations of condition* because they expressed a condition or relationship that would hold if no errors were present. The modern tendency would be to write, say, $(h - 90°) = -\beta + \alpha \sin(g - k) - \alpha \sin \theta \cos(g - k) + \mathrm{E}$, treating $h - 90°$ as the dependent variable and $-\beta$, α, and $-\alpha \sin \theta$ as the parameters in a linear regression model.

How did Mayer address his overdetermined system of equations? His approach was a simple and straightforward one, so simple and straightforward that a twentieth-century reader might arrive at the very mistaken opinion that the procedure was not remarkable at all. Mayer divided his equations into three groups of nine equations each, added each of the three groups separately, and solved the resulting three linear equations for α, β, and $\alpha \sin \theta$ (and then solved for θ). His choice of which equations belonged in which groups was based upon the coefficients of α and $\alpha \sin \theta$. The first group consisted of the nine equations with the largest positive values for the coefficient of α, namely, equations 1, 2, 3, 6, 9, 10, 11, 12, and 27. The second group were those with the nine largest negative values for this coefficient: equations 8, 18, 19, 21, 22, 23, 24, 25, and 26. The remaining nine equations formed the third group, which he described as having the largest values for the coefficient of $\alpha \sin \theta$.

3. Mayer started with $\sin(g - k - \theta) = \sin(g - k)\cos \theta - \sin \theta \cos(g - k) \cong \sin(g - k) - \sin \theta \cos(g - k)$. Then, setting $\cos \alpha = 1$ in equation (A), he had equation (B) $\sin \beta - \cos h = \sin \alpha \sin h \sin(g - k) - \sin \alpha \sin h \sin \theta \cos(g - k)$; letting $x = \beta - (90° - h)$, a small quantity, he had $\sin \beta = \cos(h - x) = \cos h \cos x + \sin h \sin x \cong \cos h + x \sin h$, which with (B) gives $x \sin h \cong \sin \alpha \sin h \sin(g - k) - \sin \alpha \sin h \sin \theta \cos(g - k)$, and setting $\alpha \cong \sin \alpha$ and dividing through by $\sin h$ gives (C).

Table 1.1. Mayer's twenty-seven equations of condition, derived from observations of the crater Manilius from 11 April 1748 through 4 March 1749.

Eq. no.	Equation	Group
1	$\beta - 13°10' = +0.8836\alpha - 0.4682\alpha \sin\theta$	I
2	$\beta - 13°8' = +0.9996\alpha - 0.0282\alpha \sin\theta$	I
3	$\beta - 13°12' = +0.9899\alpha + 0.1421\alpha \sin\theta$	I
4	$\beta - 14°15' = +0.2221\alpha + 0.9750\alpha \sin\theta$	III
5	$\beta - 14°42' = +0.0006\alpha + 1.0000\alpha \sin\theta$	III
6	$\beta - 13°1' = +0.9308\alpha - 0.3654\alpha \sin\theta$	I
7	$\beta - 14°31' = +0.0602\alpha + 0.9982\alpha \sin\theta$	III
8	$\beta - 14°57' = -0.1570\alpha + 0.9876\alpha \sin\theta$	II
9	$\beta - 13°5' = +0.9097\alpha - 0.4152\alpha \sin\theta$	I
10	$\beta - 13°2' = +1.0000\alpha + 0.0055\alpha \sin\theta$	I
11	$\beta - 13°12' = +0.9689\alpha + 0.2476\alpha \sin\theta$	I
12	$\beta - 13°11' = +0.8878\alpha + 0.4602\alpha \sin\theta$	I
13	$\beta - 13°34' = +0.7549\alpha + 0.6558\alpha \sin\theta$	III
14	$\beta - 13°53' = +0.5755\alpha + 0.8178\alpha \sin\theta$	III
15	$\beta - 13°58' = +0.3608\alpha + 0.9326\alpha \sin\theta$	III
16	$\beta - 14°14' = +0.1302\alpha + 0.9915\alpha \sin\theta$	III
17	$\beta - 14°56' = -0.1068\alpha + 0.9943\alpha \sin\theta$	III
18	$\beta - 14°47' = -0.3363\alpha + 0.9418\alpha \sin\theta$	II
19	$\beta - 15°56' = -0.8560\alpha + 0.5170\alpha \sin\theta$	II
20	$\beta - 13°29' = +0.8002\alpha + 0.5997\alpha \sin\theta$	III
21	$\beta - 15°55' = -0.9952\alpha - 0.0982\alpha \sin\theta$	II
22	$\beta - 15°39' = -0.8409\alpha + 0.5412\alpha \sin\theta$	II
23	$\beta - 16°9' = -0.9429\alpha + 0.3330\alpha \sin\theta$	II
24	$\beta - 16°22' = -0.9768\alpha + 0.2141\alpha \sin\theta$	II
25	$\beta - 15°38' = -0.6262\alpha - 0.7797\alpha \sin\theta$	II
26	$\beta - 14°54' = -0.4091\alpha - 0.9125\alpha \sin\theta$	II
27	$\beta - 13°7' = +0.9284\alpha - 0.3716\alpha \sin\theta$	I

Source: Mayer (1750, p. 153).
Note: One misprinted sign in equation 7 has been corrected.

Even though Mayer's description of the third group is not fully accurate (compare equation 8, in group II, with equation 13 in group III), his specification of the three groups shows insight into the geometry of the situation and reveals that his choice of the crater Manilius was perhaps motivated by at least dimly perceived notions of experimental design. The coefficients of α and $\alpha \sin\theta$ are $\sin(g-k)$, and $-\cos(g-k)$, which are related by $[\sin(g-k)]^2 + [-\cos(g-k)]^2 = 1$. The first group consists of those nine equations whose coefficients of α are nearest 1.0, the second group of those nearest -1.0, leaving the third group as those equations with $\sin(g-k)$ "near" zero, that is, with large (although not without exception the largest) $-\cos(g-k)$'s (which all happen, because of the choice of crater, to be positive). This way of choosing equations for aggregation tends, subject to the restriction to equal groups, to maximize the contrast among the coefficients of α and produce good estimates of α, and, with the present selection of crater, good estimates of $\alpha \sin \theta$ as well. Mayer seems to have understood this because he wrote, "These equations [Table 1.2] can take the place of the foregoing totality of equations [Table 1.1] because each of these three equations has been formed in the most advantageous manner *(die vortheilhaftigste Art).* The advantage consists in the fact that through the above division into three classes, the differences between the three sums are made as large as is possible. The greater these differences are, the more accurately *(richtiger)* one may determine the unknown values of α, β, and θ" (Mayer, 1750, p. 154).

Mayer solved the three equations he found, getting $\alpha = 89'.90 \cong 1°30'$, $\theta = -3°45'$, and $\beta = 14°33'$, and he went on to consider the accuracy of these values. He noted that even under favorable conditions an individual observation of an arc could only be counted as accurate to within 10 or 15 minutes, and he claimed that the effect of an error of this magnitude in g and h on the final determinations could be traced through the formulas he had given. He did not attempt this kind of nonstatistical error analysis, however. Instead he presented an empirical assessment of accuracy.

Earlier in his paper Mayer had illustrated how α, β, and θ were calculated on the basis of only three observational equations (equations 9, 16, and 19 of Table 1.1). The value he had found for α based on those three equations was $\alpha = 1°40'$. Now, he noted, "Because these last values [based on all twenty-seven equations] were derived from nine times as many observations, one can therefore conclude that they are nine times more correct *(neunmal richtiger)*; therefore the error *(Fehler)* in each of the constants is in inverse relationship to the number of their observations" (Mayer, 1750, p. 155). Mayer turned this statement into an interval description of the most important of the unknowns, α, as follows:

> Let the true value *(wahre Wehrt)* be $\alpha = 1°30' \pm x$; then x is the difference or the error *(der Unterschied oder Irrthum)*: how far the quantity α, determined

Table 1.2. Mayer's three equations, as derived from Table 1.1 by adding equations 1, 2, 3, 6, 9, 10, 11, 12, and 27 in group I, equations 8, 18, 19, 21, 22, 23, 24, 25, and 26 in group II, and the rest in group III.

Group	Equation
I	$9\beta - 118°8' = +8.4987\alpha - 0.7932\alpha \sin \theta$
II	$9\beta - 140°17' = -6.1404\alpha + 1.7443\alpha \sin \theta$
III	$9\beta - 127°32' = +2.7977\alpha + 7.9649\alpha \sin \theta$

Source: Mayer (1750, p. 154).

from the 27 observations, can deviate from the true value. Since from three observations we found $\alpha = 1°40'$, the error *(der Fehler)* of that determination is found to be $= 10 \pm x$; consequently we are led to conclude that

$$\pm x : \frac{1}{27} = 10 \pm x : \frac{1}{3},$$

from which we find $x = \pm 1' \frac{1}{4}$. The true value of α can therefore be about 1′ or 2′ smaller or larger than 1°30′. (Mayer, 1750, p. 155)

Thus Mayer introduced the symbol $\pm x$ for the error made in taking $\alpha = 1°30'$; and $10 \pm x$ was the error made by taking $\alpha = 1°40'$. Because the determination 1°30′ was based on nine times as many observations as 1°40′ (and was thus nine times more accurate), he supposed that $\pm x \cdot 27 = (10 \pm x) \cdot 3$. His solution to this equation makes it clear that he assumed both sides must have the same sign, for, setting $e = \pm x$, he actually solves $e \cdot 27 = (10 + e) \cdot 3$ to get $e = 30/24 = 1.25$. Thus either he misses the possibility of solving $|e| \cdot 27 = |10 + e| \cdot 3$ to get $e = -1$ (the error in taking $\alpha = 1°40'$ is nine times that of $\alpha = 1°30'$, albeit in a different direction), or he has deliberately taken the larger of the two values, to give a conservative bound to the error.

We now know that Mayer's judgment of the inverse relationship between the number of equations used and the accuracy of the determination was too optimistic; statistical accuracy at most increases only as the square root of the number of equations, subject to various assumptions on the conditions under which the observations are made. It was to be several years before that relationship emerged in the works of Laplace and, later, Gauss. Thus, rather than be surprised at Mayer's overly optimistic view of his procedure's accuracy, we should be surprised at how qualitatively correct this view was. In fact, even to attempt a numerical estimate of the accuracy of an empirical determination was remarkable for the time.

We can express Mayer's error assessment in modern notation, in the special case of determining a mean as follows: Let e be the limit of accuracy (analogous to Mayer's $\pm x$) for a mean $\overline{X}$ (analogous to his observed 1°30′), let X_1 be a single determination (analogous to his 1°40′), and let n be the ratio of the sample sizes entering into $\overline{X}$ and X_1 (analogous to his $9 = 27/3$). Then $|X_1 - \overline{X}|$ is analogous to his 10, and he would take

$$e \div (1/n) = (|X_1 - \overline{X}| + e) \div 1, \qquad \text{or} \qquad e = |X_1 - \overline{X}|/(n - 1).$$

Of course, this represents a considerable formal extrapolation of Mayer's intention, but it shows that his approach was at least qualitatively sound, even if far from the best we can do today. The point is not that he found a particularly clever method of combining his twenty-seven equations but that he found it useful to combine the equations at all, instead of, say, being content with selecting three "good" ones and solving for the unknowns from them, as he did by way of illustration. This aspect of Mayer's approach is best appreciated by comparing Mayer's work with Euler's memoir of a year earlier.

Saturn, Jupiter, and Euler

Leonhard Euler (1707–1783) was and is best known for his work in pure analysis, but he worked in nearly every area of pure and applied mathematics known at the time (or invented in the succeeding century). Euler was the most prolific mathematician of all time; his collected works now run to nearly eighty quarto volumes and are still in the process of publication. If the maxim "Publish or perish" held literally, Euler would be alive today. Yet for all this abundance, the quality of his work did not suffer; and on several occasions he was honored by foreign academies for his solutions to outstanding problems. One such instance was the prize announced by the Academy of Sciences in Paris for the year 1748, when Euler was in Berlin.

The Academy problem of 1748 concerned the second of the major scientific problems we shall consider in this chapter; entrants were invited to prepare memoirs giving "A Theory of Saturn and of Jupiter, by which one can explain the inequalities that the two planets appear to cause in each other's motion, principally near the time of their conjunction" (Euler, 1749, p. 45).

In 1676 Halley had verified an earlier suspicion of Horrocks that the motions of Jupiter and Saturn were subject to slight, apparently secular, inequalities. When the actual positions of Jupiter and Saturn were compared with the tabulated observations of many centuries, it appeared that the mean motion of Jupiter was accelerating, whereas that of Saturn was retarding. Halley was able to improve the accuracy of the tables by an empirical adjustment, and he speculated that the irregularity was somehow due to the mutual attraction of the planets. But he was unable to provide a mathematical theory that would account for this inequality.

This problem, like that of the motions of the moon, was an instance of the three-body problem. Its appearance at this time was also due to improved accuracy of astronomical observations revealing inadequacies of simple two-body theories of attraction. Unlike the problem of the moon, however, that of Jupiter and Saturn was not commercially motivated;[4]

4. The motions of Jupiter and Saturn about the sun are too slow to be useful for determining longitude at sea. In 1737 W. Whiston suggested that if reflecting telescopes were made part of a ship's navigational equipment then longitude could be determined by observing the eclipses of the moons of Jupiter. Although this method would work on land, the suggestion must have been a source of some amusement to naval astronomers who knew the instabilities of a ship's deck as an observational platform (W. R. Martin, "Navigation," in *Encyclopaedia Britannica,* 11th ed., p. 289).

rather it drew its major impetus from the philosophical implications of an unstable solar system. If the observed trends were to continue indefinitely, Jupiter would crash into the sun as Saturn receded into space! The problem posed by the Academy could be (and was) interpreted as requiring the development of an extension of existing theories of attraction to incorporate the mutual attraction of three bodies, in order to see whether such a theory could account for at least the major observed inequalities as, it was hoped, periodic in nature. Thus stability would be restored to the solar system, and Newtonian gravitational theory would have overcome another obstacle.

Euler's memoir on this difficult subject was judged the winner of the prize even though it fell far short of providing a complete resolution of the problem. This 123-page memoir, "Researches on the question of the inequalities in the movement of Saturn and of Jupiter," was published separately in Paris in 1749, and it may still be read today as a model of clear and orderly mathematical exposition. Euler focused his attention primarily on Saturn (which as the smaller of the two is subject to more pronounced perturbations) and developed an equation for the longitudinal position of Saturn that took into account the mutual attractions of the three bodies. He began by assuming that Jupiter and Saturn followed circular orbits about the sun and that the orbits lay in the same plane. Finding that the simple theory that resulted from this assumption would not admit inequalities of the size actually observed, he considered more complicated hypotheses. He first permitted Saturn's orbit to be an ellipse, he then permitted Jupiter to follow an elliptical orbit also, and he finally incorporated the fact that the planes of the two orbits are not coincident, but at a slight inclination, into the calculations.

After he had completed his mathematical analysis, it remained for Euler to check his results empirically. He wrote, "After having determined the derangements that the action of Jupiter should cause in the movement of Saturn, I now pass to an examination of the degree of precision with which they agree with the observations" (Euler, 1749, p. 111). To make this comparison, he developed a formula for the heliocentric longitude φ of Saturn in the following form (p. 121):

$$\begin{aligned}\varphi = \eta &- 23525'' \sin q + 168'' \sin 2q - 32'' \sin 2\omega \\ &- 257'' \sin(\omega - q) - 243'' \sin(2\omega - p) + m'' \\ &- x'' \sin q + y'' \sin 2q - z'' \sin(\omega - p) \\ &- u(\alpha + 360\, v + p)\cos(\omega - p) + Nn'' \\ &- 0.11405k'' \cos q + (1/600)k'' \cos 2q.\end{aligned}$$

Of this impressive array of symbols, some ($\varphi, \eta, q, \omega, p, N, v$) were given by observation and varied from observation to observation, and some ($x, y, m, z, \alpha, k, n, u$) were fixed unknown corrections whose values were not specified by the theory. A full explanation of this equation is not required here. In essence it represents the observed heliocentric longitude of Saturn (φ) as equal to Saturn's mean heliocentric longitude η (what Saturn's longitude would be if we ignored perturbations and assumed an elliptical orbit) and correction terms. These terms depend on the difference between the longitudes of the two planets (ω), the number of years since 1582 (N), and various orbital characteristics (the planet's eccentric anomalies, p and q; and the number of complete orbits by Jupiter since 1582, ν).

The problem Euler faced was this: He had available seventy-five complete sets of observations of φ, η, q, ω, p, N, and ν made in the years from 1582 through 1745. From these he first derived values of n and u in which he had confidence. It remained to determine the six unknown corrections x, y, m, z, α, and k and to check whether, when their values were substituted in the equation for φ, the values derived for the right-hand side agreed well enough with the observed values of φ to enable him to say that the theory explained the observed motions of Saturn.

The problem was an extraordinarily difficult one for the time, and Euler's attempts to grope for a solution are most revealing. Euler's work was, in comparison with Mayer's a year later, a statistical failure. After he had found values for n and u, Euler had the data needed to produce seventy-five equations, all linear in x, y, m, z, α, and k. He might have added them together in six groups and solved for the unknowns, but he attempted no such overall combination of the equations.

To see how Euler did work with his data, it is instructive to look at the way he found the first two unknowns, n and u. He noted that the coefficients of all terms except those involving n and u were, to a close approximation, periodic, with a period of fifty-nine years. He subtracted the equation for 1703 from that for 1585 (2×59 years apart) and that for 1732 from that for 1673 (fifty-nine years apart), thereby getting two linear equations in n and u alone. He solved these equations and checked his results by comparing them with another set derived from four other equations, similarly spaced in time.

Euler attempted to evaluate other correction factors by the same method, that is, by looking at small sets of equations taken under astronomically similar conditions, and thus creating a situation in which many of the coefficients would be approximately equal and the difference of two equations would annihilate most terms. But he did not succeed in finding other situations (like that for n and u) where different sets of equations gave the same results. Once he derived six inconsistent linear equations in only two unknowns but stated that, "Now, from these equations we can conclude nothing; and the reason, perhaps, is that I have tried to satisfy several observations exactly, whereas I should have only satisfied them approximately; and this error has then multiplied itself" (Euler, 1749, p.

136). Immediately after this, he presented twenty-one of the equations, involving the six unknowns other than n and u, only to throw up his hands with no real attempt at a solution. The furthest he went was to set five of the unknowns equal to zero (all except the term whose coefficient was unity in all equations, m). He adjusted this remaining term to be halfway between the largest and smallest constants in the twenty-one equations, thus making the maximum discrepancy as small as possible.

The comparison between the approaches of Euler and Mayer is dramatic. In 1750 Mayer, faced with a set of twenty-seven inconsistent equations in three unknowns, devised a sensible method of combining them into three equations and solving for the unknowns. In 1749 Euler, faced with up to seventy-five equations in up to eight unknowns, was reduced to groping for solutions. Euler worked with small sets of equations (usually as many as there were unknowns), and he only accepted numerical answers when different small sets of equations yielded essentially the same results. Euler's problem was similar to Mayer's, yet of the two only Mayer succeeded in finding a statistical solution to his "problem": a "combination of observations" that Euler could not devise (and, we shall argue, would not have accepted).

The two men brought absolutely first-rate intellects to bear on their respective problems, and both problems were in astronomy. Yet there was an essential conceptual difference in their approaches that made it impossible for Euler to adopt a statistical attitude and a subtle difference between their problems that made it extremely unlikely that Euler would overcome this conceptual barrier. The differences were these: Mayer approached his problem as a practical astronomer, dealing with observations that he himself had made under what he considered essentially similar observational conditions, despite the differing astronomical conditions. Euler, on the other hand, approached his problem as a mathematician, dealing with observations made by others over several centuries under unknown observational conditions. Mayer could regard errors or variation in his observations as random (even though no explicit probability considerations were introduced), and he could take the step of aggregating equations without fear that bad observations would contaminate good ones. In fact, he approached his problem with the conviction that a combination of observations increased the accuracy of the result in proportion to the number of equations combined. Euler could not bring himself to accept such a view with respect to his data. He distrusted the combination of equations, taking the mathematician's view that errors actually increase with aggregation rather than taking the statistician's view that random errors tend to cancel one another.

It has long been the practice of mathematicians to think in terms of the *maximum* error that could occur in a complex calculation rather than in terms of the likely error, to think in terms of absolute error bounds (which would typically increase with aggregation) rather than in terms of likely error sizes (which would not). For example, if a quantity is derived from adding together four numbers, any one of which could be in error by two units, then the sum could err by $4 \cdot 2 = 8$ units, and any attempt at an exact calculation would have to allow for ± 8 units of *possible* error. The longer the chain of calculation, the greater the maximum possible error — the more the potential error would tend to multiply. On the other hand, later statistical theory would show that under some conditions the likely error in such a sum could be much less (perhaps ± 2 units, if the likely error in one number was half the maximum possible) and the likely error in averages would actually decrease even though the mathematician's error bounds would not. Although that theory was yet to come (see Chapters 2 and 3), practicing astronomers like Mayer already had, based upon experience, at least a qualitative sense of its results in simple situations. For example, when Nevil Maskelyne wrote in 1762 that "by examining the error of the adjustment in this manner, by at least three trials, and taking a medium of the results, one can scarce err above half a minute in determining the exact error of the quadrant; whereas one may be mistaken a minute, or more, by a single trial" (Maskelyne, 1762; 1763, p. 4), it was based on his experience with his instrument, not his reading of the small amount of theory available by that time. Euler lacked that kind of direct experience. One bit of evidence supporting the idea that Euler took the more conservative mathematician's view is his previously quoted remark that the error made by supposing the equations held exactly has "multiplied itself" *(cette faute s'est ensuite augmentée)* (Euler, 1749, p. 136). An earlier statement was slightly more to the point: "By the combination of two or more equations, the errors of the observations and of the calculations can multiply themselves" (Euler, 1749, p. 135).

At one point Euler did combine two equations by averaging, but only when all corresponding coefficients were approximately equal. By subtracting three pairs of equations (and annihilating all terms except those involving x, y, z, and αu), Euler found three equations for x. He averaged the first two:[5]

$$\begin{array}{l} x = 683'' - 0.153y'' + 0.179z'' + 0.984\alpha u^\circ \\ \underline{x = 673'' - 0.153y'' + 0.187z'' + 0.983\alpha u^\circ} \\ x = 678'' - 0.153y'' + 0.183z'' + 0.983\alpha u^\circ. \end{array}$$

But the third gave

$$x = 154'' + 0.067y'' + 0.192z'' + 0.980\alpha u^\circ,$$

5. This single instance of averaging seems to be the source of the occasional mistaken attribution to Euler of Mayer's method, which was later called the "method of averages."

and he would conclude only that "the value of *y* is quite large" (Euler, 1749, pp. 130–131). At no other point did he average or add equations except to take advantage of an existing periodicity to cancel terms.

In calling attention to Euler's statistical failure, I mean to imply no criticism of Euler as a mathematical scientist. Rather by contrasting his work with Mayer's I want to highlight the extremely subtle conceptual advance that was evident in Mayer's work. Euler's memoir made a significant contribution to the mathematical theory of attraction. Even his crude empirical solution, setting most of the correction terms equal to zero, provided him with an improvement over existing tables of Saturn's motion. His inability to resolve the major inequalities in the planets' motions was in the end due to the inadequacy of his theory as well as to his lack of statistical technique. Euler himself was satisfied, correctly, that no values of his correction factors could adequately account for the planet's motions, but he suggested as a possible cause for this the failure of Newton's inverse square law of attraction to hold exactly over large distances! The problem in fact withstood successive assaults by Lagrange, Lambert, and Laplace before finally yielding to Laplace in 1787. The lesson Euler's work has for the history of statistics is that even though before 1750 mathematical astronomers were willing to average simple measurements (combining observational evidence from several days or observers into a single number), it was only after 1750 that the conceptual advance of combining observational equations (with varying coefficients for several unknowns testifying to the differing circumstances under which the observations had been made) began to appear.

To what extent was Mayer's approach a method that could be generalized and transferred to problems other than its original application? Did his contemporaries or immediate followers attempt such generalizations? Mayer himself used the approach three times in his 1750 memoir: on the twenty-seven equations for Manilius, on nine equations for the crater Dionysius, and on twelve equations for the crater Censorinus.[6] He made no attempt, however, to describe his calculation as a method that would be useful in other problems. He did not do what Legendre did, namely, abstract the method from the application where it first appeared. His work proved to be influential nonetheless.

In a widely read treatise, *Astronomie* (1771), Joseph Jérôme Lalande presented an extensive discussion of Mayer's work for the specific purpose of explaining how large numbers of observational equations could be combined to determine unknown quantities. Indeed, Lalande presented virtually the whole of Mayer's analysis of Manilius, in what amounts to an

6. We can speculate that the numbers of equations were chosen to permit an equal division into three groups, perhaps by discarding one or two equations. Mayer does not comment on this point, however.

only slightly abridged translation (Lalande, 1771, vol. 3, pp. 418–428), saying, "I report the following numbers only to serve as an example of the method that I wish to explain" (p. 419). It seems plausible that it was Lalande's exposition that called the method to Laplace's attention and that it was Laplace who first developed it into the form in which it became widely known in the nineteenth century as "Mayer's method."

Laplace's Rescue of the Solar System

Pierre Simon Laplace was born in Normandy on 23 March 1749, and his life spanned the Napoleonic era. In what most would agree was the golden age of French science, Laplace was France's most illustrious scientist. Upon his death on 5 March 1827, Poisson eulogized him as "the Newton of France," and the phrase seems apt: Laplace was Newtonian in outlook, breadth, and, at least in probability and statistics, in accomplishment. By the age of twenty his mathematical talent had won him the patronage of d'Alembert; by the end of 1773 he was a member of the Academy of Sciences. At one time or another he was professor at the Ecole Militaire (he is said to have examined, and passed, Napoleon in 1785), member of the Bureau des Longitudes, Professor at the Ecole Normale, Minister of the Interior (for six weeks, in 1799, before being displaced by Napoleon's brother), and Chancellor of the Senate. Laplace's scientific work was no less varied than his public career. His scientific memoirs constitute seven of the total of fourteen volumes of his (not quite complete) *Oeuvres complètes*. About half of them were concerned with celestial mechanics, nearly one-quarter with mathematics exclusive of probability; and the remainder were divided between probability and physics (Stigler, 1978b). He is best known in the history of science for two major treatises that were distilled from this work: the *Traité de mécanique céleste* (four volumes, 1799–1805, with a supplementary volume published in 1825) and the *Théorie analytique des probabilités* (1812). John Playfair called the first of these "the highest point to which man has yet ascended in the scale of intellectual attainment" (Playfair, 1808, pp. 277–278), Augustus De Morgan described the second as "the Mont Blanc of mathematical analysis" (De Morgan, 1837).

In 1787, in the course of a memoir on the inequalities in the motions of Saturn and Jupiter, Laplace proposed what amounts to an extension of Mayer's method of reconciling inconsistent linear equations. In this epochal work, Laplace finally laid to rest what was by then a century-old problem by showing that the inequalities were in fact periodic (with a very long period). In the course of his demonstration Laplace was confronted with equations of the type that had stalled Euler's drive toward a solution.

Laplace's success in this celebrated problem was in considerable part a statistical triumph, a model of the ways in which analyses of data suggested by theory may in turn suggest hypotheses requiring further theoretical development and then observational confirmation. In 1773 Lambert had deduced, on the basis of an empirical investigation of contemporary observations, that the retardation in Saturn's motion noticed by Halley had apparently reversed—Saturn was accelerating and apparently had been doing so at least since 1640. This observation suggested (but did not prove) that the inequality was periodic rather than secular as Halley (and even Euler and Lagrange in early work) had thought. Laplace sought to account for the motions within the constraints of Newtonian gravitational theory, to within the limits of observational accuracy. The problem was an exceedingly difficult one. Even if the inequality was periodic, which of the known planets or moons would he need to include in the theory to obtain satisfactory agreement? Even if only two planets were needed, which of the many ways of developing the equations of motion would both be tractable and permit the desired empirical check?

Building upon his own and Lagrange's earlier work on planetary motions, Laplace succeeded in first proving that a remarkably simple conservation property held for the eccentricity of planetary orbits. This property implied in particular that, given the known ratio of the masses of Jupiter and Saturn, the ratio of the maximum retardation of the mean motion of Saturn to the maximum acceleration of the mean motion of Jupiter should be nearly in the ratio of 7 to 3—if, in fact, only the mutual attractions of these two planets and the sun needed to be taken into account. Laplace found that two quantities in the correct ratio, namely, 9°16′ and 3°58′, differed from the largest values given in Halley's tables by only 9′. Encouraged by this close agreement, he embarked upon the arduous mathematical development of a theoretical formula for Saturn's motion.

It had long been known that the average annual mean motions of Jupiter (now known to be $n = 30^\circ\!.349043$) and of Saturn (now known to be $m = 12^\circ\!.221133$) were approximately in the ratio of 5 to 2; in fact $5m - 2n = 0.40758$. In *his* treatment of this problem, Euler had curtailed his expansion of the longitude of Saturn, omitting terms that contributed less than 30″ each on the principle that some of the observations he would be using could only be counted as accurate to 1′ (Euler, 1749, p. 118). Euler, as we have seen, seems not to have realized that unknown quantities could be determined to greater precision than that of the individual observations, and he was here oblivious to the possible cumulative effect of such terms. Laplace, on the other hand, followed a different tack. Although $5m - 2n$ is a very small quantity (only about 1/74 of Jupiter's mean annual motion), Laplace focused his attention on terms involving this difference because work of Lagrange had shown that such near commensurability of planetary motions could produce significant inequalities. Laplace noticed, first, that the periodic inequality due to the planets' motions corresponding to terms involving $5m - 2n$ would have a period of about 900 years (about the right order of magnitude to explain the observed inequalities) and, second, that whereas the coefficients of these terms are very small in the differential equations of the planets' motions, they become potentially significant after the successive integrations needed to derive formulas for the planets' longitudes.

Encouraged further by a sense of empirical agreement between these theoretical observations and the known characteristics of the yet unexplained inequality, Laplace undertook to develop a formula for Saturn's longitude, in effect out of Euler's discarded scraps—out of a selection of the terms Euler had omitted as neither measurable nor likely to be important. The results of this intricate analysis were assembled into a 127-page memoir, "Théorie de Jupiter et de Saturne," printed separately by the Academy of Sciences in 1787 and reprinted the following year as part of the Academy's *Mémoires* for the year 1785.

The capstone of Laplace's investigation was his comparison of his theory with observations. He made use of the best available data on the "elements" of the planets' motions, but his theory required four quantities that were not readily available with sufficient accuracy for his purposes; he had to determine them from the observations themselves. These were, in Laplace's notation $\delta\epsilon^I$, δn^I, δe^I, and $\delta\tilde{\omega}^I$; they denoted necessary "corrections" (rates of change) for, respectively, the mean longitude of Saturn in 1750, its mean annual motion, its eccentricity, and the position of its aphelion (its most distant position from the sun). Laplace selected twenty-four observations of Saturn, made at times of opposition (when the sun, earth, and Saturn were aligned) over a 200-year period as being particularly likely to be accurately made. In each case he expressed the difference between the observed longitude of Saturn and that given by his theory as an "equation of condition." For example, the equation for the year 1672 was given as

$$0 = -3'32.8'' + \delta\epsilon^I - 77.28\delta n^I - 2\delta e^I\, 0.98890$$
$$- 2e^I(\delta\tilde{\omega}^I - \delta\epsilon^I)0.14858.$$

(The eccentricity e^I was considered as known and was given elsewhere in the memoir). The entire data set is given in Table 1.3, in which $-a_i$ stands for the first term of the ith equation ($a_i = -3'32.8''$ in the equation for the year 1672), and b_i, c_i, d_i are the coefficients of the unknowns δn^I, $2\delta e^I$, and $2e^I(\delta\tilde{\omega}^I - \delta\epsilon^I)$. It should be noted that the coefficients c_i and d_i are strongly related, as were the coefficients in Mayer's investigation, by $c_i^2 + d_i^2 = 1$. In fact, $c_i = -\sin(\varphi_i - \omega_i)$ and $d_i = \cos(\varphi_i - \omega_i)$, where $\varphi_i - \omega_i$ is the difference between Saturn's observed longitude and its aphelion in the ith year; b_i is the number of years from the beginning of 1750 to the time the observation was made.

Table 1.3. Laplace's Saturn data.

Eq. no.	Year (i)	$-a_i$	b_i	c_i	d_i	Laplace residual	Halley residual	L.S. residual
1	1591	1′11.9″	−158.0	0.22041	−0.97541	+1′33″	−0′54″	+1′36″
2	1598	3′32.7″	−151.78	0.99974	−0.02278	−0.07	+0.37	+0.05
3	1660	5′12.0″	−89.67	0.79735	0.60352	−1.36	+2.58	−1.21
4	1664	3′56.7″	−85.54	0.04241	0.99910	−0.35	+3.20	−0.29
5	1667	3′31.7″	−82.45	−0.57924	0.81516	−0.21	+3.50	−0.33
6	1672	3′32.8″	−77.28	−0.98890	−0.14858	−0.58	+3.25	−1.06
7	1679	3′9.9″	−70.01	0.12591	−0.99204	−0.14	−1.57	−0.08
8	1687	4′49.2″	−62.79	0.99476	0.10222	−1.09	−4.54	−0.52
9	1690	3′26.8″	−59.66	0.72246	0.69141	+0.25	−7.59	+0.29
10	1694	2′4.9″	−55.52	−0.07303	0.99733	+1.29	−9.00	+1.23
11	1697	2′37.4″	−52.43	−0.66945	0.74285	+0.25	−9.35	+0.22
12	1701	2′41.2″	−48.29	−0.99902	−0.04435	+0.01	−8.00	−0.07
13	1731	3′31.4″	−18.27	−0.98712	−0.15998	−0.47	−4.50	−0.53
14	1738	4′9.5″	−11.01	0.13759	−0.99049	−1.02	−7.49	−0.56
15	1746	4′58.3″	−3.75	0.99348	0.11401	−1.07	−4.21	−0.50
16	1749	4′3.8″	−0.65	0.71410	0.70004	−0.12	−8.38	+0.03
17	1753	1′58.2″	3.48	−0.08518	0.99637	+1.54	−13.39	+1.41
18	1756	1′35.2″	6.58	−0.67859	0.73452	+1.37	−17.27	+1.35
19	1760	3′14.0″	10.72	−0.99838	−0.05691	−0.23	−22.17	−0.29
20	1767	1′40.2″	17.98	0.03403	−0.99942	+1.29	−13.12	+1.34
21	1775	3′46.0″	25.23	0.99994	0.01065	+0.19	+2.12	+0.26
22	1778	4′32.9″	28.33	0.78255	0.62559	−0.34	+1.21	−0.19
23	1782	4′4.4″	32.46	0.01794	0.99984	−0.23	−5.18	−0.15
24	1785	4′17.6″	35.56	−0.59930	0.80053	−0.56	−12.07	−0.57

Source: Laplace (1788).
Note: Residuals are fitted values minus observed values.

Laplace was faced with twenty-four inconsistent equations of condition, each linear in the four unknowns. It was a situation similar to one that Euler had failed to resolve and to one that Mayer had resolved by breaking the equations into disjoint groups and adding each group together. Laplace dealt with the problem by using a method that bears a superficial resemblance to Mayer's approach; yet it differed from that of Mayer in one subtle respect that marks it as an important advance toward least squares.

Laplace did not provide an algebraic description of his solution, but he did give a detailed description of the steps he followed. What he did, he explained, was to reduce his twenty-four linear equations to four equations: (i) the sum of equations 1 − 24; (ii) the difference between the sum of equations 1 – 12 and the sum of equations 13 – 24; (iii) the linear combination of equations: $-1 + 3 + 4 - 7 + 10 + 11 - 14 + 17 + 18 - 20 + 23 + 24$; and (iv) the linear combination of equations: $+2 - 5 - 6 + 8 + 9 - 12 - 13 + 15 + 16 - 19 + 21 + 22$. He then solved equations (i) – (iv) and checked the degree to which the resulting equations fit the observations by computing each residual, defined as the "excess" of the fitted value over the observed value (the negative of the modern definition of residual).

Laplace did not explain his selection of four linear combinations, but he seems to have based it upon their effect upon the coefficients of the unknowns in equations (i) – (iv). Thus (i) and (ii) are natural linear combinations to consider: (i) maximizes the coefficient of the constant term, whereas (ii) eliminates it. Nearly the reverse is true for the coefficient of δn^I. Evidently, the choice of which equations were included in (iii) and which in (iv) was made according to whether $|c_i| < |d_i|$ or $|c_i| > |d_i|$. The one exception to this rule is the reversal of it with respect to equations 3 and 5, a minor exception that may have been made to reduce the coefficient of the constant term in (iv) by 2. Once the twenty-four equations were divided between (iii) and (iv), the signs + and − were chosen according to the signs of d_i [for equation (iii)] and c_i [for equation (iv)], thus nearly maximizing the contrast between the coefficients of the last two unknowns.

The subtle advance Laplace had made was this: Where Mayer had only added his equations of condition together within *disjoint* groups, Laplace combined the *same* equations together in several different ways. The relationship between the methods of Mayer, Laplace, and Legendre and the importance of Laplace's advance can be best understood if we interpret them all in terms of a uniform notation. If we write an equation of condition involving four unknowns including a constant term as

$$0 = a_i + w + b_i x + c_i y + d_i z, \qquad i = 1, \ldots, n,$$

where a_i, b_i, c_i, and d_i are observable and w, x, y, and z are unknown, then we might write the jth aggregated equation, $j = 1, 2, 3, 4$, as

$$0 = \sum_i k_{ij}\, a_i + \sum_i k_{ij}\, w + \sum_i k_{ij}\, b_i x + \sum_i k_{ij}\, c_i y + \sum_i k_{ij}\, d_i z,$$

where $\{k_{ij}\}$ form a system of multipliers.

All three of the schemes we have discussed fit this description, although it should be borne in mind that viewing them in terms of this notation is an anachronism and that this unifying view was not to appear until later, in Laplace's 1812 work on this subject.

Mayer treated the case where $n = 27$ and $d_i = 0$ for all i (that is, only three unknowns appeared) and determined the multipliers $\{k_{ij}\}$ for ($j = 1, 2, 3$) from the coefficient of the first unknown with a nonconstant coefficient, that is, from the b_i's. If we let $\underline{b} < \bar{b}$ be the ninth largest and ninth smallest b_i, respectively, then Mayer effectively took

$$k_{i1} = \begin{cases} 1 & \text{if } b_i \leq \underline{b} \\ 0 & \text{otherwise} \end{cases}$$

$$k_{i2} = \begin{cases} 1 & \text{if } b_i \geq \bar{b} \\ 0 & \text{otherwise} \end{cases}$$

$$k_{i3} = \begin{cases} 1 & \text{if } \underline{b} < b_i < \bar{b} \\ 0 & \text{otherwise.} \end{cases}$$

Thus each equation of condition influences only one aggregated equation, and the question of which one is influenced was determined by the coefficient of a single unknown. Legendre's "method of least squares" went to another extreme, taking

$$k_{i1} = 1, \qquad k_{i2} = b_i, \qquad k_{i3} = c_i,$$

and, if the fourth unknown is present,

$$k_{i4} = d_i.$$

Here, unless a coefficient of an unknown is zero, all equations of condition influence all aggregated equations and the coefficients of all unknowns are important in the aggregation. Laplace's approach took a middle ground between these approaches and amounted to a "rounded off" version of the least squares aggregation, where the multipliers $\{k_{ij}\}$ are only allowed to take the values $-1, 0, +1$. If we let $B = \text{median}\,(b_i)$ and note that since in Laplace's case $c_i^2 + d_i^2 = 1$ (so $|c_i| > |d_i|$ if and only if $|c_i| > 2^{-1/2} \cong 0.707$), then (with the single minor exception noted earlier) Laplace's aggregation was

$$k_{i1} = 1 \quad \text{all } i$$

$$k_{i2} = \begin{cases} -1 & \text{if } b_i < B \\ 1 & \text{if } b_i > B \end{cases}$$

$$k_{i3} = \begin{cases} -1 & \text{if } -1 \leq c_i < -0.707 \\ 0 & \text{if } -0.707 < c_i < 0.707 \\ 1 & \text{if } 0.707 < c_i \leq 1 \end{cases}$$

$$k_{i4} = \begin{cases} -1 & \text{if } -1 \leq d_i < -0.707 \\ 0 & \text{if } -0.707 < d_i < 0.707 \\ 1 & \text{if } 0.707 < d_i \leq 1. \end{cases}$$

Now, the point of this comparison is not to argue that Laplace almost arrived at the method of least squares. In the first place, even describing his method in this algebraic form is a considerable formal extrapolation of what he actually did, which was to present a single worked example. In the second place, Legendre's method, unlike Laplace's, was formally derived from an explicitly stated criterion of best fit. Rather the point is that Laplace had moved forward from Mayer in treating the data set as a whole and in allowing the values of all the coefficients of the unknowns to influence the aggregation. Mayer had broken his data set into disjoint sets which, judged by the values of the coefficient of one unknown, were made under similar circumstances. In this he went beyond Euler, who would have insisted that the equations be alike in all coefficients before he would combine them. Mayer nevertheless clung to the older tradition in that he treated observations made under very different conditions separately, at least until the final stage of his analysis. Laplace went further by combining all the observational equations in the very first stage of his analysis, and more important, by letting all coefficients influence the manner of combination.

Mayer had focused on the coefficient of only one unknown. Because it was indeed the most important unknown in his application,[7] he arrived at what was, for his situation, a good solution. But if we consider his approach as a method to be applied in other situations, it has the serious shortcoming of requiring that there be a single important unknown, that the investigator be able to tell which unknown is important, and that the coefficients of the other unknowns be distributed so as not to confound the solution by, for example, producing a nearly singular set of aggregated equations. Mayer was successful because he knew what he was doing in his particular application, but another investigator mimicking his procedure in another application might not be so lucky. Laplace's generalization of this approach, on the other hand, did not suffer from this drawback. All unknowns could be determined by his method with reasonable accuracy—at least when there was sufficient information in the original equations of condition to determine them accurately by any method. Another investigator imitating Laplace's method would not require the same degree of "good luck" required by a follower of Mayer and usually would be rewarded by greater accuracy as well.

The criterion minimized by the method of least squares—the sum of squared residuals—is not an unexceptionable measure of the success of a fit, but it provides one useful way of assessing these early efforts. When we compare the square root of the sum of the squares of Mayer's residuals to the square root of the residual sum of squares given by least squares, we

7. In fact, because of the small size of θ (and thus of $\alpha \sin \theta$) and the relatively large measurement error in determining $g - k$ (caused by the small angle between the moon's orbit and the ecliptic and the consequent difficulty of estimating the node F in Figure 1.2), the term involving $\alpha \sin \theta$ does not contribute significantly to the regression. Interestingly Mayer was aware of this measurement problem and did not consider his determination of θ to be reliable.

find that for the Manilius data Mayer's method gives a value only 6 percent larger, for the Dionysius data, 5.7 percent larger, and for the Censorinus data, 66 percent larger. For Laplace's Saturn data, the same comparison has Laplace's method with a value only 5.5 percent larger than that for least squares; whereas when Halley's 1719 empirical adjustment is extrapolated to 1786, the square root of the sum of the twenty-four corresponding squared residuals is 90 percent larger than Laplace's. Note (Table 1.3) that the pattern of either Laplace's or the least squares residuals hints correctly that not all of the periodic inequalities in Saturn's motion had been accounted for, even though the most controversial one had been.

Laplace had come quite a bit closer to providing a general method than Mayer had. Indeed, it was Laplace's generalization that enjoyed popularity throughout the first half of the nineteenth century, as a method that provided some of the accuracy expected from least squares, with much less labor. Because the multipliers were all -1, 0, or 1, no multiplication, only addition, was required.

Among the writers to present Laplace's version of the method as an alternative to least squares was Mary Somerville, who described it in her 1831 book *Mechanism of the Heavens* and attributed it to Mayer. In telling how a set of equations of condition of the form

$$\text{Error} = \epsilon + 0.''3133P + 0.''2969e,$$

but with varying coefficients, could be combined to solve for three unknowns P, e, and ϵ, Somerville wrote:

> For example, in finding the value of P before the other two, the numerous equations must be so combined, as to render the coefficient of P as great as possible; and the coefficients of e and ϵ as small as may be; this may always be accomplished by changing the signs of all the equations, so as to have the terms containing P positive, and then adding them; for some of the other terms will be positive, and some negative, as they may chance to be; therefore the sum of their coefficients will be less than that of P.
>
> Having determined this equation, in which P has the greatest coefficient possible, two others must be formed on the same principle, in which the coefficients of the other two errors must be respectively as great as possible, and from these three equations values of the three errors will be easily obtained, and their accuracy will be in proportion to the number of observations employed. (Somerville, 1831, p. 409)

In this description, and in our abstraction of Laplace's extension of Mayer, there is an implicit assumption that the signs of the coefficients change, as would be the case if they were centered at their means. In fact, in these early applications this was essentially the case because the unknowns represented corrections to a mean. We note in addition that Mary Somerville follows Mayer in claiming that accuracy is proportional to the number of observations rather than to the square root of the number of observations.

Other writers to mention Laplace's version of the method, crediting it to Mayer, included Francoeur (1830; 1840, p. 432), Bowditch (1832, p. 485), Puissant (1842, vol. 2, p. 344; Puissant does not mention the method in the first edition, 1805), Wolf (1869–1872, vol. 1, p. 279; vol. 2, p. 199), and Whittaker and Robinson (1924, pp. 258–259). After about 1850 the references to the method were of a historical rather than a practical nature, and they describe the method in terms of Mayer's original formulation, with disjoint groups being added.

Roger Boscovich and the Figure of the Earth

Thus far we have seen how problems involving Jupiter, Saturn, and the moon led to the introduction and development of a method of combining inconsistent linear equations in the half-century before the appearance of the method of least squares. "Mayer's method," as it was known, was easy to use and generally led to sensible results, two properties that conspired to keep it as an actively employed tool in astronomers' and geodesists' workshops for a half-century after least squares was introduced. But Mayer's method lacked one quality that the method of least squares was found to have in abundance; and it was this quality that contributed to the eventual eclipse of Mayer's method by the method of least squares. Mayer's method, unlike least squares, was not "best" in the sense that it appeared as the solution to a mathematically posed problem of finding the "best" combination of inconsistent equations. It was an ad hoc method, and its acceptance depended upon its reputation for past successful use, its ease of application, and the investigator's intuitive feeling that by combining the equations in such a way that the coefficients of the unknowns are successively maximized, a mechanically stable (and hence reliable) solution would result. The first of these—a reputation for successful use—was widely believed. Mayer's tables of the moon's motion and his map of the face of the moon were commonly and justly seen as among the most accurate such achievements of eighteenth-century observational astronomy. Even those who knew that the development of Mayer's method into a more general tool was first found in Laplace's work would not be dissuaded from its use—who could not feel confident using the method that had reconciled the motions of Jupiter and Saturn with Newtonian gravitational theory? As the years went by, the vivid impression of these past triumphs faded, however, to be replaced by stories of the triumphs of least squares. Then, as more experience and ways of simplifying computation made least squares easier to use and as succeeding generations of mathematicians made no successful attempt to give formal statement to the vague intuitive notions of the reasonableness of Mayer's method, it faded from view—or rather moved from the workshop of the practitioner to the display case of the statistical museum.

Least squares was the most successful of the early methods of combining inconsistent equations, and the fact that it was based on and derived from an easily understood objective criterion was a major reason for its success. Nevertheless, least squares missed being the first method to be so based by nearly half a century. To trace the genesis and fate of its most famous predecessor, "Boscovich's method," we shall first consider a third major eighteenth-century scientific problem—the problem of the figure of the earth.

The first post-Columbian hint that the earth was not a perfect sphere seems to have been the discovery by Richer in 1672 that a pendulum near the equator was less affected by gravitational attraction than was the same pendulum at Paris. Newton, in the *Principia* (1687), showed how the rotation of the earth could be expected to produce a flattening of the earth at the poles and a bulging at the equator, a shape known as an oblate spheroid. Newton estimated the oblateness or ellipticity (the fraction by which a radius at the equator exceeds the radius at the pole) to be 1/230. This conclusion about the shape or figure of the earth did not go unchallenged, however. Domenico Cassini, director of the Royal Observatory in Paris, thought in fact that the earth was a prolate spheroid, flattened at the equator, not the poles. Several attempts were made over the succeeding century both to determine whether the earth was oblate or prolate and to measure the departure of its shape from spherical.

The two principal methods of determining the figure of the earth were pendulum experiments and arc measurements. We shall be concerned primarily with arc measurements in this chapter, although the statistical problems that arise in the two cases have striking similarities. The determination of the earth's figure from arc measurements required the cooperation of a team of scientists and months of labor under adverse circumstances; it was the perfect sort of challenge for the growing and increasingly adventuresome French scientific community. The idea was to measure the linear length of a degree of latitude at two (or more) widely separated latitudes. If a degree near the equator is found to be shorter than one nearer the pole, then the shape of the earth is oblate; and the difference between the two measurements can be used to calculate the oblateness.

At first thought it may seem paradoxical that a shorter degree near the equator would indicate a *bulging* at the equator, but only because there is a common misapprehension of the definition of a degree of latitude. The latitude of a point on the earth's surface is not, as might be supposed, the angle formed between two rays from the center of the earth, one to the given point and the other at the intersection of the equatorial plane and the point's meridian plane. Indeed, the operational difficulties in taking such a measurement would tax the resources of even the largest geodetical organization. Latitude in fact is measured as the angle between a ray to the zenith of the given point and the equatorial plane or, alternatively, as the complement of the angle between rays from the given point to the zenith and to the Pole Star. Figure 1.3 shows arcs of 10° latitude for an exaggeratedly oblate earth.

The relationship between arc length and latitude can be derived from the geometry of conic sections, the exact relation being given by an elliptic

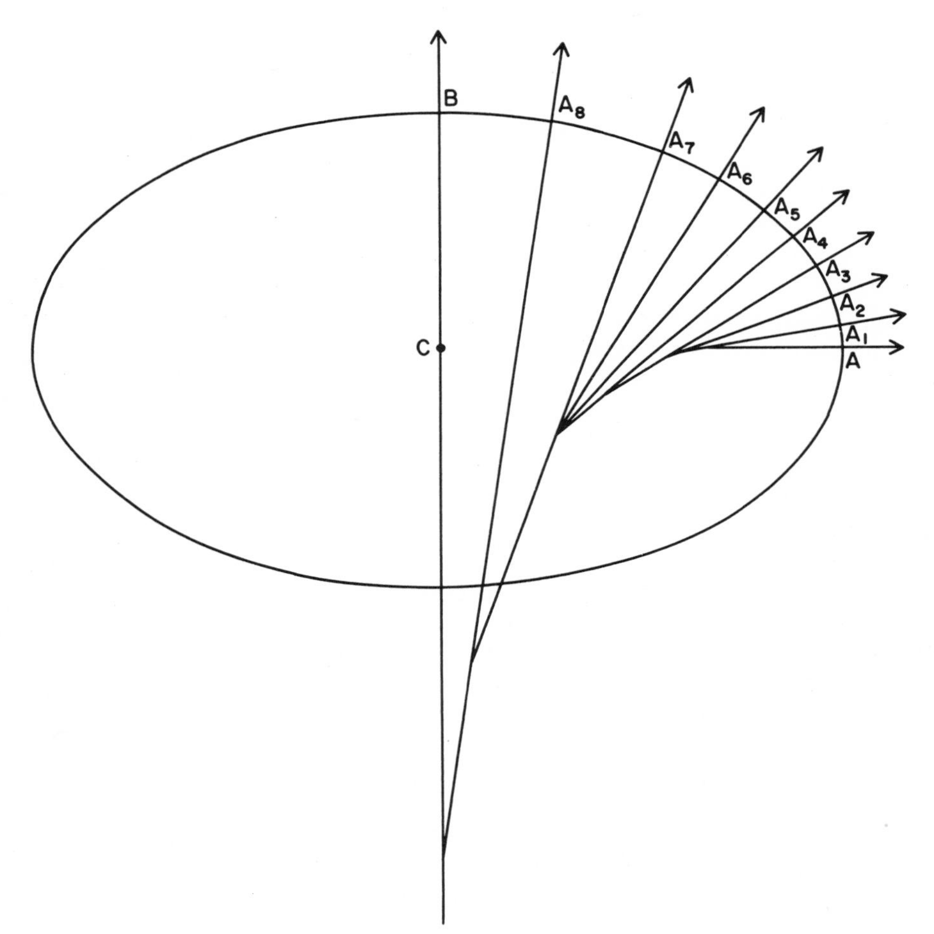

Figure 1.3. A side view of an exaggeratedly oblate earth, illustrating the lengthening of degrees of arc toward the pole. The meridian quadrant AB is broken into nine segments, each of 10° latitude. (Based upon Berry, 1898, p. 277.)

integral. For short arcs, however (the only ones it was practical to measure), a simple approximation will do: If a is the length of 1° of latitude centered at latitude θ, measured along a meridian, then to a good approximation,

$$a = z + y \sin^2 \theta,$$

where z is the length of a degree at the equator, and y is the excess (or deficiency) of a degree at the North Pole over one at the equator. Some early works expressed this as

$$a = z + y \cdot \frac{1}{2} \cdot \text{versed sine } 2\theta,$$

where "versed sine" $2\theta = 1 - \cos 2\theta$. Then the identity $1 - \cos 2\theta = 2 \sin^2 \theta$ would save the bother of squaring $\sin \theta$ in any calculation.

Measurements of a French arc made by Domenico Cassini and his son and successor Jacques before 1720 supported the hypothesis that the earth was prolate, but the narrow range of latitude (9°) and the possibility of a low accuracy in the measurements prevented this anti-Newtonian conclusion from gaining wide acceptance. In 1735 the French Academy launched expeditions by Bouguer to Peru and by Maupertuis to Lapland to measure arcs near the equator and at about 66° latitude for comparison with measurements near Paris. The results effectively refuted the Cassini hypothesis and settled the matter in Newton's favor: The earth was oblate. The only question remaining was the size of the oblateness or ellipticity, because different pairs of arcs gave different values.

In 1755 the results of measuring a length of a meridian arc near Rome were published by the English Jesuit Christopher Maire and the Dalmatian Jesuit Roger Joseph Boscovich (or Rudjer J. Bŏsković) under the title *De Litteraria Expeditione per Pontificiam ditionem ad dimetiendas duas Meridiani gradus.* In successive analyses by Boscovich of these data we find the first successful resolution of the inconsistency of the different arc measurements and the introduction of the statistical procedure that is our immediate concern.

When Boscovich first addressed this problem in 1755—in a chapter of his joint work with Maire for which he took sole responsibility—he met with only limited success. Boscovich was aware, as others before him had been, that to obtain an accurate determination of the figure of the earth it would be necessary to compare measurements widely separated in latitude, as even small errors made in proximate arc measurements would be greatly exaggerated in any pairwise combination of them. Boscovich thus focused his attention on only five determinations that were made at well-separated locations and were likely to be accurate (Table 1.4). Boscovich gave no analytic description of his handling of these data; here as else-

Table 1.4. Boscovich's data on meridian arcs.

Location	Latitude (θ)	Arc length (toises)	Boscovich's $\sin^2\theta \times 10^4$
(1) Quito	0°0′	56,751	0
(2) Cape of Good Hope	33°18′	57,037	2,987
(3) Rome	42°59′	56,979	4,648
(4) Paris	49°23′	57,074	5,762
(5) Lapland	66°19′	57,422	8,386

Source: Boscovich and Maire (1755, p. 500). Reprinted in Boscovich and Maire (1770, p. 482).

Note: Arc lengths are given as toises per degree measured, where 1 toise ≅ 6.39 feet. The value for $\sin^2\theta \times 10^4$ for the Cape of Good Hope is erroneous and is evidently based on 33°8′. The correct figure would be 3,014.

where he followed in a Newtonian tradition of giving geometric descriptions rather than analytic ones.[8] It will be easier, however, to relate Boscovich's different efforts to later work if we adopt an analytic formulation from the beginning. In analytic terms, Boscovich was faced with the equivalent of five observational equations,

$$a_i = z + y \sin^2 \theta_i,$$

where a_i and θ_i are the length of an arc (in toise per degree, 1 toise ≅ 6.39 feet) and the latitude of the midpoint of the arc, both at location i. The unknowns y and z are, respectively, the excess of a 1° arc at the pole over one at the equator and the length of a degree at the equator.

In principle any two of the five locations could be used to solve for the polar excess y and the equatorial degree z or, equivalently, for the polar

8. Isaac Todhunter is among those who have experienced frustration because Boscovich retained the cumbersome geometric apparatus of Newton instead of using the more elegantly concise analytic formulations of Clairaut and Euler. Commenting on one of Boscovich's proofs, Todhunter wrote, "Boscovich professes to use Geometry alone: but the Geometry consists chiefly in denoting the length of every straight line by two capital letters instead of a single small letter: this strange notion of Geometry has survived to our own times in the University of Cambridge" (Todhunter 1873, vol. 1, p. 309). Todhunter later took another opportunity to link sarcastically Boscovich and his own University of Cambridge, "In forming an estimate of the treatise we must remember that the author had prescribed to himself the condition of supplying *geometrical* investigations; so the Differential Calculus was not to be introduced. We must consider the treatise rather as the work of a professor for the purposes of instruction than of an investigator for the advancement of science; and then we may award the praise that the task proposed is fairly accomplished. It would have been more desirable to study Clairaut's work than to be confined to Boscovich's geometrical methods: but the experience of our own university shews us that it is possible to find the methods used for teaching occasionally some years in arrear of those used for investigation" (Todhunter, 1873, vol. 1, p. 319).

excess y and the ellipticity (Boscovich computed[9] it here as 1/ellipticity $= 3z/y$). And in fact this is exactly what Boscovich did: He calculated y and the ellipticity based upon each of the $\binom{5}{2} = 10$ pairs and presented the results shown in Table 1.5. This gave him not one, but ten solutions to his problem, and in 1755 he showed himself not quite able to deal with this embarrassment of riches. He did make a weak attempt to combine these findings: He averaged the ten values of the excess[10] and found, using the equatorial degree at Quito ($z = 56{,}751$), the value 1/155 for the ellipticity.

This value must have seemed too large, for he then recomputed the ellipticity after rejecting the pairs (2, 4) and (2, 3) as "so different from the others," possibly because of the close proximity of their degrees of latitude. The mean excess based on the remaining pairs gave, with the Quito degree, an ellipticity of 1/198, but this still seemed unsatisfactory to Boscovich. Instead of accepting either of these figures as a compromise, as an average determination of the ellipticity, Boscovich focused on the discrepancy between this average value and the ten (or eight) components that had made up the average, taking what appeared to him to be large discrepancies as evidence against an ellipsoidal shape for the earth.

> Thus it is evident that the determinations of these degrees cannot be reconciled with the ellipse of Newton, nor with any other ellipse, either more or less oblate. Five degrees, taken arbitrarily, must always give the same ellipse, and we have seen what little agreement there is between those we have chosen. The differences between them are not proportional to the versed sine of double the latitude [that is, versed sine $2\theta \equiv 2 \sin^2 \theta$]. If they were, each combination of degrees, as we have said, should give the same ellipticity. (Boscovich and Maire, 1755, p. 501; 1770, p. 484)

A modern geodesist would not quarrel with Boscovich's rejection of an ellipsoidal hypothesis, but in the context of his own time he was wrong. If the likely size of measurement errors, even as perceived by Boscovich and his contemporaries, is taken into account and the observational evidence combined in a reasonable way, then Boscovich's data is not wholly incon-

9. A slightly better local approximation would be one Laplace used later, namely 1/ellipticity $= 3z/y + 5/3$, but the difference between the two formulas is negligible in the present application. Other workers used 1/ellipticity $= 3z/y + 3/2$, or $= 3z/y + 2$. All workers of the period wrote the ellipticity in reciprocal form (for example, 1/230), even when it was first calculated in decimal form. We would describe this practice as reparametrization.

10. Actually the text (Boscovich and Maire, 1755, p. 501; 1770, p. 484) gives the average excess as 222 (just one-third of the correct value) even though the ellipticities were correctly calculated. Evidently Boscovich inadvertently inserted the average of the ten values of $y/3$ into the text, a number he would perhaps have found as an intermediate step toward calculating 1/ellipticity $= z/(y/3)$.

Table 1.5. Boscovich's pairwise solutions, based on the data of Table 1.4, for the polar excess y (the amount by which a degree at the pole exceeds a degree at the equator) and the ellipticity (found from the formula $1/\text{ellipticity} = 3z/y$, where z is the length of a degree at the equator as found from the pair of equations).

Pair	Polar excess (y, in toises)	Ellipticity	Pair	Polar excess (y, in toises)	Ellipticity
1, 5	800	1/213	2, 4	133	1/128
2, 5	713	1/239	3, 4	853	1/200
3, 5	1,185	1/144	1, 3	491	1/347
4, 5	1,327	1/128	2, 3	−350	−1/486
1, 4	542	1/314	1, 2	957	1/78

Source: Boscovich and Maire (1755, p. 501). Reprinted in Boscovich and Maire (1770, p. 483).

Note: The ellipticities for pairs (2, 4) and (1, 2) were evidently misprinted in the original; they should be 1/1282 and 1/178. The figures for the pair (1, 4) are erroneous; they should be 560 and 1/304.

sistent with an ellipsoidal hypothesis. The arc at Paris was widely seen as the most accurate of those measured before 1755, and even it was susceptible to large changes whenever it was carefully rescrutinized: From 1738 to 1740, estimates of the Paris degree changed from Picard's original (1671) figure of 57,060 toises to 56,926 toises (Maupertuis in 1738) to 57,183 toises (Maupertuis in 1740) to 57,074 toises (Cassini de Thury in 1740), see Todhunter (1873, vol. 1, p. 127). These successive changes of over 100 toises in the most-studied arc would not justify confidence in much greater accuracy than 100 toises, although individual investigators evidently had higher opinions of their own arcs. Even forty years later, Laplace (1799, vol. 2, p. 448) felt that an error of 97.20 toises "is exactly on the least limit of those which might be considered as possible." If Boscovich's data (Table 1.4) is fit by least squares, the residuals are 13, 83, −95, −80, 78. A slightly better fit is achieved if Boscovich's error in the $\sin^2 \theta$ for the Cape is corrected, that is, 15, 82, −94, −80, 78. The accuracy of the Lapland arc is discussed by Todhunter (1879).

Boscovich himself must have felt some uneasiness at his own conclusion, for he did not let the matter rest with his 1755 analysis. Two years later he published a synopsis of the 1755 volume that included a brief statement of a radically new principle for the combination of inconsistent arc measurements. And in 1760, in a prose supplement to a versified treatise on natural philosophy by Benedict Stay, Boscovich gave a full description of his principle, an explanation of how it could be used in practice and a worked example based on the five degrees he had considered in 1755. This 1760 version was later translated into French and appended to Boscovich and Maire (1770) as a part of a note[11] (pp. 501–510).

The principal novelty in Boscovich's approach was its novelty of principle. Where Mayer had proceeded ad hoc, his underlying motivation remaining unformulated, Boscovich began with a generalizable principle, a list of properties that the mean based on a combination of arc measurements should have; and he went on to derive an ingenious geometric algorithm that would find such a mean. He introduced his principle as follows (the italics are Boscovich's):

> The mean we will take will not be a simple arithmetic mean, rather it will be one tied by a certain law to the rules of fortuitous combination and the calculus of probabilities. We are faced here with a problem I have discussed toward the end of a Dissertation inserted in the proceedings of the Institute of *Bologna, volume 4* [that is, Boscovich's 1757 summary], where I contented myself with giving the result of its solution. Here is the problem: *Being given a certain number of degrees, find the correction that must be made to each of them, supposing these three conditions are complied with: the first, that their differences shall be proportional to the differences between the versed sines of twice their latitudes; the second, that the sum of the positive corrections shall be equal to the sum of the negative ones; the third, that the sum of all the corrections, positive as well as negative, shall be the least possible, for the case where the first two conditions will be fulfilled.* The first condition is called for by the law of equilibrium, which requires an elliptical shape; the second, from the fact that deviations of a pendulum, or errors by observers, that augment or diminish degrees have the same degree of probability; the third is necessary in order to approximate the observations as closely as possible, for, as we have observed above, it is clearly very probable that the deviations are quite small, because the scrupulous exactitude of the observers would not permit any suspicion of large errors in their observations. (Boscovich, 1760, as translated from the French of Boscovich and Maire, 1770, p. 501)

If we introduce the symbols $a_1, a_2, \ldots$ for the measured lengths, in toises per degree, of the measured arcs at latitudes $\theta_1, \theta_2, \ldots$, and let δa_i stand for the "correction" that Boscovich would make to the degree a_i, then his first condition is that the corrected degrees, $a_i + \delta a_i$, satisfy

$$a_i + \delta a_i - (a_j + \delta a_j) \propto \text{versed sine } 2\theta_i - \text{versed sine } 2\theta_j.$$

Since versed sine $2\theta = 2 \sin^2 \theta$, this is equivalent to supposing

$$a_i + \delta a_i = z + y \sin^2 \theta_i,$$

11. The 1770 translation also included an additional worked example based on nine arc measurements. Boscovich's original manuscript copy of this additional material is preserved in the Boscovich Archives at the Bancroft Library of the University of California, Berkeley (folder 28).

or

$$\delta a_i = z + y \sin^2 \theta_i - a_i,$$

for some choice of z and y. Boscovich's second and third conditions were intended to lead to a best choice of z and y. The second was justified by an appeal to the intuitively plausible notion that positive and negative errors are equally likely, and it stated that the sum of positive corrections should equal the sum of negative corrections. In our notation this is compactly expressed by

$$\sum_{\text{all } i} \delta a_i = 0.$$

Boscovich's third condition was that the sum of the corrections, taken without regard to sign, was to be a minimum; that is,

$$\sum_{\text{all } i} |\delta a_i| \text{ is minimized.}$$

Now, it should be emphasized that this analytic formulation is not found in Boscovich's work. His statement of the conditions was only given in the verbal version we have quoted, and at no place in his work did he give other than a geometric or mechanical description of his solution to the problem. But even though this restriction to a geometric approach was costly in that it limited the generalizability of the approach, there was an important compensating benefit: Boscovich's geometric approach suggested a solution to the problem that would have been far less apparent in an analytic formulation.

Boscovich began by noting that his first condition *could* be expressed analytically—as meaning that the corrected arcs are expressible as an equation of the first degree involving the versed sines—and that if this were done, the second condition would be expressible as a sum involving the unknown coefficients. His statements again were verbal, not symbolic, but he clearly had the equation

$$\sum (z + y \sin^2 \theta_i - a_i) = 0$$

in mind. He mistakenly felt, however, that an analytic solution could only be had by differentiating this expression with respect to the unspecified coefficients; and he saw that this would be absurd, writing: "Thus supposing $dz = 0$ we will have nothing: the formula will vanish completely, together with the hopes of the mathematician" (Boscovich, 1760; Boscovich and Maire, 1770, p. 502; Boscovich used the letter x where I use z). Therefore, he moved on to present a simple geometric and mechanical solution.

Boscovich's discussion was accompanied by a diagram in the 1770 translation, reproduced here as Figure 1.4. We may consider AF as being one

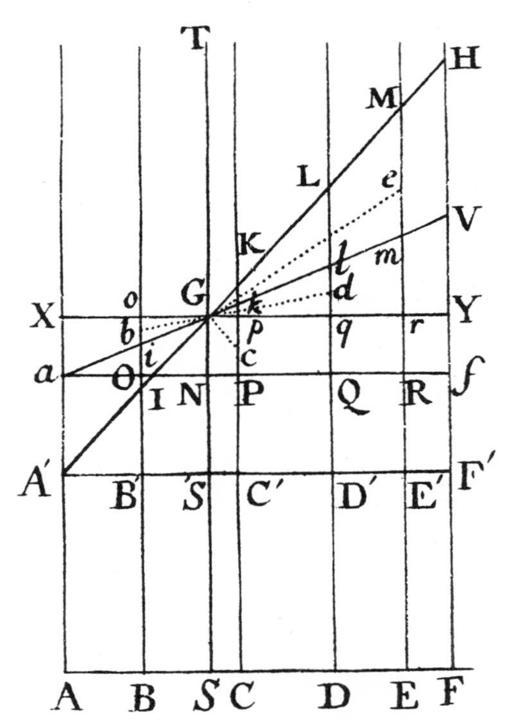

Figure 1.4. Boscovich's algorithm. The horizontal axis AF gives $\sin^2 \theta$*, where* θ *is the latitude of the midpoint of the arc; the vertical axis AX gives the arc length in toises per degree. The five arcs are indicated by* a, b, c, d, *and* e; *G is the center of gravity. (From Boscovich and Maire, 1770, plate I.)*

unit in length; A as the origin; and A, B, C, D, and E as representing the five values of the $\sin^2 \theta_i$ as marked off on the unit interval. The lengths A*a*, B*b*, C*c*, D*d*, and E*e* may be considered as representing the lengths (in toises per degree) of the corresponding measured arcs. The problem was to find a straight line A′H such that the corrections *a*A′, *b*O, *c*K, *d*L, and *e*M satisfied the second and third conditions. Boscovich began by observing that the second condition, interpreted as describing a condition of mechanical equilibrium, implied that the line must pass through the center of gravity (G) of the points. This reduced the problem to one of finding that line through G such that the third condition was satisfied.

To find the solution, Boscovich imagined a straight line SGT that was rotated clockwise, kept anchored at G. As the line rotated, the sum of the corrections (taken without regard to sign) would decrease until a minimum was achieved, then increase again. Since the corrections for the individual arcs would change as the line rotated in proportion to the distances AS, BS, SC, SD, and SE, it was only necessary to continue until the line had passed (or just reached) a sufficient number of the five points so that at least half the sum of these five distances, AS + BS + SC + SD + SE, was accounted for by the distances corresponding to the points passed (or just reached). Before such a point was reached, the sum of the corrections would decrease as the line rotated; after that point the sum would increase. He further simplified the solution by noting that the rotating line would encounter the five points in the inverse order to the slopes of the five lines from G to the five points: aX/AS, bo/BS, and so forth.

Boscovich clarified his geometric description of this algorithm by working through the details for the data of Table 1.4. He first located G by calculating the mean of the values for $10^4 \cdot \sin^2 \theta_i$ (giving 10^4 AS = 4,356.6) and the mean of the arcs a_i (giving SG = 57,052.6). He then calculated the differences $10^4(\sin^2 \theta_i - \text{AS})$ (that is, AS, BS, . . . ; given in column 2 of Table 1.6), the differences $a_i - \text{SG}$ (that is, aX, bo, . . . ; given in column 3 of Table 1.6), and the ratios $10^4(\sin^2 \theta_i - \text{AS})/(a_i - \text{SG})$ (that is, AS/aX, BS/bo, . . . ; given in column 4 of Table 1.6). These ratios are the inverses of the slopes described earlier. Thus the numbers in column 4 of Table 1.6 told Boscovich that the rotating line would encounter the five points in the order e, a, d, b, and, finally, c. How far through this list should Boscovich travel? The total of the distances AS, BS, SC, SD, SE ($\times 10^4$, taken without regard to sign) was found from the numbers in column 2 to be twice 5,726.2; thus he should continue until he reached a point such that the sum of the values in column 2 (without regard for sign) corresponding to the passed points reached 5,726.2. Since SE = 4,029.4 $<$ 5,726.2 and SE + SA = 8,386.0 $>$ 5,726.2, the line should be allowed to rotate until it reached a. While the line rotated from e to a, the correction to the point e increased by an amount proportional to SE = 4,029.4 and the total decrease in the other corrections was proportional to SA + SD + SB + SC = 7,423.0. Once a was passed, the total increase was

Table 1.6. Intermediate calculations for Boscovich's 1760 solution.

Arc	$10^4(\sin^2 \theta - \text{AS})$	Arc length − SG (toises)	Ratio
(1) = a	−4,356.6	−301.6	14
(2) = b	−1,369.6	−15.6	88
(3) = c	291.4	−73.6	−4
(4) = d	1,405.4	21.4	66
(5) = e	4,029.4	369.4	11

Source: Boscovich and Maire (1770, pp. 505–506).

proportional to SE + SA = 8,386.0, and thus more than counterbalanced the decreasing corrections, proportional to SD + SB + SC = 3,066.4. The solution then was the line passing through G and *a*! This he found corresponded to $z = 56{,}751$ and $y = 692$, from which he calculated the ellipticity as 1/248. Here, in 1760, he used the formula 1/ellipticity = $3z/y + 2$. Perhaps encouraged by the close proximity of this value to Newton's 1/230 and by the fact that this new value emerged from his own new method of combining observations, Boscovich did not repeat his earlier doubts about the elliptical shape of the earth.

The verbal description quoted earlier and the worked example are really all Boscovich ever wrote about his method. He added an updated discussion of arc measurements to the 1770 translation of his work with Maire. This discussion used his method on nine measured arcs and, after successive reanalyses omitting the three most discordant of the measurements, came to the conclusion that the evidence supported a hypothesis that the earth was somewhat irregular but was formed around an ellipsoidal core. Boscovich gave no further development of the method, no study of its properties, no analytic formulation, and no application of the method to problems other than the figure of the earth. He did briefly indicate the possibility of other applications in 1760, in these words: "Now we see that the method is generally for the correction of any terms which must be in a given ratio, because in substituting this ratio for that of the versed sines, all remains the same" (Boscovich, 1760; Boscovich and Maire, 1770, p. 505). But he appears never to have followed up on this statement by applying the method elsewhere, and he made no further statements regarding its generalizability. In fact, it is tempting to suppose that the method might have faded into obscurity had not a brief reference to its existence, in a 1772 review of the 1770 translation, caught the eye of Laplace.[12]

Laplace and the Method of Situation

In 1789 Laplace took up the question of the figure of the earth for the second time. He had previously approached the question of comparing arc measurements with an ellipsoidal hypothesis in 1783, but on that earlier occasion he had ignored Boscovich's work. He had considered only four of the five arcs studied by Boscovich (omitting the one at Rome measured by Boscovich and Maire), and he had been content to determine the best-fitting ellipsoidal figure as that which minimized the maximum correction needed. For that purpose he introduced an algorithm of his own invention

12. The review was in Jean Bernoulli III's *Recueil pour les astronomes* (Tome II, 1772, pp. 245–249). The relevant portions are quoted, together with evidence that Laplace read the review soon after it appeared, in Stigler (1978b).

(Laplace, 1786a). In his 1789 return to this subject, however, he had Boscovich very much in mind.

After admitting that his earlier approach of minimizing the maximum correction would become too laborious *(très pénible)* to use when many degrees were to be considered, Laplace turned to other approaches to the problem. He first derived an improved version of the earlier algorithm, and he illustrated its use with a set of nine arc measurements (which now included that at Rome); he then turned to Boscovich's method. Laplace felt that the elliptical figure that minimized the maximum correction was

> not the one that the measurements indicated with the greatest likelihood. This latter ellipse must, it appears to me, fulfil the following two conditions: 1° that the sum of the errors be zero; 2° that the sum of the errors taken all with the sign + be a *minimum.* Boscovich has given an ingenious method of achieving this; it is explained at the end of the French edition of his *Voyage astronomique et géographique* [that is, Boscovich and Maire, 1770]. But since he has unnecessarily complicated the method by the consideration of diagrams, I shall present it here in its simplest analytical form. (Laplace, 1793, p. 32)

Laplace proceeded to do just that, giving a precise algebraic statement to Boscovich's algorithm and accompanying his description with a rigorous analytic demonstration that the algorithm solved the problem as stated. He also gave two numerical examples, one involving nine measured arcs, the other involving thirteen observed lengths of seconds pendulums at various latitudes.

In this 1789 treatment Laplace added only an analytic formulation to Boscovich's earlier presentation. Even though this step alone was to prove crucial to Laplace's (and others') later studies of the method's statistical properties (he was to give it the name "Method of Situation"), it was nonetheless only a small conceptual advance—a translation from the language of Newton to the language of Euler. Ten years later, however, Laplace took an additional step, a subtle development of Boscovich's idea that was a symptom perhaps of Laplace's increasingly sharp statistical intuition.

The occasion for Laplace's renewed interest in the figure of the earth was the preparation of the second volume of the *Mécanique céleste.* After developing a mathematical theory of the figures of the heavenly bodies, Laplace returned to the problem he had faced a decade before—that of comparing measured arcs of the meridian with the hypothesis that the earth's figure was ellipsoidal. His analysis now considered only seven arcs; he omitted an ancient measurement in Holland and substituted a French arc measured in 1795 by Delambre and Méchain for two earlier French arcs. Now, all earlier analyses of arc measurements had treated all measurements as equally reliable; that is, all measurements considered good enough to be used at all were allowed an equal opportunity to influence the result of the calculation. Of course, the mechanics of the problem dictated that measurements made at extremes in latitude exerted a greater weight on the combined result than did those made at middle latitudes, but as far as the intrinsic accuracy of the measurement was concerned it was an all-or-nothing proposition. In 1799 Laplace evidently thought this was inappropriate.

In the preceding discussion, I introduced Laplace's notation (but with subscripts for superscripts) for the equations of condition describing the relationship to be tested:

$$a_k - z - p_k y = x_k,$$

where a_k is the arc length of the kth arc (in toises per degree), z the length of a degree at the equator, y the polar excess, p_k the square of the sine of the kth latitude, and x_k the error, due to measurement or failure of the ellipsoidal hypothesis. Now a_k was in fact determined by measuring the length of the arc both on the ground[13] and by astronomical observation of the Pole Star from the arc's two extreme points; the ratio of these two measurements would then give the length a_k in toise per degree. The actual lengths of the measured arcs varied considerably, from just under 1° in Lapland to nearly 10° in France. Surely these lengths would affect accuracy, and they should be incorporated into the analysis. In the *Mécanique céleste* (vol. 2, bk. 3, §40) Laplace therefore modified Boscovich's earlier conditions to the following form:

> First, that the sum of the errors committed in the measures of the whole arcs, ought to be zero. Second, that the sum of all these errors, taken positively, ought to be a *minimum*. By considering, in this manner, the whole arcs, instead of the degrees which have been deduced from them, we shall give to each of these degrees so much more influence, in the computation of the ellipticity of the earth, as the corresponding arc is of greater extent, which ought to be the case. (Laplace, 1799–1805, vol. 2, p. 134; Bowditch, 1832, pp. 434–437)

Following Laplace, let i_k represent the length, in degrees, of the kth arc (so $i_k a_k$ is the arc's length in toises). Then the whole arcs satisfy $i_k a_k - i_k z - i_k p_k y = i_k x_k$, and Laplace's two new conditions become

$$\text{(i)} \quad \sum_k i_k x_k = 0$$

$$\text{(ii)} \quad \sum_k |i_k x_k| = \text{minimum.}$$

13. A baseline (perhaps one-fifth to one-tenth the total distance surveyed) was measured directly by chains. Then the entire length was surveyed by triangulation.

Laplace derived the solution to the problem by a simple modification of the earlier algorithm. The condition (i) was equivalent to

$$A - z - y\,P = 0$$

where

$$A = \frac{\sum i_k a_k}{\sum i_k}, \qquad P = \frac{\sum i_k p_k}{\sum i_k}.$$

Subtracting this equation from each of the equations of condition gave

$$b_k - y q_k = x_k,$$

where

$$b_k = a_k - A, \qquad q_k = P - p_k.$$

Thus Laplace began, following Boscovich, by tying the line to the one point $G^* = (P, A)$, effectively using condition (i) to reduce the problem to one involving a single unknown, the slope y of the line through the weighted center of gravity, G^*.

The next step in the algorithm was to suppose that the equations are labeled to correspond to a decreasing sequence b_k/q_k; that is, so that

$$\frac{b_1}{q_1} \geq \frac{b_2}{q_2} \geq \ . \ . \ . \ \geq \frac{b_n}{q_n}.$$

Again, this is an algebraic statement of Boscovich's ordering of the points according to their encounter with the moving line. Finally, let $h_k = |i_k q_k|$, and let $F = h_1 + \ . \ . \ . \ + h_n$. Then if r is that integer such that

$$h_1 + \ . \ . \ . \ + h_{r-1} < \frac{1}{2}\,F \quad \text{and} \quad h_1 + \ . \ . \ . \ + h_r > \frac{1}{2}\,F,$$

the solution to the problem was to take $y = b_r/q_r$ and $z = A - Py$. The algorithm is of course just an analytic version of Boscovich's geometric procedure, where the points (p_k, a_k) are replaced by $(i_k p_k, i_k a_k)$. Laplace provided an analytic proof that this was indeed the solution to the stated problem, and he applied the method to seven arc measurements. On the basis of this calculation he concluded that an ellipticity of 1/312 was indicated; but he thought the large error (172.52 toises) this implied in the Lapland arc was evidence that the earth was not ellipsoidal (Laplace, 1799–1805, vol. 2, pp. 138–141; Bowditch, 1832, pp. 443–450). [Ironically, a remeasurement of this arc in 1803 by Svanberg *did* show an error of this magnitude (in fact, larger than 200 toises; Svanberg, 1805, p. 192). Later analyses, however, called Svanberg's work into question (Todhunter, 1879).]

Laplace's weighted analysis may now be seen as a small but significant advance in statistical technique. Previous workers had weighted different measurements differently depending on their realized values. James Short, for example, had in 1763 averaged determinations of the parallax of the sun in such a way as to discount those whose distance from the arithmetic mean was large (Stigler, 1973b). Other workers had discarded discordant measurements. In addition, analyses such as those of Mayer and Boscovich had had the effect of giving greater weight, or greater leverage, to measurements taken under extreme conditions; Cotes's rule, quoted earlier, may be most plausibly read as a statement of this principle. Laplace differed from all of these in weighting the measurements according to an intrinsic measure of the measurements' perceived accuracies, the length of the measured arcs.

Was Laplace's weighting correct from a modern perspective? Two sources of error enter into each a_k: the error in measurement on the ground and the error in the astronomical observations. Only two sets of astronomical determinations are required for each arc, regardless of the length of the arc, so we shall ignore this source of error in evaluating the weighting scheme. (Actually, this is not quite correct, because the astronomically measured arc affects a_k as a divisor and errors in small arcs will have a greater effect than the same error for a large arc. However, the errors from this source were at this time likely to be small relative to those from other sources, and we are not likely to be greatly misled by ignoring them in the present analysis.) The error in the measurement on the ground would have had a variance roughly proportional to the arc length i_k (assuming a constant baseline to arc ratio), so we might then expect the variance of a_k to be *inversely* proportional to i_k. Now Boscovich's original method weighted the measurements as if the a_k had equal variances, and Laplace weighted them as if the a_k had variances inversely proportional to i_k^2. Thus it would seem that Laplace's scheme gave too much weight to long arcs. And in particular, it gave much too much weight to the French arc, a circumstance that would probably not have been deplored by Laplace's colleagues at the Academy. If the errors in the astronomical determinations were not negligible, however, Laplace's weights may have been nearly appropriate. In any event, the fact that he attempted any weighting at all at this early date is most interesting.

The method of Boscovich, as formalized by Laplace, has continued to enjoy occasional use since the publication of the *Mécanique céleste.* Prony described it in detail and applied it to problems of water flow in 1804 (Prony, 1804, pp. xxi–xxxii). Three years later Puissant (1807, p. 63) presented it, again in full analytic detail, and recommended its use in surveying. And in 1809 and 1815 Bowditch published applications of a generalization of the method to cometary data. Bowditch's generalizations

(1809, 1815) were particularly interesting in that they combined Mayer's and Boscovich's methods. For example, faced with fifty-six equations of condition involving five unknowns, Bowditch applied the condition that the equations sum to zero separately to four different (but partially overlapping) subsets of equations; he thus eliminated four of the unknowns and solved for the fifth using Laplace's algorithm. As late as 1832 Bowditch was recommending Boscovich's method over least squares because it gave less weight to defective observations than did least squares (1832, p. 434).

Legendre and the Invention of Least Squares

We have now almost arrived at the method of least squares, both in chronological and conceptual terms. We have seen how by 1800 the principle of combining observational equations had evolved, through work of Mayer and Laplace, to produce a convenient ad hoc procedure for quite general situations. We have also seen how the idea of starting with a mathematical criterion had led, in work of Boscovich and Laplace, to an elegant solution suitable for simple linear relationships involving only two unknowns. The first of these approaches developed through problems in astronomy; the second was (at least in these early years) exclusively employed in connection with attempts to determine the figure of the earth. These two lines came together in the work of a man who, like Laplace, was an excellent mathematician working on problems in both arenas—Adrien Marie Legendre.

Legendre came to deal with empirical problems in astronomy and geodesy at a time when the methods we have discussed had been developed separately in the two fields. It was also a time when a half-century's successful use of these methods had seen a change in the view scientists took of them—from Euler's early belief that combination of observations made under different conditions would be detrimental to the later view of Laplace that such combination was essential to the comparison of theory and experience. Legendre brought a fresh view to these problems; and it was Legendre, and not Laplace, who took the next important step.

Legendre did not hit upon the idea of least squares in his first exposure to observational data. From 1792 on he was associated with the French commission charged with measuring the length of a meridian quadrant (the distance from the equator to the North Pole) through Paris. One of the major projects initiated by the National Convention in the early years after the French Revolution had been the decision in 1792 to change the ancient system of measurement by introducing the metric system as a new order, toppling existing standards of measurement in an action symbolic of the French Revolution itself. The basis of the new system was to be the meter, defined to be 1/10,000,000 of a meridian quadrant. It remained for French science to come up with a new determination of the length of this arc. In keeping with the nationalism that inspired the enterprise, the determination was to be based only on new measurements made on French lands. To this end an arc of nearly 10°, extending from Montjouy (near Barcelona) in the south to Dunkirk in the north, was measured in 1795. By 1799 the complex task of reducing the multitude of angular measurements to arc lengths had been completed by J. B. J. Delambre and P. F. A. Méchain. Although the official reports did not appear until after 1805, a summary of the data was widely circulated by 1799.[14]

In early 1799, before the appearance of the first two volumes of the *Mécanique céleste,* Delambre published an extensive discussion of the theoretical results underlying the reduction of the raw data on this arc. This volume (Delambre, 1799) is prefaced by a short memoir by Legendre that is dated 9 Nivôse, an VII (30 December 1798) and indicates that Legendre did not have the method of least squares at that time. He wrote, in a theoretical discussion of the reduction of arc lengths:

> In this way we obtain four equations of the form
>
> $$0 = fx - gy + hz$$
> $$0 = f'x - g'y + h'z$$
> $$e'' = f''x - g''y + h''z$$
> $$e''' = f'''x - g'''y + h'''z,$$
>
> from which we need to find the values of x, y, z. In this type of analysis, of which astronomical questions offer many examples, it is not necessary to seek to satisfy three of the equations exactly; that would force all the error onto the fourth equation. Rather, we need to try to balance the errors in such a way that they are borne nearly equally by all four equations; this will not be difficult when numerical values have been substituted in the equations. (Legendre, 1798, pp. 9–10)

This little-known comment of Legendre's is revealing: It shows that as early as 1798 he had accepted the notion, evolved from Mayer's early writings, that a balance should be struck between measurements, that all should contribute to the final result. But the comment shows no sense of a need for a general method of striking this balance. Rather it suggests that it will not be difficult to proceed ad hoc: After the numerical values have been substituted in the equations, a balance could be found acceptable for the specific case. Evidently Legendre was to change his mind on this in the next five years.

The occasion for Legendre's reconsideration of observational equa-

14. For example, in France it was published in Laplace's *Mécanique céleste,* vol. 2., bk. 3, §41 (1799), and in Germany in *Allgemeine Geographische Ephemeridenz* vol. 4, p. xxxv (1799).

tions, and for the appearance of the method of least squares, was the preparation in 1805 of a memoir on the determination of cometary orbits (Figure 1.5). The memoir is a scant seventy-one pages (excluding the appendix); and, aside from a few brief remarks at the end of the preface, the method of least squares makes no appearance before page 64. Even this first mention of least squares seems to be an afterthought because, after presenting an arbitrary solution to five linear equations in four unknowns (one that assumed that two equations held exactly and two of the unknowns were zero), Legendre wrote that the resulting errors were of a size "quite tolerable in the theory of comets. But it is possible to reduce them further by seeking the *minimum* of the sum of the squares of the quantities E′, E″, E‴ " (1805, p. 64). He then reworked the solution in line with this principle. It seems plausible that Legendre hit on the method of least squares while his memoir was in the later stages of preparation, a guess that is consistent with the fact that the method is not employed earlier in the memoir, despite several opportunities.

It is clear, however, that Legendre immediately realized the method's potential and that it was not merely applications to the orbits of comets he had in mind. On pages 68 and 69 he explained the method in more detail (with the word *minimum* making five italicized appearances, an emphasis reflecting his apparent excitement), and the memoir is followed on pages 72–80 by the elegant appendix from which the quotation near the beginning of this chapter was taken. The example that concludes the appendix reveals Legendre's depth of understanding of his method (notwithstanding the lack of a formal probabilistic framework). It also suggests that it was because Legendre saw these problems of the orbits of comets as similar to those he had encountered in geodesy that he was inspired to introduce his principle and was able to abstract it from the particular problem he faced. Indeed, the example he chose to discuss was not just given as an illustration, it was a serious return to what must have been the most expensive set of data in France—the 1795 measurements of the French meridian arc from Montjouy to Dunkirk.

To determine the figure of the earth from these data (Table 1.7), Legendre developed the relationship between arc length in degrees and in toises in a form slightly different from that we encountered earlier. Letting L and L' be the astronomically determined latitudes of the end points of an arc (from column 2 of Table 1.7) and S the measured length of the arc (given in column 3 of Table 1.7 in modules, where a module is just 2 toises), Legendre wrote

$$\begin{aligned} L' - L &= \frac{S}{D} + \frac{3}{2} \cdot \alpha \cdot \frac{180}{\pi} \sin (L' - L)\cos(L' + L) \\ &= \frac{S}{28{,}500} + \mathscr{C} \cdot \frac{S}{28{,}500} + \alpha \frac{270}{\pi} \sin(L' - L)\cos(L' + L). \end{aligned}$$

APPENDICE.

Sur la Méthode des moindres quarrés.

DANS la plupart des questions où il s'agit de tirer des mesures données par l'observation, les résultats les plus exacts qu'elles peuvent offrir, on est presque toujours conduit à un système d'équations de la forme

$$E = a + bx + cy + fz + \&c.$$

dans lesquelles a, b, c, f, &c. sont des coëfficiens connus, qui varient d'une équation à l'autre, et x, y, z, &c. sont des inconnues qu'il faut déterminer par la condition que la valeur de E se réduise, pour chaque équation, à une quantité ou nulle ou très-petite.

Si l'on a autant d'équations que d'inconnues x, y, z, &c., il n'y a aucune difficulté pour la détermination de ces inconnues, et on peut rendre les erreurs E absolument nulles. Mais le plus souvent, le nombre des équations est supérieur à celui des inconnues, et il est impossible d'anéantir toutes les erreurs.

Dans cette circonstance, qui est celle de la plupart des problèmes physiques et astronomiques, où l'on cherche à déterminer quelques élémens importans, il entre nécessairement de l'arbitraire dans la distribution des erreurs, et on ne doit pas s'attendre que toutes les hypothèses conduiront exactement aux mêmes résultats; mais il faut sur-tout faire en sorte que les erreurs extrêmes, sans avoir égard à leurs signes, soient renfermées dans les limites les plus étroites qu'il est possible.

De tous les principes qu'on peut proposer pour cet objet, je pense qu'il n'en est pas de plus général, de plus exact, ni d'une application plus facile que celui dont nous avons fait usage dans les recherches précédentes, et qui consiste à rendre

(73)

minimum la somme des quarrés des erreurs. Par ce moyen, il s'établit entre les erreurs une sorte d'équilibre qui empêchant les extrêmes de prévaloir, est très-propre à faire connoître l'état du système le plus proche de la vérité.

La somme des quarrés des erreurs $E^2 + E'^2 + E''^2 + \&c.$ étant

$$\begin{aligned} &(a + bx + cy + fz + \&c.)^2 \\ &+ (a' + b'x + c'y + f'z + \&c.)^2 \\ &+ (a'' + b''x + c''y + f''z + \&c.)^2 \\ &+ \&c.; \end{aligned}$$

si l'on cherche son *minimum*, en faisant varier x seule, on aura l'équation

$$o = \int ab + x\int b^2 + y\int bc + z\int bf + \&c.,$$

dans laquelle par $\int ab$ on entend la somme des produits semblables $ab + a'b' + a''b'' + \&c.$; par $\int b^2$ la somme des quarrés des coëfficiens de x, savoir $b^2 + b'^2 + b''^2 + \&c.$, ainsi de suite.

Le *minimum*, par rapport à y, donnera semblablement

$$o = \int ac + x\int bc + y\int c^2 + z\int fc + \&c.,$$

et le *minimum* par rapport à z,

$$o = \int af + x\int bf + y\int cf + z\int f^2 + \&c.,$$

où l'on voit que les mêmes coëfficiens $\int bc$, $\int bf$, &c. sont communs à deux équations, ce qui contribue à faciliter le calcul.

En général, *pour former l'équation du* minimum *par rapport à l'une des inconnues, il faut multiplier tous les termes de chaque équation proposée par le coëfficient de l'inconnue dans cette équation, pris avec son signe, et faire une somme de tous ces produits.*

On obtiendra de cette manière autant d'équations du *minimum*, qu'il y a d'inconnues, et il faudra résoudre ces équations par les méthodes ordinaires. Mais on aura soin d'abréger tous les calculs, tant des multiplications que de la résolution, en n'admettant dans chaque opération que le nombre de chiffres

10

(74)

entiers ou décimaux que peut exiger le degré d'approximation dont la question est susceptible.

Si par un hasard singulier, il étoit possible de satisfaire à toutes les équations en rendant toutes les erreurs nulles, on obtiendroit également ce résultat par les équations du *minimum*; car si après avoir trouvé les valeurs de x, y, z, &c. qui rendent nulles E, E′, &c., on fait varier x, y, z, &c. de δx, δy, δz, &c., il est évident que E^2 qui étoit zéro deviendra par cette variation $(a\delta x + b\delta y + c\delta z, \&c.)^2$. Il en sera de même de E'^2, E''^2, &c. D'où l'on voit que la somme des quarrés des erreurs aura pour variation une quantité du second ordre par rapport à δx, δy, &c.; ce qui s'accorde avec la nature du *minimum*.

Si après avoir déterminé toutes les inconnues x, y, z, &c., on substitue leurs valeurs dans les équations proposées, on connoîtra les diverses erreurs E, E′, E″, &c. auxquelles ce système donne lieu, et qui ne peuvent être réduites sans augmenter la somme de leurs quarrés. Si parmi ces erreurs il s'en trouve que l'on juge trop grandes pour être admises, alors on rejettera les équations qui ont produit ces erreurs, comme venant d'expériences trop défectueuses, et on déterminera les inconnues par le moyen des équations restantes, qui alors donneront des erreurs beaucoup moindres. Et il est à observer qu'on ne sera pas obligé alors de recommencer tous les calculs; car comme les équations du *minimum* se forment par l'addition des produits faits dans chacune des équations proposées, il suffira d'écarter de l'addition les produits donnés par les équations qui auront conduit à des erreurs trop considérables.

La règle par laquelle on prend le milieu entre les résultats de différentes observations, n'est qu'une conséquence très-simple de notre méthode générale, que nous appellerons *Méthode des moindres quarrés.*

En effet, si l'expérience a donné diverses valeurs a', a'', a''', &c.

(75)

pour une certaine quantité x, la somme des quarrés des erreurs sera $(a' - x)^2 + (a'' - x)^2 + (a''' - x)^2 + \&c.$, et en égalant cette somme à un *minimum*, on a

$$o = (a' - x) + (a'' - x) + (a''' - x) + \&c.;$$

d'où résulte $x = \dfrac{a' + a'' + a''' + \&c.}{n}$, n étant le nombre des observations.

Pareillement, si pour déterminer la position d'un point dans l'espace, on a trouvé, par une première expérience, les coordonnées a', b', c'; par une seconde, les coordonnées a'', b'', c'', &c. ainsi de suite; soient x, y, z, les véritables coordonnées de ce point: alors l'erreur de la première expérience sera la distance du point (a', b', c') au point (x, y, z); le quarré de cette distance est

$$(a' - x)^2 + (b' - y)^2 + (c' - z)^2;$$

et la somme des quarrés semblables étant égalée à un *minimum*, on en tire trois équations qui donnent $x = \dfrac{\int a}{n}$, $y = \dfrac{\int b}{n}$, $z = \dfrac{\int c}{n}$, n étant le nombre des points donnés par l'expérience. Ces formules sont les mêmes par lesquelles on trouveroit le centre de gravité commun de plusieurs masses égales, situées dans les points donnés; d'où l'on voit que le centre de gravité d'un corps quelconque jouit de cette propriété générale.

Si on divise la masse d'un corps en molécules égales et assez petites pour être considérées comme des points, la somme des quarrés des distances des molécules au centre de gravité sera un minimum.

On voit donc que la méthode des moindres quarrés fait connoître, en quelque sorte, le centre autour duquel viennent se ranger tous les résultats fournis par l'expérience, de manière à s'en écarter le moins qu'il est possible. L'application que nous allons faire de cette méthode à la mesure de la méridienne, achèvera de mettre dans tout son jour sa simplicité et sa fécondité.

Figure 1.5. Legendre's 1805 appendix, introducing the method of least squares. (From Legendre, 1805, pp. 72–75.)

Table 1.7. Measurements of the French meridian arc, made in 1795 between Montjouy (near Barcelona) and Dunkirk.

Place of observation	Latitude, L	Arc length S	$L' - L$	$L' + L$
Dunkirk	51°2′10.″50			
		62,472.59	2°11′20.″75	99°53′0″
Pantheon (Paris)	48°50′49.″75			
		76,145.74	2°40′7.″25	95°1′32″
Evaux	46°10′42.″50			
		84,424.55	2°57′48.″10	89°23′37″
Carcassonne	43°12′54.″40			
		52,749.48	1°51′9.″60	84°34′39″
Montjouy	41°21′44.″80			

Source: Legendre (1805, p. 76). Reprinted in Harvey (1822); given in a slightly different form in Laplace (1799–1805, vol. 2, bk. 3, §41) and Bowditch (1832, p. 453).
Note: Arc lengths (S) are in modules; 1 module = 2 toises ≅ 12.78 feet.

Here D is the length in modules of 1° centered at 45° latitude, α is the ellipticity of the earth, and $\mathscr{C}$ is defined by the relationship $D^{-1} = (1 + \mathscr{C})/28{,}500$. At first glance this appears to be quite a change from the earlier relationship $a = z + y \sin^2\theta$; but it is not. Just note that, since $\sin^2 45° = 0.5$, $D = z + y/2$; and, since $\theta = (L' + L)/2$, $2 \sin^2 \theta = 1 - \cos 2\theta = 1 - \cos(L' + L)$. Then since $3\alpha = y/D$ and $a = S/(L' - L)$, we can easily see that the formulas are equivalent, save that Legendre employs $180 \sin(L' - L)/\pi$ instead of its local approximation, $L' - L$. Actually, Legendre's formulation is [except for the local approximation of $\sin(L' - L)$] exactly the same as that used by Laplace for his weighted analysis: In our previous notation, Legendre's equation is equivalent to $i_k a_k = i_k z + i_k y p_k$. The introduction of $\mathscr{C}$ instead of D is just a reparametrization, based on the fact that D is known to be near 28,500; it is both easier and more accurate to work with smaller numbers.

Legendre then let E^i represent the error made in the determination of the ith latitude; and, on the basis of data of Table 1.7, he obtained four equations:

$$E^{I} - E^{II} = 0.002923 + \mathscr{C}(2.192) - \alpha(0.563)$$

$$E^{II} - E^{III} = 0.003100 + \mathscr{C}(2.672) - \alpha(0.351)$$

$$E^{III} - E^{IV} = -0.001096 + \mathscr{C}(2.962) + \alpha(0.047)$$

$$E^{IV} - E^{V} = -0.001808 + \mathscr{C}(1.851) + \alpha(0.263).$$

In forming these equations, Legendre assumed that the effect of these errors on $\sin(L' - L)\cos(L' + L)$ was negligible. Now, he might have applied the method of least squares to these equations directly, but instead he noted that "it is necessary to consider the errors separately." I take this to mean that, despite his lack of any formulation of any probability model, he correctly feared one of the consequences of the correlation of the equation's left-hand sides, namely, that treating these differences as four errors would restrict his choice of solutions. Therefore he introduced a fifth equation, $E^{III} = E^{III}$, which permitted him to reexpress the equations as

$$E^{I} = E^{III} + 0.006023 + \mathscr{C}(4.864) - \alpha(0.914)$$

$$E^{II} = E^{III} + 0.003100 + \mathscr{C}(2.672) - \alpha(0.351)$$

$$E^{III} = E^{III}$$

$$E^{IV} = E^{III} + 0.001096 - \mathscr{C}(2.962) - \alpha(0.047)$$

$$E^{V} = E^{III} + 0.002904 - \mathscr{C}(4.813) - \alpha(0.310).$$

He then solved these equations by the method of least squares, treating E^{III} on the right-hand side as an unknown, together with $\mathscr{C}$ and α. He found $\alpha = 0.00675 = 1/148$ and $\mathscr{C} = 0.0000778$. Thus $D = 28{,}500/(1 + \mathscr{C}) = 28{,}497.78$, and the corresponding length of the meridian quadrant would be $90 \cdot D = 2{,}564{,}800.20$ modules, a value leading to a meter of 0.256480 modules = 0.512960 toises ≅ 3.280 feet.

The actual meter was based upon the value found for D by Laplace in the *Mécanique céleste* (vol. 2, bk. 3, §41; see Bowditch, 1832, p. 465), namely (expressed in terms of a standard degree), $D = 28{,}504.11$, which gave the meridian quadrant as 2,565,370 modules and the meter as 0.256537 modules = 0.513074 toises ≅ 3.281 feet. Laplace's determination incorporated the measurement of the arc at Peru into the calculation of the ellipticity; then he found the value of D from the French arc based on this predetermined ellipticity using his own algorithm to minimize the maximum error. We note that the use of only the French arc would not, because it extended only about 10°, permit a very accurate determination of the ellipticity. The same was not true, however, with respect to D and thus of the meridian quadrant, $90D$. Thus the restriction to French data was made at less cost in efficiency than might be feared, at least as far as the determination of the meter was concerned.

Two observations on Legendre's procedure are in order: Viewed from a later perspective it was not a correct way of dealing with the type of correlation he encountered, and it was not original with Legendre. In fact, Laplace had employed the same sort of approach attempting to untangle errors that appear as differences by treating one as an unknown to be estimated—in his handling of these same data in the *Mécanique céleste* (vol. 2, bk. 3, §41; see Bowditch, 1832, p. 459). It is likely that it was Laplace's analysis that suggested the approach, even though Laplace minimized the maximum error, not the sum of squared errors. This approach is incorrect from a postcorrelation point of view. It does not introduce a bias into the results; but, because it ignores the variability of the artificially designated "unknown," it produces a weighting of the observations that is not of maximum efficiency for any reasonable specification of the errors' stochastic structure. This observation should not be construed as criticism of Legendre or Laplace, however, for it was to be more than a century before efficient methods were developed for dealing with the type of correlation they faced, and in the present case the difference in result is negligible. Rather, it is remarkable that, lacking any explicit probabilistic formulation, they made any attempt at all to deal with the problem. The attempt they did make was a limited one and seems to have been based on a rough intuitive notion of dependence and tied to the explicit notational appearance of the same errors in different equations. It is nonetheless surprising to find even this crude recognition of the dependence at this early time. We shall see that even a half-century after probability was introduced formally into the analysis of such problems, little more understanding of dependence was evident than is found in this, the first published example of the method of least squares.

With Legendre's introduction of least squares, we reach the end of the first stage of the lines of development begun separately by Mayer and Boscovich about a half-century before. The idea of combining different observational equations evolved slowly from Mayer's astronomical work, the idea of an objective criterion of fit was born in Boscovich's geodetic work, and their inspired synthesis was signaled by Legendre's geodetic example, appended to an astronomical memoir. But a key element was missing: There was, in all of this work, no formal appeal to probability and, more to the point, no move to quantify the uncertainty in the derived estimates (save only Mayer's weak attempt in 1750). All of this is the more puzzling because Laplace, who had been writing extensively on probability since 1774, had played a key role in all of this development. In Chapter 4 we shall see how, in the two decades following Legendre's analyses, the next stage was completed by Laplace, with Carl Friedrich Gauss providing a key catalytic agent. To explain this properly, however, we must first examine the major currents of eighteenth-century probability, and it is to this topic I now turn, starting with the work of Jacob Bernoulli.

2. *Probabilists and the Measurement of Uncertainty*

Abraham De Moivre (1667–1754), as portrayed in 1736

Jacob Bernoulli (1654–1705), as portrayed in 1687

EARLY WORK in mathematical probability was dominated by a consideration of equally likely cases. The problems considered were in a loose sense motivated by other problems, problems in the social sciences, annuities, insurance, meteorology, and medicine; but the paradigm for the mathematical development of the field was the analysis of games of chance. Why men of broad vision and wide interests chose such a narrow focus as the dicing table and why the concepts that were developed there were applied in astronomy before they were returned to the fields that originally motivated them, are both interesting questions and central to our main argument; however, a discussion of their answers will be deferred until we consider the introduction of statistical methods in the social sciences. In this chapter we shall restrict our attention to the way in which the consideration of games of chance led to the first mathematical treatment of the quantification of uncertainty.

By the end of the seventeenth century the mathematics of many simple (and some not-so-simple) games of chance was well understood and widely known. Fermat, Pascal, Huygens, Leibniz, Jacob Bernoulli, and Arbuthnot all had examined the ways in which the mathematics of permutations and combinations could be employed in the enumeration of favorable cases in a variety of games of known properties. But this early work did not extend to the consideration of the problem: How, from the outcome of a game (or several outcomes of the same game), could one learn about the properties of the game and how could one quantify the uncertainty of our inferred knowledge of these properties? The early works had been concerned with a priori computations. Given an urn known to contain r red balls and s black balls, the chance of a red ball being drawn is computed to be $r/(r+s)$. The a posteriori question of determining r and s based on observations of the game had not yet been addressed. The eventual successful treatment of this question is closely associated with the family Bernoulli.

Jacob Bernoulli

The Bernoullis are surely the most renowned family in the history of the mathematical sciences. Perhaps as many as twelve Bernoullis have contributed to some branch of mathematics or physics, and at least five have written on probability. So large is the set of Bernoullis that chance alone may have made it inevitable that a Bernoulli should be designated father of the quantification of uncertainty. The individual in question is Jacob Bernoulli (1654–1705),[1] professor at the University of Basel from 1687, contemporary and occasional rival to Isaac Newton.

Bernoulli (I shall take the risky course of allowing the context to indicate which of the clan is meant) was born in Basel, Switzerland, on 27 December 1654. His earliest biographer, Fontenelle, tells us that he was originally destined by his father for the ministry and it was only over the latter's objections that Jacob studied mathematics. Fontenelle's choice of words is apposite: Bernoulli was "instructed in Latin, Greek, and the Philosophy of the Schools, but nothing of Geometry; however, having seen by chance some Geometrical Figures, he was struck with those Charms of which few

1. The size of the Bernoulli clan has sometimes seemed larger than in fact it actually was because several members were known by different names in different languages. Jacob (or Jakob) Bernoulli, James Bernoulli, and Jacques Bernoulli were one and the same, even if his total contribution to our subject was enough for at least three men. Similarly Jean Bernoulli I through III, Johann Bernoulli I through III, and John Bernoulli I through III, were three (and only three) other people.

Men in the World are sensible. He had hardly any one Book of the Mathematicks, nor durst he make use of those he had but by stealth . . . He likewise apply'd himself even to Astronomy" (Fontenelle, 1717, p. 37). His rebellion against his father's wishes was significant in his choice of a motto: *Invito patre sidera verso* (I am among the stars in spite of my father).

By 1684 Jacob Bernoulli and his brother John had developed the differential calculus from hints and solutions published by Leibniz, and they were widely recognized as mathematicians of the first rank. They subsequently worked on integral calculus and studied curves and several minimization problems that would later evolve into the calculus of variations in the hands of Euler and Lagrange. The Bernoulli brothers were not often collaborators—they were more often rivals. Jacob, professor at Basel, would pose a question in a journal, inviting solutions. John, professor at Groningen, responded in the same journal (their sole mode of communication in later years), only to be told by his older brother (again in print) that he had erred. Such disputes (evidence supporting Fontenelle's description of Jacob as being "of a bilious and melancholly temper") help explain why the appearance of Jacob's posthumous papers was delayed until eight years after his death, when they were edited by his nephew Nicholas, not by his brother John.

When Jacob Bernoulli died of a "slow fever" on 16 August 1705, he left behind a legacy of unpublished (and some uncompleted) works on many topics in mathematics. The most important of these concerned probability. Bernoulli had worried over problems of the a posteriori determination of chances for twenty years before he died, and it was the fruits of these labors that were the focus of the major treatise his nephew finally produced in 1713, the *Ars Conjectandi.*

Bernoulli's book has variously been regarded as the beginning of the mathematical theory of probability and as the end of the emergence of the concept of probability (Hacking, 1975). Gouraud (1848, p. 38) wrote that "his *Ars Conjectandi* changed the face of the Calculus of Probabilities." The book is remarkable in many aspects, from its advances in combinatorics (including the "Bernoulli numbers") to its pathbreaking analysis of the interpretation of evidence (Shafer, 1978). The relevant point for our analysis is his introduction in the fourth part of *Ars Conjectandi* of what has come to be regarded as the first law of large numbers. Bernoulli began the discussion[2] leading up to his theorem by noting that, in games employing homogeneous dice with similar faces or urns with equally accessible tickets of different colors, the a priori determination of chances was straightforward. One would simply enumerate the possible cases and take the ratio of the number of "fertile" cases to the total number of cases, whether "fer-

2. English translations of large extracts of this discussion can be found in Uspensky (1937, pp. 105–107), Adams (1974, pp. 10–15), and Shafer (1978).

tile" or "sterile." But, Bernoulli asked, what about problems such as those involving disease, weather, or games of skill, where the causes are hidden and the enumeration of equally likely cases impossible? In such situations, Bernoulli wrote, "It would be a sign of insanity to attempt to learn anything in this manner."

Instead, Bernoulli proposed to determine the probability of a fertile case a posteriori: "For it should be presumed that a particular thing will occur or not occur in the future as many times as it has been observed, in similar circumstances, to have occurred or not occurred in the past" (1713, p. 224). The proportion of favorable or fertile cases could thus be determined empirically. Now this empirical approach to the determination of chances was not new with Bernoulli, nor did he consider it to be new. What was new was Bernoulli's attempt to give formal treatment to the vague notion that the greater the accumulation of evidence about the unknown proportion of cases, the closer we are to certain knowledge about that proportion.

Bernoulli took it as commonly known that uncertainty decreased as the number of observations increased: "For even the most stupid of men, by some instinct of nature, by himself and without any instruction (which is a remarkable thing), is convinced that the more observations have been made, the less danger there is of wandering from one's goal" (1713, p. 225). Bernoulli sought both to provide a proof of this principle and to show that there was no natural lower bound to the residual uncertainty: By multiplying the observations, "moral certainty" about the unknown proportion could be approached arbitrarily closely.

> To illustrate this by an example, I suppose that without your knowledge there are concealed in an urn 3000 white pebbles and 2000 black pebbles, and in trying to determine the numbers of these pebbles you take out one pebble after another (each time replacing the pebble you have drawn before choosing the next, in order not to decrease the number of pebbles in the urn), and that you observe how often a white and how often a black pebble is withdrawn. The question is, can you do this so often that it becomes ten times, one hundred times, one thousand times, etc., more probable (that is, it be morally certain) that the numbers of whites and blacks chosen are in the same 3:2 ratio as the pebbles in the urn, rather than in any other different ratio? (Bernoulli, 1713, pp. 225–226)

Bernoulli recognized that we could not count on determining the ratio exactly but would have to content ourselves with an approximation to the true ratio:

> To avoid misunderstanding, we must note that the ratio between the number of cases, which we are trying to determine by experiment, should not be taken as precise and indivisible (for then just the contrary would happen, and it would become less probable that the true ratio would be found the more

numerous were the observations). Rather, it is a ratio taken with some latitude, that is, included within two limits which can be made as narrow as one might wish. For instance, if in the example of the pebbles alluded to above we take two ratios 301/200 and 299/200 or 3001/2000 and 2999/2000, etc., of which one is immediately greater and the other immediately less than the ratio 3 : 2, it will be shown that it can be made more probable, that the ratio found by often repeated experiments will fall within these limits of the 3 : 2 ratio rather than outside them. (Bernoulli, 1713, pp. 226–227)

The temptation to restate Bernoulli's problem and his formal solution in modern notation is strong—and I shall not resist it. It must be borne in mind, however, that here, as always, modern notation brings with it modern concepts that distort a proper understanding of the way Bernoulli viewed the result. Let us take X to be the number of *observed* successes, favorable cases, or fertile cases out of a total of N observations, and let p be the unknown proportion. Then a modern statement of Bernoulli's solution is that for any given small positive number ϵ and any given large positive number c (say, $c = 10$, 100, or 1,000), N may be specified so that

$$\mathrm{P}\left(\left|\frac{X}{N} - p\right| \leq \epsilon\right) > c\mathrm{P}\left(\left|\frac{X}{N} - p\right| > \epsilon\right).$$

This statement can be easily converted into what is now known as Bernoulli's weak law of large numbers. By simple algebra this becomes

$$\text{(1)} \qquad \mathrm{P}\left(\left|\frac{X}{N} - p\right| > \epsilon\right) < \frac{1}{(c+1)}.$$

Thus, since we recognize that c is arbitrary, we have that given any $\epsilon > 0$ and any c (however large) N can be specified large enough that (1) holds—and Bernoulli's law is proved. It is remarkable that Bernoulli's proof may be viewed as a fully rigorous proof of this result. In fact, Uspensky (1937, chap. 6) presented Bernoulli's original proof, despite its length, as more natural than later, simpler proofs.

This modern synopsis is inaccurate in several respects, however, as is the occasional claim that Bernoulli presented the first example of an interval estimate of a probability. Bernoulli's proof can indeed be bent in modern hands to prove the theorem that bears his name, but in fact he did both less and more than this synopsis implies. His actual result was deeper, subtler, more precise, more difficult, and more ambitious than the simple and elementary statement of the weak law of large numbers given above. Yet, the investigation was in a sense a failure, both in fact and in Bernoulli's eyes.

In the first place Bernoulli did less than is often claimed in that he dealt only with the case where the numbers of fertile cases (r) and sterile cases (s) were integers, not with the modern situation in which the proportion $p = r/(r + s)$ is allowed to range over all real numbers in the interval [0, 1]. His aim was to show that, in essence, the exact ratio $r/(r+s)$ could be recovered with "moral certainty" for a sufficiently large N. Bernoulli realized that in seeking this goal some latitude was needed. In fact, unless N is a multiple of $r+s$, it is not even possible to obtain $X/N = r/(r+s)$ exactly. Even when N is a multiple of $r+s$ the probability that this exact equality is observed decreases as N increases. He spoke to this point in the previously quoted passage (Bernoulli, 1713, pp. 226–227). He did view the ratio $r/(r+s)$ as possibly an approximation to the real state of affairs, and he knew that r and s were not identifiable ($r' = 10r$ and $s' = 10s$ would give the same ratio as r and s). But up to the order of approximation determined by a given $r+s$ he sought to determine the ratio exactly, as his statements and examples make clear. In Bernoulli's work the limits within which he sought to include the empirical ratio were not arbitrary. Bernoulli's ϵ was not an arbitrary positive number; it was always taken as $1/(r+s)$. When he wrote that the limits "can be made as narrow as one might wish," the clear implication of his examples was that this was accomplished by taking r and s in the same ratio but larger by a factor of a power of 10, so that he was actually identifying $r/(r+s)$ exactly from among a larger class of possible ratios based on $10(r+s)$ cases or $100(r+s)$ cases. That is, within the context of a closer approximation he still sought to determine the ratio as exactly as the approximation permitted—within $\pm 1/(r+s)$. His estimation was essentially an attempt to identify a discrete r and s with "moral certainty."

I hasten to add that the limitation in Bernoulli's formulation and statement was not intrinsic in his proof; the proof was rigorous even for a later formulation. But from two other points of view, this limitation was critical. First, it signaled the lack of a conceptual advance that was, from a practical point of view, to doom the whole enterprise. Second, the fine distinction I draw between Bernoulli's attempt to hit an integral ratio as exactly as possible and a modern formulation in which an unknown proportion p is to be captured with high probability in an arbitrarily small interval is important for the evaluation of later developments in mathematical statistics.

When we come to the work of De Moivre I shall return to a discussion of what Bernoulli did not successfully accomplish and why he did not publish the work, but first I shall expand upon what was truly remarkable in the work. Bernoulli in fact did more than our synopsis implies. Let us consider in outline his ingenious proof, again in modern notation. Bernoulli wished to prove that

$$\mathrm{P}\left(\left|\frac{X}{N}-p\right|\leq\epsilon\right)>c\mathrm{P}\left(\left|\frac{X}{N}-p\right|>\epsilon\right),$$

or, equivalently,

$$\mathrm{P}(|X-Np|\leq N\epsilon)>c\mathrm{P}(|X-Np|>N\epsilon).$$

Bernoulli took N as a multiple of $t = r + s$, namely, $N = nt$, so that $Np = nr$ and $N\epsilon = n$ are integers; and he broke the range of values of X into equal portions of length $N\epsilon$ each. Focusing attention first on values of X above Np, he had

$$A_0 = P(Np < X \leq Np + N\epsilon)$$
$$A_1 = P(Np + N\epsilon < X \leq Np + 2N\epsilon)$$
$$A_2 = P(Np + 2N\epsilon < X \leq Np + 3N\epsilon)$$
$$\cdot$$
$$\cdot$$
$$\cdot$$

His intention was to show that, given c, an $N_0(c)$ could be found so that for $N \geq N_0(c)$, $A_0 > c(A_1 + A_2 + \ldots)$. A similar inequality for values of X below Np would complete the proof.

To demonstrate this Bernoulli showed that for an η depending only on N and p,

$$A_{k+1} < \eta A_k \quad \text{for } k = 0, 1, \ldots,$$

and that for a given p, η decreased toward zero as N increased. This gave him his proof because then

$$A_k < \eta^k A_0,$$

and

$$A_1 + A_2 + \ldots < A_0(\eta + \eta^2 + \eta^3 + \ldots) = A_0 \cdot \frac{\eta}{1 - \eta},$$

so

$$A_0 > \frac{(1 - \eta)}{\eta}(A_1 + A_2 + \ldots),$$

and he had $(1 - \eta)/\eta \geq c$ for N sufficiently large.

It remained to prove $A_{k+1} < \eta A_k$. Now if we let $b_k = Np + kN\epsilon + 1 = n(r + k) + 1$,

$$\frac{A_{k+1}}{A_k} = \frac{P(X = b_{k+1}) + P(X = b_{k+1} + 1) + \ldots + P(X = b_{k+1} + N\epsilon - 1)}{P(X = b_k) + P(X = b_k + 1) + \ldots + P(X = b_k + N\epsilon - 1)}$$
$$< \max\left\{\frac{P(X = b_{k+1})}{P(X = b_k)}, \ldots, \frac{P(X = b_{k+1} + N\epsilon - 1)}{P(X = b_k + N\epsilon - 1)}\right\}$$

(since for positive $F, G, H, \ldots, P, Q, R, \ldots,$

$$\frac{F + G + H + \ldots}{P + Q + R + \ldots} \leq \max\left\{\frac{F}{P}, \frac{G}{Q}, \frac{H}{R}, \ldots\right\},$$

with equality only when all ratios are equal). By a detailed analysis of the probabilities Bernoulli was able to show that

$$\frac{P(X = a)}{P(X = b)} > \frac{P(X = a + l)}{P(X = b + l)},$$

as long as $l > 0$ and $b < a$. Thus he had

$$\frac{A_{k+1}}{A_k} < \frac{P(X = b_{k+1})}{P(X = b_k)} \leq \frac{P(X = b_1)}{P(X = b_0)} = \eta.$$

By further detailed analysis he had $\eta \downarrow 0$ as $N \uparrow \infty$.

Thus far the proof flows quite naturally, but it is certainly longer and more involved than a two-line appeal to Chebychev's inequality, the proof most common in today's texts. Modern proofs, however, would not permit the final refinement Bernoulli was looking for and actually achieved. Bernoulli had announced earlier that it was no mere limit theorem he sought: "I would consider that I had done too little if I only gave a demonstration of this one thing which is known to all." Rather, he wanted to find a value for N for which he could actually achieve, with "moral certainty," the recovery of the ratio $r/(r + s)$. He accomplished this by returning to a detailed analysis of the ratio of probabilities $\eta = P(X = b_1)/P(X = b_0)$ and showing that he could force $(1 - \eta)/\eta > c$ by taking N to be the next integer above (or equal to) the larger of

$$mt + \frac{st(m - 1)}{r + 1},$$

where $m \geq \log[c(s - 1)]/[\log(r + 1) - \log r]$, ($m$ an integer), and

$$mt + \frac{rt(m - 1)}{s + 1},$$

where $m \geq \log[c(r - 1)]/[\log(s + 1) - \log s]$. (Here, as above, $t = r + s$.)

After completing this analysis Bernoulli worked an example. Taking $r = 30$ and $s = 20$, he found the second of the above expressions to be the larger, giving $N = 25{,}550$ for $c = 1{,}000$, $N = 31{,}258$ for $c = 10{,}000$, and $N = 36{,}966$ for $c = 100{,}000$. There the book ends (Figure 2.1).

Bernoulli's statement is capable of improvement; nonetheless we should view it as a remarkable achievement for the time. For perhaps the first time a mathematical approach to the measurement of uncertainty had been developed. Bernoulli had not shown merely that, qualitatively, the greater the number of observations the less the uncertainty in the result, he had shown how this statement could be quantified. Bernoulli could guarantee that, with a chance exceeding 1,000/1,001, $N = 25{,}550$ observations would produce a relative frequency of fertile cases that fell within 1/50 of

238 *ARTIS CONJECTANDI*

$\frac{nr-n}{ns}$, ſeu $\frac{r+1}{s}$ & $\frac{r-1}{s}$, pluribus quàm c vicibus ſuperet ſumam caſuum reliquorum; h. e. ut pluribus quàm c vicibus probabilius reddatur, rationem numeri obſervationum fertilium ad numerum omnium intra hos limites $\frac{r+1}{s}$ & $\frac{r-1}{s}$, quàm extra caſuram eſſe. Quod demonſtrandum erat.

In ſpeciali autem horum applicatione ad numeros ſatis per ſe patet, quòd quo majores in eadem ratione aſſumuntur numeri r, s & t, eo arctius quoque conſtringi poſſunt limites $\frac{r+1}{s}$ & $\frac{r-1}{s}$ rationis $\frac{r}{s}$. Idcirco si ratio inter numeros caſuum $\frac{r}{s}$, per experimenta determinanda, ſit ex. gr. ſeſquialtera, pro r & s non pono 3 & 2, ſed 30 & 20, vel 300 & 200 &c. ſufficiat poſuiſſe $r \infty 30$, $s \infty 20$, & $t \infty r+s \infty 50$, ut limites fiant $\frac{r+1}{t} \infty \frac{31}{50}$, & $\frac{r-1}{t} \infty \frac{29}{50}$; & ſtatuatur inſuper $c \infty 1000$: sic fiet ex Scholii præſcripto, pro terminis ad

ſiniſtram:

$$m > \frac{Lc.\overline{s-1}}{L\overline{r+1}-Lr} \infty \frac{4.2787536}{142405} < 301$$

$$nt \infty mt + \frac{mst-st}{r+1} < 24728$$

dextram:

$$m > \frac{Lc.\overline{r-1}}{L\overline{s+1}-Ls} \infty \frac{4.4623980}{211893} < 211.$$

$$nt \infty mt + \frac{mrt-rt}{s+1} \infty 25550.$$

Unde per ibi demonſtrata infertur, quòd inſtitutis 25550 experimentis multo plus millies veriſimilius ſit, rationem quam numerus fertilium obſervationum obtinebit ad numerum omnium, intra hos limites $\frac{31}{50}$ & $\frac{29}{50}$ caſuram, quàm extra. Atque eodem pacto, poſita $c \infty 10000$, aut $c \infty 100000$ &c. cognoſcetur, idem plus decies millies probabilius fore, ſi fiant experimenta 31258; & plus quàm centies millies,

PARS QUARTA. 239

millies, ſi capiantur 36966, &c. & ſic porrò in infinitum, additis nempe continuo ad 25550 aliis 5708 experimentis. Unde tandem hoc ſingulare ſequi videtur, quòd si eventuum omnium obſervationes per totam æternitatem continuarentur, (probabilitate ultimo in perfectam certitudinem abeunte) omnia in mundo certis rationibus & conſtanti viciſſitudinis lege contingere deprehenderentur; adeo ut etiam in maximè caſualibus atque fortuitis quandam quaſi neceſſitatem, &, ut ſic dicam, fatalitatem agnoſcere teneamur; quam neſcio annon ipſe jam Plato intendere voluerit, ſuo de univerſali rerum apocataſtaſi dogmate, ſecundum quod omnia poſt innumerabilium ſeculorum decurſum in priſtinum reverſura ſtatum prædixit.

Figure 2.1. The final two pages from Jacob Bernoulli's Ars Conjectandi. *(From Bernoulli, 1713, pp. 238–239.)*

a true proportion of 30/50. Bernoulli's bound was not sharp (and it depended upon the unknowns r and s), but it was explicit.[3] Mathematical scientists, Bernoulli had shown, not only could learn about nature a posteriori, they also could use mathematics to measure the extent of their knowledge. The conflicting philosophical interpretations of Bernoulli's quantitative statement would come much later, but he had taken the first step on what was to be a long and eventful journey.

De Moivre and the Expanded Binomial

Jacob Bernoulli had begun the journey toward a mathematical quantification of uncertainty, but the next major step was to come from a different quarter, from a French expatriate living in London. Abraham De Moivre

3. Bernoulli's result is conservative. In his example, for $c = 1000$, the approximation given by Feller (1968, p. 195, problem 16) suggests that $N = 6600$ will do. But Bernoulli's result is much sharper than Chebychev's inequality, which gives $N = 600,600$ in this case!

(born 26 May 1667, died 27 November 1754) was born a Protestant in Vitry, France. At the age of 21 he went to London—and freedom from persecution—after more than two years of imprisonment in France (Walker, 1934). In England he continued his study of mathematics while working as a tutor, principally by reading Newton; and in 1697 he was elected to the Royal Society at the age of 30. But it is not just because of his hazardous early emigration that De Moivre's scientific immortality rests on chance, for his fame is securely founded in three books upon differing mathematical aspects of that subject, the *Doctrine of Chances* (1718, 1738, 1756), *Annuities upon Lives* (1725, and several later editions), and *Miscellanea Analytica de Seriebus et Quadraturis* (1730).

De Moivre's earliest book on probability, the first edition of the *Doctrine of Chances,* was an expansion of a long (fifty-two pages) memoir he had published in Latin in the *Philosophical Transactions of the Royal Society* in 1711 under the title "De Mensura Sortis" (literally, "On the measurement of lots"). De Moivre tells us that in 1711 he had read only Huygens's 1657 tract *De Ratiociniis in Ludo Aleae* and an anonymous English 1692 tract based on Huygens's work (now known to have been written by John Arbuthnot). By 1718 he had encountered both Montmort's *Essay d'analyse sur les jeux de hasard* (2nd ed., 1713) and Bernoulli's *Ars Conjectandi* (1713), although the latter had no pronounced effect upon De Moivre at that early date. Indeed, he wrote in the preface in regard to Bernoulli's work, "I wish I were capable of carrying on a Project he had begun, of applying the Doctrine of Chances to *Oeconomical* and *Political* Uses, to which I have been invited [by Nicholas Bernoulli], but I willingly resign my share of that Task into better Hands" (De Moivre, 1718, p. xiv).[4] All indications are that by 1721 De Moivre's reluctance to pursue Bernoulli had disappeared for by that time De Moivre began to make progress in approximating the terms of a binomial expansion, in work that was to culminate in 1733 with the publication of what we now call the normal approximation to the binomial distribution.

De Moivre's first published comments on this topic appeared in 1730 in his *Miscellanea Analytica,* by which time he had nearly achieved a complete solution. His work was completed in 1733 in a short, separately printed note in Latin (De Moivre, 1733; Daw and Pearson, 1972). Later this note was translated into English and included in successively more expanded forms in the second and third editions of the *Doctrine of Chances* (De Moivre, 1738, pp. 235–243; 1756, pp. 243–254).

Book V of the *Miscellanea Analytica* ("On the Binomial $a + b$ raised to high powers") began by quoting extensively from that portion of *Ars Conjectandi* where Bernoulli had first come to grips with the problem of

4. This passage was omitted from De Moivre's subsequent republications of this preface in 1738 and 1756.

specifying the number of experiments needed to determine the actual ratio of cases, within a given approximation. De Moivre repeated Bernoulli's example (see earlier) and also rephrased one of Nicholas Bernoulli's examples that Montmort had published in 1713 (Montmort, 1713, pp. 388–393; Todhunter, 1865, pp. 130–131; Sheynin, 1968; Hald, 1984a). In this second example Nicholas, Jacob's nephew, had developed a different approximation for the same probability Jacob had considered. Nicholas's approach had differed from Jacob's in one important aspect: Nicholas had taken the number of observations as given, and he had sought then to determine a bound for the probability rather than following Jacob by specifying the probability and using the inequality to bound the number of observations needed. In Nicholas's actual example he had shown that if the chances of male and female births are in the ratio of 18 to 17, then the odds are at least 43.58 to 1 that of 14,000 births the number of male births will be within 163 of its expected value of 7200.[5]

De Moivre's approach followed more in the spirit of the approach of Nicholas Bernoulli (given N, find the probability) than of that of Jacob, but the mathematical treatment was his own. De Moivre began by looking at the approximate behavior for large n of the terms of the symmetric binomial $(1 + 1)^n$. In his 1730 treatment (p. 102)[6] he presented the results of two mathematical derivations. First, he gave the ratio of the maximum term of $(1 + 1)^n$—that is, $\binom{n}{n/2}$, in modern notation—to the sum of all terms (that is, 2^n) to be approximately

$$2\,\frac{21}{125}\cdot\frac{\left(1-\dfrac{1}{n}\right)^n}{\sqrt{n-1}}.$$

This is essentially the approximation now available from Stirling's formula. Indeed, it is clear from De Moivre's account that he had the practical equivalent to "Stirling's formula" before Stirling turned to this problem and that it was in response to De Moivre's investigation that Stirling took up this question. De Moivre had found the constant $A = 2\frac{21}{125} = 2.168$ through the numerical evaluation of the first four terms of the series

$$\ln\left(\frac{A}{2}\right) = \frac{1}{12} - \frac{1}{360} + \frac{1}{1260} - \frac{1}{1680} + \cdots.$$

Stirling's contribution (which was presented by De Moivre in a supplement to *Miscellanea Analytica*) consisted of his discovery that

$$\ln\sqrt{2\pi} = 1 - \frac{1}{12} + \frac{1}{360} - \frac{1}{1260} + \cdots;$$

5. A normal approximation would place the odds at 170 to 1.

6. De Moivre later indicated that these results had been found by him by 1721 (1738, p. 235; 1756, p. 243).

and so De Moivre's constant A was in fact

$$A = \frac{2e}{\sqrt{2\pi}}.$$

De Moivre noted in 1733 that, even though Stirling's discovery was unnecessary (because the numerical value of the constant was available from the series), it "has spread a singular Elegancy on the Solution" (De Moivre, 1738, p. 236; 1756, p. 244). With Stirling's constant—and approximating

$$\left(1 - \frac{1}{n}\right)^n \text{ by } e^{-1}, \quad \sqrt{n-1} \text{ by } \sqrt{n},$$

De Moivre in 1733 wrote the ratio of the maximum term of $(1 + 1)^n$ to 2^n as

$$\frac{2}{\sqrt{nc}}, \quad \text{where } c = 2\pi.$$

What De Moivre had found (written in modern notation) was a large sample approximation to $\mathrm{P}(X = n/2)$, where X has a symmetric binomial distribution (n trials, n even). He next sought to approximate the relationship between other terms in the series and this maximum one. In 1730 this took the form of writing the ratio of the maximum term M and a term Q a distance p from the maximum as approximately

$$\frac{M}{Q} = \frac{(m + p - 1)^{m+p-1/2}(m - p + 1)^{m-p+1/2}(m + p)/m}{m^{2m}}$$

where $m = n/2$ (De Moivre, 1730, p. 103). In 1733 he carried this a step further, writing l for p and noting "that if m or $\frac{1}{2}n$ be a Quantity infinitely great, then the Logarithm of the Ratio, which a Term distant from the middle by the Interval l, has to the middle Term, is $-2ll/n$" (De Moivre, 1738, p. 237; 1756, p. 245; see Figure 2.2). What De Moivre had shown (written in modern notation) was that if n is large relative to l, then

$$\ln\left[\mathrm{P}\left(X = \frac{n}{2} + l\right) \Big/ \mathrm{P}\left(X = \frac{n}{2}\right)\right] \cong -\frac{2l^2}{n},$$

which we may also write as

$$\mathrm{P}\left(X = \frac{n}{2} + l\right) \cong \mathrm{P}\left(X = \frac{n}{2}\right) \exp\left\{-\frac{2l^2}{n}\right\}$$

$$\cong \frac{2}{\sqrt{nc}} \exp\left\{-\frac{2l^2}{n}\right\}, \quad \text{where } c = 2\pi.$$

to every body: in order thereto, I ſhall here tranſlate a Paper of mine which was printed *November* 12, 1733, and communicated to ſome Friends, but never yet made public, reſerving to myſelf the right of enlarging my own Thoughts, as occaſion ſhall require.

Novemb. 12. 1733.

A Method of approximating the Sum of the Terms of the Binomial $\overline{a+b}^n$ *expanded into a Series, from whence are deduced ſome practical Rules to eſtimate the Degree of Aſſent which is to be given to Experiments.*

ALTHO' the Solution of Problems of Chance often require that ſeveral Terms of the Binomial $\overline{a+b}^n$ be added together, nevertheleſs in very high Powers the thing appears ſo laborious, and of ſo great a difficulty, that few people have undertaken that Taſk; for beſides *James* and *Nicolas Bernoulli*, two great Mathematicians, I know of no body that has attempted it; in which, tho' they have ſhewn very great ſkill, and have the praiſe which is due to their Induſtry, yet ſome things were farther required; for what they have done is not ſo much an Approximation as the determining very wide limits, within which they demonſtrated that the Sum of the Terms was contained. Now the Method which they have followed has been briefly deſcribed in my *Miſcellanea Analytica*, which the Reader may conſult if he pleaſes, unleſs they rather chuſe, which perhaps would be the beſt, to conſult what they themſelves have writ upon that Subject: for my part, what made me apply myſelf to that Inquiry was not out of opinion that I ſhould excel others, in which however I might have been forgiven; but what I did was in compliance to the deſire of a very worthy Gentleman, and good Mathematician, who encouraged me to it: I now add ſome new thoughts to the former; but in order to make their connexion the clearer, it is neceſſary for me to reſume ſome few things that have been delivered by me a pretty while ago.

I. It is now a dozen years or more ſince I had found what follows; If the Binomial 1 + 1 be raiſed to a very high Power de-

Hh 2 noted

2

Figure 2.2. De Moivre's 1738 translation of his 1733 paper on approximating the binomial, giving the first appearance in English of what was to become known as the normal curve, in Corollaries 1 and 2. (From De Moivre, 1738, pp. 235, 237.)

tion, (ſuppoſing $m = \frac{1}{2}n$) by the Quantities $\overline{m + l - \frac{1}{2}} \times \log. \overline{m + l - 1} + \overline{m - l + \frac{1}{2}} \times \log. \overline{m - l + 1} - 2m \times \log. m + \log. \frac{m+l}{m}$.

COROLLARY I.

This being admitted, I conclude, that if m or $\frac{1}{2}n$ be a Quantity infinitely great, then the Logarithm of the Ratio, which a Term diſtant from the middle by the Interval l, has to the middle Term, is $-\frac{2ll}{n}$.

COROLLARY 2.

The Number, which anſwers to the Hyperbolic Logarithm $-\frac{2ll}{n}$, being

$$1 - \frac{2ll}{n} + \frac{4l^4}{2nn} - \frac{8l^6}{6n^3} + \frac{16l^8}{24n^4} - \frac{32l^{10}}{120n^5} + \frac{64l^{12}}{720n^6}, \text{ \&c.}$$

it follows, that the Sum of the Terms intercepted between the Middle, and that whoſe diſtance from it is denoted by l, will be

$$\frac{2}{\sqrt{nc}} \text{ into } l - \frac{2l^3}{1 \times 3n} + \frac{4l^5}{2 \times 5nn} - \frac{8l^7}{6 \times 7n^3} + \frac{16l^9}{24 \times 9n^4} - \frac{32l^{11}}{120 \times 11n^5}, \text{ \&c.}$$

Let now l be ſuppoſed $= s\sqrt{n}$, then the ſaid Sum will be expreſſed by the Series

$$\frac{2}{\sqrt{c}} \text{ into } s - \frac{2s^3}{3} + \frac{4s^5}{2 \times 5} - \frac{8s^7}{6 \times 7} + \frac{16s^9}{24 \times 9} - \frac{32s^{11}}{120 \times 11}, \text{ \&c.}$$

Moreover, if s be interpreted by $\frac{1}{2}$, then the Series will become

$$\frac{2}{\sqrt{c}} \text{ into } \frac{1}{2} - \frac{1}{3 \times 4} + \frac{1}{2 \times 5 \times 8} - \frac{1}{6 \times 7 \times 16} + \frac{1}{24 \times 9 \times 32} - \frac{1}{120 \times 11 \times 64}, \text{ \&c.}$$

which converges ſo faſt, that by help of no more than ſeven or eight Terms, the Sum required may be carried to ſix or ſeven places of Decimals: Now that Sum will be found to be 0.427812, independently from the common Multiplicator $\frac{2}{\sqrt{c}}$, and therefore to the Tabular Logarithm of 0.427812, which is $\overline{9}.6312529$, adding the Logarithm of $\frac{2}{\sqrt{c}}$, *viz.* $\overline{9}.9019400$, the Sum will be $\overline{19}.5331929$, to which anſwers the number 0.341344.

LEMMA.

If an Event be ſo dependent on Chance, as that the Probabilities of its happening or failing be equal, and that a certain given number n of

Even though he gave no derivation, De Moivre had evidently followed a route such as

$$\begin{aligned}
\ln\left(\frac{M}{Q}\right) &= \left(m+l-\frac{1}{2}\right)\ln(m+l-1) \\
&\quad + \left(m-l+\frac{1}{2}\right)\ln(m-l+1) \\
&\quad - 2m\ln m + \ln\left(\frac{m+l}{m}\right) \\
&= \left(m+l-\frac{1}{2}\right)\ln\left(1+\frac{l-1}{m}\right) \\
&\quad + \left(m-l+\frac{1}{2}\right)\ln\left(1-\frac{l-1}{m}\right) \\
&\quad + \ln\left(1+\frac{l}{m}\right) \\
&\cong (m+l)\ln\left(1+\frac{l}{m}\right) + (m-l)\ln\left(1-\frac{l}{m}\right) \\
&\cong (m+l)\left(\frac{l}{m}-\frac{l^2}{2m^2}\right) + (m-l)\left(-\frac{l}{m}-\frac{l^2}{2m^2}\right) \\
&= \frac{l^2}{m} \\
&= \frac{2l^2}{n}.
\end{aligned}$$

So $\ln(Q/M) \cong -2l^2/n$, as long as terms of smaller order than l^2/m are neglected.

This approximation has been taken by Karl Pearson (1926) and others as the original appearance of the "normal curve." There is some justice to this view. Although De Moivre could not be said to have developed the concept of a probability density function—and he attached no importance to this exponential function other than as an approximation to the binomial distribution—he most definitely did think of it as a "curve." In fact, in 1730, even before he had completed the derivation of the approximation $\ln(Q/M) \cong -2l^2/n$, he had written, "Si termini omnes Binomii intelligantur normalitar erigi super lineam rectam ad intervalla aequalia, ducaturque Curva per extremitates omnium terminorum seu ordinatarum; Curva sic descripta habebit duplex punctum inflexus, unum ab utraque parte termini maximi." ("If the terms of the binomial are thought of as set upright, equally spaced at right angles to and above a straight line, the extremities of the terms follow a curve. The curve so described has two inflection points, one on each side of the maximal term.") He went on to find these two points of inflection, both for general binomial distributions and for the symmetric binomial; he showed that they came at a distance $\frac{1}{2}(n + 2)^{1/2}$ from the maximum term, which he noted was essentially equal to $\frac{1}{2}n^{1/2}$ when n was large (De Moivre, 1730, pp. 109–110).

Bernoulli's Failure

The more important aspect of De Moivre's work from our point of view is the *use* he made of this curve, first in a limited way in 1730 and then more extensively in 1733 and thereafter. De Moivre was not merely playing with mathematical symbols in an impressive display of his considerable talents; rather, he was attempting a fundamental refinement of Jacob Bernoulli's first attempt at the quantification of uncertainty. Bernoulli's upper bound was a start but it must have been a disappointing one, both to Bernoulli and to his contemporaries. To find that 25,550 experiments are needed to learn the proportion of fertile to sterile cases within one part in fifty is to find that nothing reliable can be learned in a reasonable number of experiments. The entire population of Basel was then smaller than 25,550; Flamsteed's 1725 catalogue listed only 3,000 stars. The number 25,550 was more than astronomical; for all practical purposes it was infinite. I suspect that Jacob Bernoulli's reluctance to publish his deliberations of twenty years on the subject was due more to the magnitude of the number yielded by his first and only calculation than to any philosophical reservations he may have had about making inferences based upon experimental data.

The abrupt ending of the *Ars Conjectandi* (Figure 2.1) would seem to support this—Bernoulli literally quit when he saw the number 25,550, mustering strength only to add one further sentence, a weak but flowery description of the limit theorem he had scorned earlier:

> If all events are observed for all of eternity . . . all will occur in certain ratios . . . Plato himself may have predicted this. (Bernoulli, 1713, p. 239)

Indeed, it is entirely possible that the editor–nephew Nicholas added this sentence to avoid too abrupt an ending. Jacob, as well as Nicholas Bernoulli and others, must have felt that something was wrong, that the mathematical steps of the approximation had given up too much. An experienced gambler could—or so he would believe—guess his chances of success to within one part in fifty based upon less experience than 25,550 games; surely a mathematical treatment of the problem should produce an answer more in accord with intuition than Bernoulli's, or be counted a failure. From this viewpoint Bernoulli's noble start *was* a failure, but one that held out hope for De Moivre.

There were two components to Bernoulli's failure. One was his insistence upon such a high standard of certainty—"moral certainty" that the empirical ratio was within one part in fifty of the true ratio. Bernoulli's insistence upon a moral certainty guaranteed by odds of 1,000 to 1 was, as we know now, more than one can expect from a small experiment. But at that time there was only a loose, intuitive notion of what could be achieved. It is understandable that a conservative standard should be entertained at first. Until a more precise manner of determining the probabilities was found, the type of numerical exploration needed to determine an acceptable standard of certainty (such as the now common 20 to 1) was not possible. Even if Bernoulli had relaxed his standard to immoral certainty and only insisted that it be an even bet (1 to 1 odds) that the empirical ratio be within one part in fifty of the true ratio 3 : 2, his inequality would still have implied that more than 8,400 experiments were needed!

De Moivre's Approximation

Either Bernoulli's bound had given up too much, or the situation was hopeless. De Moivre must have guessed the former was closer to the truth. With a more precise mathematical grip on the nature of the binomial coefficients, Bernoulli's bound could be replaced by a more precise and, De Moivre hoped, a more satisfactory estimate of the chances involved. It was at this goal that De Moivre aimed his analysis of the "Binomial curve."

Recall that in 1730 De Moivre had found that for large n,

$$\frac{M}{2^n} \cong \frac{2\frac{21}{125} \cdot \left(1 - \frac{1}{n}\right)^n}{\sqrt{n-1}}, \quad \text{and} \tag{2}$$

$$\frac{M}{Q} \cong \frac{(m+p-1)^{m+p-1/2}(m-p+1)^{m-p+1/2}(m+p)/m}{m^{2m}} \tag{3}$$

where $m = n/2$, and that in 1733 he had, with the aid of Stirling's number $c = 2\pi$ and using l for p, refined these results to

$$\frac{M}{2^n} \cong \frac{2}{\sqrt{nc}}, \quad \text{and} \quad \ln\left(\frac{Q}{M}\right) \cong -\frac{2l^2}{n}.$$

(In modern notation, $M/2^n = P(X = n/2)$ and $Q/M = P[X = (n/2) + l]/P(X = n/2)$.) If he was to improve upon Bernoulli's result (and he would *have* to improve on the result if it was to be of practical use), then he needed a way of summing the terms of the binomial near the maximum term. That is, he needed a practical and accurate way of summing Q for l near zero, or equivalently, summing $Q/2^n$ [$= P(X = \frac{1}{2}n + l)$, in modern notation]. In 1730 he *almost* fulfilled this goal, and in 1733 he finally succeeded.

De Moivre's 1730 attempt was just on the borderline of feasibility. He began by showing how the expressions (2) and (3) could be used, together with a table of logarithms, to compute M and Q. He worked an example and found $M/2^n \cong 0.026585$ for $n = 900$ from (2), rather than the value 0.026588 that he determined directly using a table of $\Sigma_{k=2}^{n} \log_{10} k = \log_{10}(n!)$ he had constructed.[7] Also, from (3) with $n = 900$, $m = 450$, and $p = 30$, he found $\log_{10}(M/Q) \cong 0.8682662628$, which he considered an adequate approximation to the actual value, found [from his table of $\log_{10}(n!)$] to be 0.8682669779.

Having shown the feasibility of calculating $M/2^n$ and $Q/2^n$, there remained the task of summing these terms, say, from $p = -\frac{1}{2}\sqrt{n}$ to $p = +\frac{1}{2}\sqrt{n}$. In his first attempt to deal with this in 1730, De Moivre had contented himself with stating a quadrature formula, without giving any numerical examples. He had shown that the binomial curve had inflection points at $p = \pm\frac{1}{2}\sqrt{n}$. Therefore, between these points the curve could be expected to be reasonably well approximated by a quadratic function of p, and the area under the curve (or the sum of the terms) could be found from a three-point quadrature formula appropriate to quadratic functions. In 1730, in a short corollary presented without proof, De Moivre stated such a formula (De Moivre, 1730, p. 110). If A, B, and C are three equidistant binomial coefficients, A being a distance p from B and B a distance p from C, then the sum of the terms from A through C is approximately

$$\text{(4)} \qquad \frac{(2p + 1)}{6p}[(p + 1)(A + C) + (4p - 2)B].$$

Presumably De Moivre had in mind applying this formula with the maximum term $M/2^n$ as B and (in the symmetric case) $Q/2^n$ as $A = C$, with $p \leq \frac{1}{2}\sqrt{n}$ and M and Q found from (2) and (3). Indeed, for the example he had given earlier, where $n = 900$, he had found $M/2^n \cong 0.026585$; (3) with $p = \frac{1}{2}\sqrt{n} = 15$ would have led to $Q/2^n = 0.016132$, and then (4) would give (in modern notation) $P(|X - 450| \leq 15) = 0.70892$. This is not very different from the value given by a normal approximation, namely, 0.68269, and De Moivre's formula would have been serviceable within the limited[8] range $\pm\frac{1}{2}\sqrt{n}$. But he presented no calculations; and in 1730 he went no further than simply stating the formula. Whether this lack of persistence was due to pessimistic uncertainty about the quality of this approximation for $p \leq \frac{1}{2}\sqrt{n}$ or to the knowledge that it would not extend to large p (hence limiting the applicability to chances far below "moral certainty") is not known.

7. His table (De Moivre, 1730, pp. 103–104) presented $\log_{10}(n!)$ for $n = 10, 20, \ldots, 900$, to fourteen decimal places, but the entries were inaccurate in the fifth and later places. In an undated supplement (probably 1730 or 1731) he presented a corrected table.

8. But the formula was hopeless beyond this range. For example, this approach would give $P(|X - 450| \leq 30) \cong 1.139$.

The 1730 attempt had only gone to the edge of feasibility, not because the expressions (2) and (3) left anything to be desired as approximations to the individual terms of a binomial, but because the lack of a compact analytic form for (3) precluded any but the crudest attempts at quadrature. A successful approximation to the individual *terms* of the binomial had been found, but not in a form that permitted the easy summation of large numbers of terms. Hence the questions Bernoulli had asked before 1705 were still frustratingly unanswerable. With the simple step De Moivre took in 1733, this changed. Instead of (2) and (3) he now had

$$(2') \qquad \frac{M}{2^n} \cong \frac{2}{\sqrt{nc}}, \qquad c = 2\pi$$

$$(3') \qquad \ln(Q/M) \cong -2l^2/n,$$

or, in modern notation,

$$P\left(X = \frac{n}{2}\right) \cong \frac{2}{\sqrt{nc}},$$

$$\ln\left[P\left(X = \frac{n}{2} + l\right) \Big/ P\left(X = \frac{n}{2}\right)\right] \cong -2l^2/n.$$

What De Moivre did in 1733 was to express the exponential $e^{-2l^2/n}$ as the series

$$1 - \frac{2l^2}{n} + \frac{4l^4}{2n^2} - \frac{8l^6}{6n^3} + \frac{16l^8}{24n^4} - \frac{32l^{10}}{120n^5} + \frac{64l^{12}}{720n^6} + \&c.$$

and integrate this series term by term between 0 and l to get

$$l - \frac{2l^3}{3n} + \frac{4l^5}{2 \cdot 5n^2} - \frac{8l^7}{6 \cdot 7n^3} + \frac{16l^9}{24 \cdot 9n^4} - \frac{32l^{11}}{120 \cdot 11n^5} + \&c.$$

This latter series times $2\sqrt{nc}$ would give the desired sum of terms from 0 to l (Figure 2.2).

He noticed that the series depended only on $l/\sqrt{n}$, and he thus evaluated it at $l = s\sqrt{n}$, giving the sum of terms from $l = 0$ to $l = s\sqrt{n}$ as

$$(5) \qquad \frac{2}{\sqrt{c}}\left(s - \frac{2s^3}{3} + \frac{4s^5}{2 \cdot 5} - \frac{8s^7}{6 \cdot 7} + \frac{16s^9}{24 \cdot 9} - \frac{32s^{11}}{120 \cdot 11} + \&c.\right).$$

De Moivre's calculations can be expressed in modern notation as follows: He desired the sum of $P(X = n/2 + l)$ from $l = 0$ to $l = s\sqrt{n}$. He had the approximation

$$\mathrm{P}\left(X=\frac{n}{2}+l\right)=\mathrm{P}\left(X=\frac{n}{2}\right)\mathrm{P}\left(X=\frac{n}{2}+l\right)\Big/\mathrm{P}\left(X=\frac{n}{2}\right)$$
$$\cong\frac{2}{\sqrt{nc}}e^{-2l^2/n}.$$

He approximated the sum

$$\sum_{l=0}^{s\sqrt{n}}\mathrm{P}\left(X=\frac{n}{2}+l\right)$$

by the integral of the right-hand side:

$$\frac{2}{\sqrt{nc}}\int_0^{s\sqrt{n}}e^{-2l^2/n}\,dl.$$

He evaluated this integral by expanding the integrand in a series and integrating term by term, then substituting in $l=s\sqrt{n}$ and arriving at (5) as an approximation to

$$\sum_{l=0}^{s\sqrt{n}}\mathrm{P}\left(X=\frac{n}{2}+l\right).$$

He felt this series converged sufficiently rapidly to be of use for $s\leq\frac{1}{2}$, but he noted that for $s=1$ "no less than 12 or 13 terms of the series will afford a tolerable approximation." For this and larger s he proposed evaluating the integral with the aid of standard quadrature formulas. Thus to find the sum from $l=0$ to $l=\sqrt{n}$ he would find the sum from $l=0$ to $l=\frac{1}{2}\sqrt{n}$ from (5) and add to it

$$\frac{2}{\sqrt{nc}}\cdot\frac{(A+D)+3(B+C)}{8}\cdot\frac{1}{2}\sqrt{n},$$

where A, B, C, D are $\exp\{-2l^2/n\}$ evaluated at, respectively, $l=(3/6)\sqrt{n}$, $(4/6)\sqrt{n}$, $(5/6)\sqrt{n}$, $(6/6)\sqrt{n}$.

At last De Moivre had found an effective, feasible way of summing the terms of the binomial. Whether by integrating a series or by using a simple four-point quadrature, the approximation $\exp\{-2l^2/n\}$ was accessible in a way the earlier expression (3) had not been, and De Moivre proceeded to demonstrate this accessibility in a series of corollaries that stand, effectively, as the first table of the normal distribution. After calculating the value of (5) to be 0.341344 for $s=\frac{1}{2}$ [in modern notation, $\mathrm{P}(0\leq X-n/2\leq\frac{1}{2}\sqrt{n})\cong 0.341344$], he found the chance that the number of occurrences of a symmetric binomial experiment would fall between

$\frac{1}{2}n - l$ and $\frac{1}{2}n + l$, for $l = \frac{1}{2}\sqrt{n}$, $\sqrt{n}$, $\frac{3}{2}\sqrt{n}$. His numerical results are spread throughout his text, but they can be conveniently summarized in tabular form (Table 2.1). In later versions of this discussion he added a paragraph in which he found, approximately, the value of l such that the probability that the number of occurrences departed from $n/2$ by more than l was $\frac{1}{2}$: $P(\frac{1}{2}n - l \leq X \leq \frac{1}{2}n + l) = 0.5$. He gave this value as "expressed by $\frac{1}{4}\sqrt{2n}$ very near" (1738, p. 239; 1756, p. 247), which is equivalent to solving

$$\int_{-a}^{a} \frac{1}{\sqrt{2\pi}} e^{-x^2/2}\, dx = 0.5$$

to find $a = 1/\sqrt{2} = 0.707$, a value that is not far from the exact solution $a = 0.674$ and that represents an interesting early appearance of what was to come to be known as the "probable error"—the median deviation from the mean.

De Moivre expressed confidence in his approximation, even when n was not "immensely great; for supposing it not to reach beyond the 900th Power, nay not even beyond the 100th, the Rule here given will be tolerably accurate, which I have had confirmed by Trials" (1738, p. 239; 1756, p. 247). He did not explain what "confirmed by Trials" meant; but, based on the precedent set by his examination of his 1730 approximation, it is likely that he had computed exact values by brute force for comparison in a few cases.

We can now recognize De Moivre's calculations as evaluations of the chance a normally distributed random variable falls within one, two, and

Table 2.1. Values of the normal integral $\int_{-a}^{a}(1/\sqrt{2\pi})e^{-x^2/2}\,dx$, computed by De Moivre in 1733 as approximations to binomial probabilities $P(\frac{1}{2}n - \frac{1}{2}a\sqrt{n} \leq X \leq \frac{1}{2}n + \frac{1}{2}a\sqrt{n})$ and computed exactly by modern methods.

a	De Moivre	Exact
1	0.682688	0.682689
2	0.95428	0.95450
3	0.99874	0.99730

Sources: De Moivre (1733, pp. 4–6; 1738, pp. 238–241; 1756, pp. 246–249) and modern tables.

Note: As this table shows, De Moivre's calculations were quite accurate. Surprisingly, though, he tripped up at one point in converting these probabilities to odds. In the original 1733 pamphlet he gave the odds of exceeding the given limits correctly, as, respectively, 28 to 13, 21 to 1, and 792 to 1. In later publications, however, (for example, De Moivre, 1756, pp. 248, 251) this last odds ratio appeared as 369 to 1. Apparently De Moivre had mistakenly convinced himself that he had forgotten to double the probability 0.00126 of exceeding $(3/2)\sqrt{n}$, and he doubled it to get 0.00252 or odds of 396 to 1, then reversed the last two digits giving 369 to 1.

three standard deviations of its mean. Our recognition of today's standardized units in these quarter-millennium-old calculations would not be entirely anachronistic either. In his evaluation of the series at values that were multiples of $\sqrt{n}$, De Moivre was in effect recognizing that this was the scale upon which deviations from the center should be judged. In a passage he expanded in the 1738 version, he went further:

> To apply this to particular Examples, it will be necessary to estimate the frequency of an Event's happening or failing by the Square-root of the number which denotes how many Experiments have been, or are designed to be taken; and this Square-root, according as it has been already hinted at in the fourth Corollary, will be as it were the *Modulus* by which we are to regulate our Estimation; and therefore suppose the number of Experiments to be taken is 3600, and that it were required to assign the Probability of the Event's neither happening oftner than [1850] times, nor more rarely than 1750, which two numbers may be varied at pleasure, provided they be equally distant from the middle Sum 1800, then make the half difference between the two numbers 1850 and 1750, that is, in this Case, $50 = s\sqrt{n}$; now having supposed $3600 = n$, then $\sqrt{n}$ will be $=60$, which will make it that 50 will be $=60s$, and consequently $s = 50/60 = 5/6$; and therefore if we take the proportion, which in an infinite power [that is, in the limit], the double Sum of the Terms corresponding to the Interval $\frac{5}{6}\sqrt{n}$, bears to the Sum of all the Terms, we shall have the Probability required exceeding near. (De Moivre, 1738, pp. 240–241; 1756, pp. 248–249)

This passage stands as a clear announcement that distances from the center are to be measured as multiples of the *square root* of the number of trials, that, if $\sqrt{n} = 60$, a difference of 50 is 5/6 units from the center. De Moivre even introduces a term for the unit $\sqrt{n}$—the *Modulus* (the italics were added in 1756; the term appears for the first time[9] in the 1738 version). We can now see in this work the beginning of the realization that in the aggregation of independent measurements, accuracy increases as the square root of the sample size. This concept was not evident in prestatistical ages (Stigler, 1977b). As Bernoulli had announced in *Ars Conjectandi,* even "the most stupid of men . . . is convinced that the more observations have been made, the less danger there is of wandering from one's aim" (Bernoulli, 1713, p. 225). To go from this qualitative intuitive judgment to a quantitative one was a large step. Bernoulli had attempted it and failed. De Moivre, with his more finely tuned analytic techniques, had reached the first success. The significance of $\sqrt{n}$, which had first appeared in De Moivre's 1730 calculation of the inflection points of the Binomial,

9. The term *Modulus* did not appear a second time in De Moivre's work, but it was to recur in later treatments. Bravais (1846) used modulus as one term for the scale parameter of a normal distribution, and Edgeworth later used it extensively as his term for the square root of twice the variance. Thus Edgeworth's modulus was De Moivre's *Modulus* divided by $\sqrt{2}$.

was announced emphatically in the 1733 series approximation. Its role was limited to binomial experiments; and the extension of this quantitative measure of accuracy to more general problems was neither trivial nor soon accomplished. Indeed, we have already seen in Chapter 1 how in 1750 Tobias Mayer had taken accuracy as increasing in direct proportion to n. But De Moivre's quantitative step was an important beginning and led to the eventual extension of these ideas by others, principally by Laplace.

De Moivre had achieved a singular mathematical success, but not one that he pursued very far. Most of his discussion and all of his examples had been concerned with the symmetric case, where the probabilities were proportional to the terms of $(1 + 1)^n$. To this he added a page of text in which he stated the results for the general case, the expansion of $(a + b)^n$, in a quick succession of two lemmas and three corollaries. As parallels to (2′) and (3′) he stated that in the general case one would have

$$(2'') \qquad \frac{M}{(a+b)^n} \cong \frac{a+b}{\sqrt{abnc}}, \qquad c = 2\pi$$

and

$$(3'') \qquad \ln(Q/M) \cong -\frac{(a+b)^2}{2abn} \cdot l^2,$$

where I have adopted our earlier notation, letting M stand for the maximum term in the expansion, and Q a term a distance l from M. In modern notation, these are

$$\mathrm{P}(X = np) \cong \frac{1}{\sqrt{p(1-p)nc}}, \qquad c = 2\pi$$

and

$$\ln[\mathrm{P}(X = np + l)/\mathrm{P}(X = np)] \cong -\frac{l^2}{2np(1-p)},$$

where X has a binomial (n, p) distribution, $p = a/(a + b)$, and np is supposed an integer. He then added as a corollary a statement that has echoes in modern mathematics as "it can easily be shown" or "it follows similarly that." He wrote

Corollary 10.

> If the Probabilities of happening and failing be in any given Ratio of inequality, the Problems relating to the Sum of the Terms of the Binomial $(a + b)^n$ will be solved with the same facility as those in which the Probabilities of happening and failing are in a Ratio of Equality. (De Moivre, 1738, p. 242; 1756, p. 250)

He provided no illustration of how this would be accomplished.

De Moivre had solved Bernoulli's mathematical problem; he had overcome one obstacle that had thwarted Bernoulli's attempt at a quantification of uncertainty. But despite the greater precision of De Moivre's result, he did not advance beyond Bernoulli's conclusion in his application of the result. From the relative precision of his normal limit to the binomial and from his signal achievement in his recognition of the significance of $\sqrt{n}$ as a standard unit of measurement, he chose to emphasize only the law of large numbers. In a statement added to the 1738 version and italicized in the 1756 edition, De Moivre summarized the importance of his results in these words: "*altho' Chance produces Irregularities, still the Odds will be infinitely great, that in process of Time, those Irregularities will bear no proportion to the recurrency of that Order which naturally results from* ORIGINAL DESIGN" (De Moivre, 1738, p. 243; 1756, p. 251). That he emphasized the disappearance of irregularity rather than its measurement in this summary can be explained either as due to the mathematical character of the work (as opposed to applied) or as a wish to prove that his mathematical argument could usefully contribute to contemporary philosophical or theological debates. But the complete absence of applications of this method in his work on annuities and the indifference with which his contemporaries greeted these advances cannot be so easily dismissed. Nonetheless, De Moivre's work was to have a profound influence upon later mathematical developments in the eighteenth century.

De Moivre's Deficiency

De Moivre's work on the doctrine of chances was widely circulated. Most encyclopedias of the last half of the eighteenth century, including the *Encyclopaedia Britannica* (1771) and Chambers's *Cyclopaedia* (1779–1791), cite De Moivre's work as the definitive treatment of the subject. Whether his solution to Bernoulli's problem was equally well known is a very different question. Looking at his work from a perspective of 250 years later it is easy to recognize the success he achieved and to see the promise his solution offered for the quantification of uncertainty in a wide variety of scientific problems. It may appear surprising at first glance that his contemporaries could have missed the potential in this masterful work; yet this is exactly what happened. Despite the three separate publications of this material, in 1733, 1738, and 1756, I know of no application or extension of these ideas before the late 1760s, in Lagrange's long memoir on choosing a best mean. Thomas Simpson (1740, pp. 74–85) presented slight variations on De Moivre's results, but he did not go beyond De Moivre.

Why was De Moivre's work on the binomial not immediately applied? One primary reason, I believe, is that it did not answer the question that is most often and most naturally asked in applications. It did not tell empirical scientists what, given an available empirical fact, they could say about the process that produced that fact. It did not provide an answer to the fundamental question of statistical inference. For example, if De Moivre or someone else of his era knew that of 346 men of age fifty, only 142 survived to age seventy,[10] he might ask, "How much credence can I give to the ratio 142/346 as an estimate of the chance of surviving from fifty to seventy? What, for example, is the probability this number is below 1/2?" De Moivre's result did not permit a direct answer to these questions. It *did* permit an answer to the question, "If the true chance were 1/2, what is the probability a ratio as small as 142/346 or smaller should occur?" But the relevance of such questions to the general inference problem would not be argued until more than a century after De Moivre's death in 1754. This question, raised in the testing of specified hypotheses, did occur naturally in a few early examples (for example, is the sex ratio or ratio of males to females at birth 1 : 1?), but even there the notion of testing the hypothesis by using De Moivre's results did not arise. Only in extreme cases, where *all* of the observed cases were "fertile" or "male" and the probability calculation a trivial one, did the assessment of hypotheses by the doctrine of chances occur in De Moivre's era. The more prevalent inference questions, those involving mixed evidence or no specified hypothesis, did not receive mathematical airing.

In two respects De Moivre's result failed to provide usable answers to the inference questions being asked at that time. Given a ratio $a/(a+b)$ of equally likely cases divided into two classes and a number n of independent trials, De Moivre's corollary 9 could be used (with some additional work) to evaluate the chance that the observed ratio would fall within a specified distance of $a/(a+b)$. But it could not be turned around to give the chance that an unknown $a/(a+b)$ would fall within the same specified distance of a given observed ratio. His calculations did not contain even a germ of inverse probability; they were totally tied to one conditional view of the random nature of the phenomenon studied: For De Moivre chance lay in the data, not in the underlying probabilities. The successful mathematical treatment of the inference problem was to require the abandonment of this restrictive view.

De Moivre himself came close to claiming that his results provided at least a qualitative answer to the inference problem. In 1730 (p. 101) and again in 1738 when introducing his binomial approximation he stated (in essence, and in modern notation) that if np is an integer,

$$\mathrm{E}|X-np| = n2pq\ \mathrm{P}\{X=np\} \cong \sqrt{2npq/\pi},$$

10. The ratio 142/346 in this example is taken from Halley's 1693 life table, as reproduced by De Moivre (1725).

noting that this last term increases much more slowly than n (in fact, $E|(X/n) - p| \cong \sqrt{2pq/n\pi}$ decreases with n). As a corollary, he added:

> From this it follows, that if after taking a great number of Experiments, it should be perceived that the happenings and failings have been nearly in a certain proportion, such as of 2 to 1, it may safely be concluded that the Probabilities of happening or failing at any one time assigned will be very near in that proportion, and that the greater the number of Experiments has been, so much nearer the Truth will the conjectures be that are derived from them. (1738, p. 234; 1756, p. 242)

This passage was as close as De Moivre came to inverse probability, and he immediately retreated to his approximation theorem. Even though the smallness of $E|(X/n) - p|$ for large n might have tempted an inversion (forgetting that the analysis fixed p, not X/n), his approximation theorem did not offer the same temptation, although it did to Neyman two centuries later.

The second shortcoming of De Moivre's approach involved his total focus upon the binomial distribution. In mathematics new concepts are usually most easily explored, and new approaches most fruitfully developed, by starting with the simplest situation. In this way both conceptual and mathematical difficulties are minimized, and an adroit analyst can more easily gauge the maximum extent to which his techniques would admit development. What, we might ask, could be simpler than analyzing experiments with but two possible outcomes? If investigators cannot learn how to quantify uncertainty in such simple situations, surely there is no hope for more complex problems. Jacob Bernoulli, Nicholas Bernoulli, and Abraham De Moivre all must have considered the problem in these terms, but all were mistaken. Their plausible but erroneous choice of a direction in which to examine the measurement of uncertainty delayed the critical breakthrough by a half-century because the simplicity with which the binomial experiment can be described is not matched by a corresponding simplicity in the associated inference problem. The binomial is in fact a very complex case, and it is no reflection upon the Bernoullis and De Moivre that their prodigious mathematical efforts met with so little applicable success.

The major development of quantified uncertainty and the first major successes in turning the approach of De Moivre on its head and creating a mathematical theory of inference are to be found in Laplace's works. Laplace was not alone in his interest in such matters, however; nor was he the first to meet with some successes on some of these problems. In fact, the literature of the period 1755 to 1770 suggests that a different approach to the inference problem, a different "easiest case," occurred to many men in many countries, apparently independently of one another. Simpson in

England, Lambert in Berlin, and Daniel Bernoulli in Basel all contributed memoirs on this new "case," and their work at least tangentially inspired Bayes, Lagrange, and finally Laplace to attack the problem and provide a framework for its eventual solution.

The new problem that excited these men was superficially unconnected to De Moivre's work on the expanded binomial, although the techniques they developed for its mathematical treatment were rooted in that work. The problem was to combine discordant observations of some astronomical body. If five observers record five different times for the passage of a star past a crosshair in a telescope, how are these numbers to be reconciled? Astronomers had long before devised practical solutions to the problem, for example, taking the arithmetic mean after discarding observations thought to be suspect. But it was only after the middle of the eighteenth century that it occurred to the more mathematically minded of their number that this problem might admit to a successful treatment that had escaped the binomial.

Simpson and Bayes

The first attempts to deal with the inference problem took place in England and were made by two Thomases, Simpson and Bayes. Their works were printed separately, first Simpson's in 1755 and then Bayes's in 1764. But for reasons partly speculative and partly documentable, these two works should be considered as part of a loosely connected line of development, despite a superficial dissimilarity. Neither work dealt with its topic with total success, but they both shared a degree of inventiveness and insight that we can appreciate from our present vantage point. They provide a natural introduction to the more successful later treatments of both topics by Laplace.

Simpson's Crucial Step toward Error

Thomas Simpson (born 20 August 1710, died 14 May 1761) was a talented self-taught mathematician who followed De Moivre both chronologically and intellectually. As a youth he became interested in astrology, which led him in turn to astronomy and mathematics. Simpson's early livelihood was earned as a weaver, but a different future loomed ahead. At the age of nineteen he married a fifty-year-old widow with two children; and at twenty-five moved to London, where he supported his family by weaving during the day and teaching mathematics in the evening.[11] By 1740 he had

11. Details of Simpson's personal life are not plentiful (Clarke, 1929). Simpson's successor as professor of mathematics at the Woolwich Academy, Charles Hutton, offered an interesting, if apologetic, hint: "It has been said that Mr. Simpson frequented low company, with

found time to write a short treatise *The Nature and Laws of Chance* on the mathematics of games of chance; and in 1742 he published another short book, *The Doctrine of Annuities and Reversions.* Both books were based on the earlier works of De Moivre and attempted to present much the same material in a brief format and at a level a wider public could afford. Simpson put this claim rather plainly in his 1740 book, referring to De Moivre's *Doctrine of Chances: "tho' it neither wants* Matter *nor* Elegance *to recommend it, yet the* Price *must, I am sensible, have put it out of the Power of many to purchase it*" (Simpson, 1740, p. i). In both books Simpson cited De Moivre's works admiringly, and he did not claim much originality beyond the form of presentation and, in the book on annuities, more accurate tables. De Moivre's 1725 work was primarily mathematical, but it included Halley's Breslau tables as a basis for some of the computations. Simpson's 1742 work included tables based upon ten years of the London Bills of Mortality and presented several other tables of the values of annuities derived from this London experience. Simpson's relations with De Moivre were initially quite good [Lalande (1765) recounts their first meeting and quotes a warm testimonial from De Moivre on that occasion.] Simpson probably would have described these books as attempts to make the works of a master more up to date and to bring them to a wider audience. De Moivre, on the other hand, felt his own income threatened and reacted angrily in 1743 by bringing out a second edition of the *Annuities upon Lives* (the first edition was then eighteen years old) that contained a scathing preface:

> After the pains I have taken to perfect this Second Edition, it may happen, that a certain Person, whom I need not name, out of *Compassion to the Public,* will publish a Second Edition of his Book on the same Subject, which he will afford at a *very moderate Price,* not regarding whether he mutilates my Propositions, obscures what is clear, makes a Shew of new Rules, and works by mine; in short, confounds, in his usual way, every thing with a croud of useless Symbols; if this be the Case, I must forgive the indigent Author, and his disappointed Bookseller. (De Moivre, 1743, p. xii)

Simpson quickly printed a spirited reply, in a sixteen-page "Appendix containing Some Remarks on a Late Book on the same Subject, with Answers to some Personal and Malignant Misrepresentations, in the Preface thereof." Although De Moivre was not mentioned by name, the target was unmistakable. Simpson defended his own work, pointed out several

whom he used to guzzle porter and gin: but it must be observed that the misconduct of his family put it out of his power to keep the company of gentlemen, as well as to procure better liquor" (Hutton, 1795, vol. 2, p. 456). On the other hand, Lalande (1765, p. 203) tells us Simpson's conduct was in all regards "irreproachable."

errors in both editions of De Moivre's book, and added some expletives of his own, accusing De Moivre of displaying "an air of self-sufficiency, ill-nature, and inveteracy, unbecoming a gentleman" (Simpson, 1743, p. 16). De Moivre prepared a rejoinder, but his friends prevailed upon him not to publish it (Lalande, 1765, p. 202). When De Moivre issued a third edition of his *Annuities* in 1750, the only significant change was the removal of the offending paragraph from the preface.

The work of Simpson's with which we are immediately concerned was read to the Royal Society on 10 April 1755, barely four months after De Moivre's death at the age of eighty-seven. By this time Simpson was a professor at the Royal Military Academy at Woolwich and a Fellow of the Royal Society, and he had become concerned with a wide variety of problems in applied mathematics. His work took the form of a letter to the Earl of Macclesfield "On the Advantage of Taking the Mean of a Number of Observations, in practical Astronomy." Simpson began his letter with a statement of his object:

> My Lord,
>
> It is well known to your Lordship, that the method practised by astronomers, in order to diminish the errors arising from the imperfections of instruments, and of the organs of sense, by taking the Mean of several observations, has not been so generally received, but that some persons, of considerable note, have been of opinion, and even publickly maintained, that one single observation, taken with due care, was as much to be relied on as the Mean of a great number.
>
> As this appeared to me to be a matter of much importance, I had a strong inclination to try whether, by the application of mathematical principles, it might not receive some new light; from whence the utility and advantage of the method in practice might appear with a greater degree of evidence. (Simpson, 1755, pp. 82–83)

Simpson's treatment of this problem was quite limited—an uncharitable but truthful characterization of his result would state that he showed only that a mean is better than a single observation when the mean is based upon six measurements and a very specific and limited hypothesis is adopted about the probabilistic character of the observations. Nevertheless his treatment was distinguished by two key features, one conceptual and one technical, that were to recur in later, more fully developed treatments of the subject.

The conceptual development in Simpson's paper was his decision to focus, not on the observations themselves or on the astronomical body being observed, but on the errors made in the observations, on the differences between the recorded observations and the actual position of the body being observed. To those of us who are now used to dealing with such matters, this may seem like a trivial, even semantic, difference. To those in the mid-eighteenth century, who lacked theories of inference or inverse probability, it was the critical step that was to open the door to an applicable quantification of uncertainty. Simpson began by assuming a specific hypothesis for the distribution of the errors. He was able to focus his attention on the mean *error* rather than on the mean *observation*. Even though the position of the body observed might be considered unknown, the distribution of errors was, for Simpson, known. By basing his analysis upon this known distribution, he was able to come to grips with the problem without having to come to grips with a stochastic structure for the unknown position. By choosing an inference problem that was susceptible to what R. A. Fisher was to call the fiducial argument,[12] Simpson limited all questions to ones involving the fixed, known (by hypothesis) error distribution. What we now call the location invariance of the problem guaranteed that the error of the mean was equal to the mean of the errors; and the uncertainty of the inference (the error of estimation using the mean) could be measured in terms of one given distribution.

This was the critical point that distinguished Simpson's problem from De Moivre's and made it tractable even without a theory of inference. For all its attractive properties, the binomial distribution has the unfortunate feature that the distribution of the difference between an empirical relative frequency and the unknown true proportion depends upon the unknown true proportion. Thus binomial "error" distributions are not fixed; they cannot be taken as known (even by hypothesis) unless the true proportion is taken as known. Even for large numbers of trials the exponent in De Moivre's corollary 9 $[-l^2(a+b)^2/2abn]$ depends upon the unknown a and b. Putting the empirical relative frequency in for $a/(a+b)$ would (in those days) have been tantamount to accepting the unknown as known and vitiating the entire point of the calculation.

Specifically, Simpson supposed that each of n independent observations was susceptible to errors of possible sizes $-v, -v+1, \ldots, -3, -2, -1, 0, 1, 2, 3, \ldots, v$, with probabilities proportional to either $r^{-v}, \ldots, r^{-3}, r^{-2}, r^{-1}, r^0, r^1, r^2, \ldots, r^v$ (a possibility Simpson considered in a Proposition I) or $r^{-v}, 2r^{-v+1}, 3r^{-v+2}, \ldots, (v+1)r^0, \ldots, 3r^{v-2}, 2r^{v-1}, r^v$. Each choice of $r > 0$ provided a distribution for the errors. At first glance these seem to be an odd choice for families of error distributions. As Simpson's analysis makes clear, however, the choice was dictated by mathematical expediency; and therein lies Simpson's key technical development.

Simpson's choice of error distributions must have been made with one eye on De Moivre's earlier use of generating functions to find the distribution of the sum of the faces of several dice. In the *Miscellanea Analytica*

12. Fisher's usage comes from the term *fiducial point*, which was used in surveying and astronomy and meant a fixed point, Fisher's idea being that a distance is the same no matter which end point is fixed.

(1730, pp. 196–197) and again in the second edition of the *Doctrine of Chances* (1738, p. 37; reproduced in 1756, p. 41), De Moivre had shown that the distribution of the sum of the faces of n rolls of a fair die with f faces numbered 1 to f was given by the coefficients of the expansion of

$$(1 + r + r^2 + r^3 + \ . \ . \ . \ + r^{f-1})^n = \frac{(1 - r^f)^n}{(1 - r)^n}.$$

De Moivre's result may be regarded as an extension of Bernoulli's presentation of the binomial distribution, which in turn was derived from Newton's binomial theorem and before that from Pascal's arithmetic triangle — this approach may have the most impressive provenance of any in probability theory. De Moivre had proved that this approach worked by imagining a large die with one face numbered 1, r faces numbered 2, r^2 faces numbered 3, ... By expanding the multinomial $(1 + r + r^2 + \ . \ . \ . \ + r^{f-1})^n$ by a generalization of the binomial theorem he concluded that, because the kth *term* itself gave the chance the sum was k for the large die, the kth *coefficient* would give the same chance for the smaller die. Simpson had taken De Moivre's large die, even retaining the symbol r, and shifted the range so that the possibilities ranged equally below and above zero. Simpson's second distribution,

$$r^{-v}, 2r^{-v+1}, \ . \ . \ . \ , (v + 1)r^0, \ . \ . \ . \ , r^v$$

came from the first by adding two independent errors together, each with distribution

$$r^{-v/2}, r^{-(v+1)/2}, \ . \ . \ . \ , r^{-1}, r^0, r^1, \ . \ . \ . \ , r^{v/2}.$$

Thus he obtained two propositions for the price of one, merely doubling n and halving v to obtain the second.

Simpson's choice of error distributions was derived from De Moivre, but the use to which he put them was new. It seems fair to say that Simpson was interested in only those cases where $r = 1$ (symmetric error distributions), because his only calculation was for the case where the error distribution was proportional to 1, 2, 3, 4, 5, 6, 5, 4, 3, 2, 1. He did recognize the advantage of considering the more general case because of the ease with which the geometric series could be summed and the separate coefficients located when $r \neq 1$, after multiplying two series in r together. His work was thus an adumbration of the use of generating functions in statistical problems, a technique that was to reach full flower in the later works of Lagrange and Laplace.

Simpson's mathematical development was both enterprising and limited. He developed expressions for the chance that the mean error would not exceed a given quantity m/n for each of his cases—and here the inelegancy of the multinomial reasserted itself. His hard-won general expression for this chance was not, in his hands, amenable to a general analysis, and he fell back upon a calculation for a special case to verify the thesis of the title of his "letter." For the error distribution that is symmetric about zero and proportional to 1, 2, 3, 4, 5, 6, 5, 4, 3, 2, 1, thinking of a unit as one second of time (and thus taking the maximum error as five seconds), Simpson computed the probability that the mean of six such errors did not exceed one, and did not exceed two seconds. He compared these results to the same probabilities for a single measurement (Table 2.2). In addition to these numbers he stated that for this example "the chance for an error exceeding 3 seconds, will not be $\frac{1}{1000}$ part so great from the Mean of six, as from one single observation" (Simpson, 1755, p. 92).

Simpson was conscious that his hypothesis about a distribution of errors was quite restrictive, but he was quite (even overly) optimistic about the generalizability of his result:

> Should not the assumption, which I have made use of, appear to your Lordship so well chosen as some others might be, it will, however, be sufficient to answer the intended purpose: and your Lordship will find, on calculation, that, whatever series is assumed for the chances of the happening of the different errors, the result will turn out greatly in favour of the method now practised, by taking a mean value. (Simpson, 1755, p. 83)

His conclusion, the earliest statistical advice from mathematician to experimental scientist of which I am aware, was sweeping:

> Upon the whole . . . it appears, that the taking of the Mean of a number of observations, greatly diminishes the chances for all the smaller errors, and

Table 2.2. Simpson's calculated values of the probability that the mean of six errors does not exceed k seconds $[P(|\overline{X}| \leq k)]$ and the probability that a single error does not exceed k seconds $[P(|X_1| \leq k)]$.

Seconds k	$P(\lvert\overline{X}\rvert \leq k)$	$P(\lvert X_1\rvert \leq k)$	$\frac{P(\lvert\overline{X}\rvert > k)}{P(\lvert X_1\rvert > k)}$
1	$\frac{788814800}{1088391168}$ (=0.725)	(0.444)	(0.495)
	or $2\frac{2}{3}$ to 1	16 to 20	
	(2.63 to 1)	or $\frac{8}{10}$ to 1	
2	$\frac{1052311761}{1088391168}$ (=0.967)	(0.667)	(0.0994)
	or 29 to 1	2 to 1	$\frac{1}{10}$
	(29.2 to 1)		

Source: Based on Simpson (1755, p. 92; 1757, pp. 70–71).
Note: Numbers in parentheses are decimal equivalents not given by Simpson.

> cuts off almost all possibility of any great ones: which last consideration, alone, seems sufficient to recommend the use of the method, not only to astronomers, but to all others concerned in making of experiments of any kind (to which the above reasoning is equally applicable). And the more observations or experiments there are made, the less will the conclusion be liable to err, provided they admit of being repeated under the same circumstances. (Simpson, 1755, pp. 92–93)

This statement of a law of large numbers was unsupported by proof, and Simpson's advice was long anticipated by at least the better astronomers. Nevertheless, the idea of basing this law upon mathematical calculation was new. Simpson had seen that the concept of error distributions permitted a back-door access to the measurement of uncertainty. Later Laplace was to slip in this same back door and come around to open the front (only to find later that Bayes's key was already in the lock).

A Bayesian Critique

There is one bit of evidence that suggests that it may have been the publication by this first Thomas (Simpson) that awakened the interest in probability in a second Thomas (Bayes). In a letter from Bayes to John Canton that must date from 1755,[13] Bayes expressed skepticism about Simpson's result, stating that he agreed that the mean

> was undoubtedly the best upon the whole and perfectly adapted to prevent any considerable error, which might possibly be committed in a single observation. But I really think Mr. Simpson has not justly represented its advantage: neither is it by far so great as he seems to make it. According to him by multiplying our observations and taking the mean we always diminish the probability of any given error, and that very fast . . . Now that the errors arising from the imperfection of instruments and the organs of sense should be thus reduced to nothing or next to nothing only by multiplying the number of observations seems to me extremely incredible. On the contrary the more observations you make with an imperfect instrument the more certain it seems to be that the error in your conclusion will be proportional to the imperfection of the instrument made use of . . . [A]s I see no mistakes in Mr. Simpson's calculations I will venture to say that there is one in the Hypothesis upon which he proceeds. And I think it is manifestly this, when we observe with imperfect instruments or organs; he supposes that the chances for the same error in excess or defect are exactly the same, and upon this

13. The letter was located in the Canton papers (vol. 2, p. 32, coded Ca. 2.32) by John D. Holland in the early 1960s; I am grateful to Churchill Eisenhart for providing me with a copy. The letter apparently bears no date, but the opening ("You may remember a few days ago we were speaking of Mr. Simpson's attempt to show the great advantage of taking a mean") would seem to fix it soon after Simpson's letter was read or published.

hypothesis only has he shown the incredible advantage, which he would prove arises from taking the mean of a great many observations. (Canton Papers)

Bayes's observation was acutely perceptive—he went on to note that Simpson's claimed generalization to "whatever series [that is, error distribution] is assumed" (Simpson, 1755, p. 83) was "only true where the chances for the errors of the same magnitude, in excess or defect, are upon an average nearly equal." Bayes's comments would plausibly have been relayed to Simpson; if so Simpson must have recognized their merit because two years after the letter appeared in the *Philosophical Transactions*, Simpson published a small volume, *Miscellaneous Tracts on Some curious, and very interesting Subjects* (1757). These tracts included a revised edition of the letter that dropped the extravagant claims of the introduction to the original, replacing them with two suppositions:

1. That there is nothing in the construction, or position of the instrument whereby the errors are constantly made to tend the same way, but that the respective chances for their happening in excess, and in defect, are either accurately, or nearly, the same.
2. That there are certain assignable limits between which all these errors may be supposed to fall; which limits depend on the goodness of the instrument and the skill of the observer. (Simpson, 1757, p. 64)

These principles were to stand as implicit hypotheses for work on this topic for the next century and beyond. But if they may have found their earliest expression in Bayes's letter to Canton, and their earliest publication in Simpson's revision, there seems little reason to doubt that their later appearances on the continent were independent of Simpson. Citation was an uncertain art in the eighteenth century, but an explicit statement written about 1770 by the widely read Jean Bernoulli III that the only published works on the subject he knew of were those of Lambert and Boscovich (Bernoulli, 1785, p. 405) suggests that Simpson's letter was not widely read.[14]

Simpson's 1757 republication of the letter incorporated one other change—three and a half pages of additional material that is frequently cited now as the first publication of a continuous error distribution. Beginning with the discrete distribution that he had considered earlier (that distribution proportional to 1, 2, . . . , $v+1$, . . . , 2, 1), Simpson (1757, p. 71) proposed to consider the situation "when the error admits of any value whatever, whole or broken, within the proposed limits, or when the result of each observation is supposed to be *accurately* known." His analysis, in the spirit of the time, consisted of viewing the interval of

14. It was briefly referred to, however, by the erudite astronomer J. Lalande in the second edition of his major treatise *Astronomie* (1771, vol. 3, p. 760). I know of no other early citation.

possible errors (AB in Figure 2.3) as being divided into an indefinitely large number of indefinitely small subintervals, and effectively passing to the limit in his earlier expression for the discrete case. He wrote:

> Let, then, the line AB represent the whole extent of the given interval, within which all the observations are supposed to fall; and conceive the same to be divided into an exceeding great number of very small, equal particles, by perpendiculars terminating in the sides AD, BD of an isosceles triangle ABD formed upon the base AB: and let the probability or chance whereby the result of any observation tends to fall within any of these very small intervals N*n*, be proportional to the corresponding area NM*mn*, or to the perpendicular NM; then, since these chances (or areas) reckoning from the extremes A and B, increase according to the terms of the arithmetical progression 1, 2, 3, 4, &c. it is evident that the case is here the same with that in the latter part of Prop. II [where the discrete triangular distribution was studied]; only, as the number v (expressing the particles in AC or BC) is indefinitely great, all (finite) quantities joined to v, or its multiples, with the signs of addition or subtraction, will here vanish, as being nothing in comparison to v. (Simpson, 1757, p. 72)

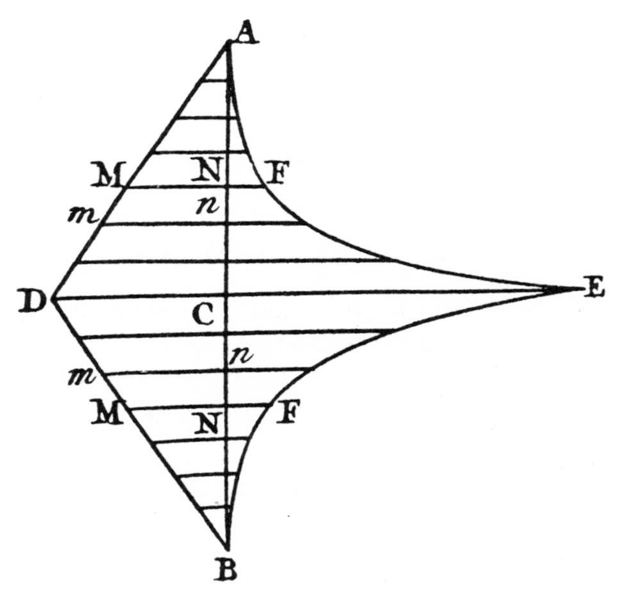

Figure 2.3. Thomas Simpson's 1757 graph of a triangular error density ADB and the density of the mean of t *errors AFEFB. (From Simpson, 1757.)*

In modern terminology, Simpson's idea is that we may consider the triangle ABD as a probability density along the axis AB, such that, for example, the area of the trapezoid NM*mn* gives the probability that the error falls in the interval N*n*. Then because when N*n* is small this probability is approximately the height NM times the length of the interval N*n* and because the heights NM form an arithmetical progression such as he had studied earlier, the earlier results can be applied by letting v increase to an infinite limit. This he proceeded to do; the curve AFEFB represented the probability density for the mean error. He integrated the resulting density to find that the probability that the mean of t errors did not exceed $(1 - y/t) \times$ CA in absolute value (that is, did not exceed the fraction $1 - y/t$ of the maximum possible error CA) to be:

$$1 - \frac{2}{1 \cdot 2 \cdot 3 \cdot \ . \ . \ . \ \cdot n} \left\{ y^n - n(y-1)^n + \frac{n}{1} \cdot \frac{(n-1)}{2} \cdot (y-2)^n - \frac{n}{1} \cdot \frac{(n-1)}{2} \cdot \frac{(n-2)}{3} (y-3)^n + \&\text{c.} \right\},$$

where here $n = 2t$ and the "series is to be continued till the quantities y, $y - 1$, $y - 2$, &c. become negative" (Simpson, 1757, p. 73). Simpson's derivation of the distribution of a sum of triangularly and continuously distributed errors anticipated later derivations (using different methods) of essentially the same formula by Lagrange and Laplace by more than ten years. It was an impressive achievement for the time, but there is no evidence that it received the audience it deserved. Simpson himself made a few calculations for the case $t = 6$ and then dropped the matter, never to take it up again. He died four years later, and the only further work he did on a statistical problem appears to have been an unpublished manuscript fragment treating Boscovich's problem analytically (Stigler, 1984).

The second Thomas I have mentioned, Thomas Bayes, was a minor figure in the history of science whose published works show a spark of intelligence few of his contemporaries possessed. The date of Bayes's birth is unknown, but it was probably in 1701; he died 7 April 1761, close to a month before Simpson died. Bayes was an ordained Nonconformist minister in Tunbridge Wells (about 35 miles southeast of London), but his mathematical interests must have been known to his contemporaries despite a lack of technical publications because he was elected a Fellow of the Royal Society in 1742. Not even one of his works on mathematics was published during his lifetime, but one of two posthumous works of his served to embed his name in what has become, two centuries after his death, one of the most widely known eponyms in all of science, Bayesian inference. It is ironic that the work responsible for this fame, the posthumously published "An Essay toward solving a Problem in the Doctrine of

Chances" (Bayes, 1764), was ignored by his contemporaries (save Richard Price) and seems to have had little or no impact upon the early development of statistics. The ideas this essay contains have been of vast influence; their association with Bayes antedates their rediscovery and development by Laplace.

The origin of Bayes's "Essay" is a mystery. Richard Price, who found this work in Bayes's papers, read it to the Royal Society on 23 December 1763, more than two and a half years after Bayes's death. Price himself contributed an introduction and an appendix that added to the "Essay" in nontrivial ways, but Price tells us Bayes had himself prepared an introduction, and we have no way of telling exactly what is Bayes and what is Price. Bayes's motive for taking up the question has been the cause of some speculation; some have even guessed that an attempt to provide a mathematical proof of the existence of a First Cause lay behind it. Another, perhaps more plausible explanation is that the question arose after Bayes read Simpson's letter in 1755, because he was drawn back to Simpson's 1740 book and De Moivre's 1738 edition of the *Doctrine of Chances* (both cited by Price in his introduction to the "Essay"). Bayes then saw a common element in De Moivre's unanswered inference problem and Simpson's evaluation of the Mean. But this scenario is conjectural; the "Essay" itself gives no clues.[15]

Whatever its origin, Bayes's "Essay" was a tentative but direct attack upon the inference problem for the binomial that De Moivre had left hanging. His approach was systematic, in the manner of De Moivre's treatises; and it adopted the geometric approach of Newton rather than the analytic approach that was emerging on the continent in the works of Euler and others. For all the clarity of insight that a determined modern reader can find in Bayes's "Essay" today, it must have seemed a curious and obscure effort to any early continental eyes that chanced upon it, a factor that no doubt contributed to the small amount of attention it received. In the line of development we are pursuing it would fit most neatly about 1780, which is also the time it seems to have come to European notice. I shall discuss Bayes's approach after analyzing Laplace's rediscovery, following this two-century-old historical precedent.

15. Another tantalizing possibility is that Bayes had seen an earlier hint of the result in David Hartley's 1749 *Observations on Man*. Still another possibility is that Bayes himself had done this work before 1749 and been in turn the inspiration for Hartley's hint. These possibilities are discussed in Stigler (1983).

3. Inverse Probability

Pierre Simon Laplace (1749–1827), as portrayed in 1799

MANY EIGHTEENTH-CENTURY scientists had at least a vague feeling that probability would underlie an eventual successful treatment of social data. From Jacob Bernoulli on we find repeated expressions of hope that this would be true, statements that help explain why many of the greatest mathematical scientists of the century expended their efforts on the study of games of chance that might otherwise have seemed a frivolous scientific pursuit. Yet these feelings remained vague, and only at an instinctive level can it be said that these works were inspired by social concerns. Some successes in the application of probability to the valuation of annuities, notably by De Moivre, had helped spur on this work, but as a tool for the reduction and measurement of uncertainty in data, the calculus of probability had proved largely sterile.

The chief conceptual step taken in the eighteenth century toward the application of probability to quantitative inference involved the inversion of the probability analyses of Jacob Bernoulli and De Moivre. We have seen how elusive this concept had proved to be in early work and how it was only with Simpson and Bayes that some measure of success was attained. In those works the first key to the successful step was to get away from games of coins and dice, where some mathematical steps were simple but the conceptual step to inverse inference was not, and to address problems of astronomical observation where this conceptual step could more easily be taken. Simpson had done this by focusing on the errors of the observations rather than on the observations themselves; and I have presented at least a tentative case that it was this step of Simpson that had encouraged Bayes's lone effort in inference. But as suggestive as Simpson's and Bayes's works are with our present historical hindsight, they did not stimulate these workers' contemporaries. In fact, Bayes's failure to publish his own paper may well have been due more to his own view of the work as analytically incomplete and of little or no practical use than to any philosophical misgivings. Laplace was able to overcome that difficulty.

Laplace and Inverse Probability

Laplace's early work on the application of probability to inference came in four bursts between 1772 and 1781. Two of these efforts were published in his lifetime—the second and the fourth. But it is only by keeping in mind all four that we can recover the order in which his ideas developed and see how his initial triumphs over the barriers that had stymied Jacob Bernoulli and De Moivre were inspired by much the same currents of thought that had spurred on Simpson and Bayes and how Laplace's analytic superiority had allowed him to open the door Bayes had tried before him. Laplace's printed works of this period were two important memoirs, his "Memoir on the Probability of the Causes of Events," probably mostly written between March and June of 1773 (Laplace, 1774; Stigler, 1978b), and his "Memoir on Probabilities," read to the Academy on 31 May 1780 (Laplace, 1781). These are among the most important and most difficult works in the early history of mathematical probability, and together they may be taken as the most influential eighteenth-century work on the use of probability in inference.

The inversion of probability statements involving the binomial distribution had proved to be a difficult step, but the same was far less true for problems of astronomical observation. Once a random distribution of errors was conceived, as was done, perhaps independently, by Simpson in 1755, Lambert in 1760, and Lagrange in about 1769, the inversion could follow almost inadvertently. If e represents the error, O the observation, and P the point observed, then $O = P + e$ implies equally well that $P =$

$O - e$. If e is taken as randomly and symmetrically distributed, then supposing P fixed gives a distribution for O; and conversely, taking O as given leads to a distribution for P. By treating only the *difference* $e = O - P$ as random, a symmetrical situation is created where inversion becomes most natural. R. A. Fisher was to call this a fiducial argument, borrowing a term for a fixed point from surveying and astronomy that suggested that a distance is the same regardless of which end point is fixed. Statisticians now know that this argument is not so simple as a naive view might have it, that deep and subtle philosophical and mathematical points must be dealt with if this most natural approach is to lead to a coherent theory of inference. But in the mid-eighteenth century the argument was a conceptual liberation. Whether expressed in mathematical symbols or in words, it led naturally to a view in which one distribution—the distribution of errors—provided the random element for both "forward" (in time) and "inverse" probability statements. And it suggested an idea of inverse inference, reasoning *probabilistically* from effect (O) to cause (P), that was to bear fruit in other applications as well.

Laplace's memoirs of 1774 and 1781 are similar in format, but this format is deceptive. They begin with general discussions of probability, statements of its applicability to "civil life," and examinations of problems involving games. Only later do they move on to the problem of choosing the mean of a set of astronomical observations. Yet the evidence now available suggests that the ideas were developed in an order that was quite the reverse of this. Both, in fact, were expansions of earlier unpublished works on the handling of astronomical data.

In his 1774 memoir, Laplace begins by boldly announcing a "Principe," a principle of inverse probability, and gives four examples of its application, the first three described in terms of drawing tickets from urns and games of chance. Only with the fourth does he move to a problem with a clear practical motivation: *Determine the mean that one should take among three given observations of the same phenomenon.* In introducing this problem, however, Laplace tells us that this portion of the work had remained in his drawer for some time. He wrote:

> We can, by means of the preceding theory, solve the problem of determining the mean that one should take among many given observations of the same phenomenon. Two years ago I presented such a solution to the Academy, as a sequel to the Memoir "Sur les Séries récurrorécurrentes" printed in this volume, but it appeared to me to be of such little usefulness that I suppressed it before it was printed. I have since learned from Jean Bernoulli's astronomical journal that Daniel Bernoulli and Lagrange have considered the same problem in two manuscript memoirs that I have not seen. This announcement both added to the usefulness of the material and reminded me of my ideas on this topic. I have no doubt that these two illustrious geometers have

> treated the subject more successfully than I; however, I shall present my reflections here, persuaded as I am that through the consideration of different approaches, we may produce a less hypothetical and more certain method for determining the mean that one should take among many observations. (Laplace, 1774, p. 634)

In fact, Laplace was not correct in his guess about how successful Daniel Bernoulli and Lagrange had been in treating the problem of choosing a mean, as he would learn later (the Bernoulli and Lagrange manuscripts were only published in 1778 and 1776, respectively, and Bernoulli's had undergone considerable change and enlargement by the time it was published). But this passage is quite revealing for what it suggests regarding the motivation and order of development of Laplace's early work on inference. The Academy's unpublished proceedings tell us that the memoir "Sur les Séries récurrorécurrentes" was read on 5 February 1772; and the journal of Jean Bernoulli that is referred to was published in May of 1772 (Stigler, 1978b). This suggests that (1) February–April 1772 was the time of Laplace's first work on inference, (2) the motivation for this early work was the problem of reconciling discrepant observations, and (3) it was the social spur of learning of related work by Lagrange and Daniel Bernoulli that drove Laplace on to develop his early groping toward a "best mean" into a general theory of reasoning from events to their causes.

The principle with which Laplace began his 1774 memoir was stated thusly:

> If an event can be produced by a number n of different causes, then the probabilities of these causes given the event are to each other as the probabilities of the event given the causes, and the probability of the existence of each of these is equal to the probability of the event given that cause, divided by the sum of all the probabilities of the event given each of these causes. (Laplace, 1774, p. 623)

This is a statement of what is now called Bayes's theorem, with the probability of causes taken a priori equal. To facilitate our discussion of this principle and its genesis, we shall introduce some notation not found in either Laplace or Bayes. Let $A_1, A_2, \ldots, A_n$ be the n causes and let E denote the event we have focused on for discussion. Then the principle is

$$\frac{\mathrm{P}(A_i|E)}{\mathrm{P}(A_j|E)} = \frac{\mathrm{P}(E|A_i)}{\mathrm{P}(E|A_j)},$$

or

$$\mathrm{P}(A_i|E) = \frac{\mathrm{P}(E|A_i)}{\sum_j \mathrm{P}(E|A_j)}.$$

Because the full modern statement of the result would be

$$P(A_i|E) = \frac{P(E|A_i)P(A_i)}{\sum_j P(E|A_j)P(A_j)},$$

we can see that Laplace's principle amounts to Bayes's theorem (modern version) with a tacit adoption of equal a priori probabilities, $P(A_i) = 1/n$. To a modern eye this may seem a severe limitation, but it would have been less so to Laplace. His basic approach to the calculation of chances, borrowed from Jacob Bernoulli and De Moivre, was to reduce the formulation of all problems to an enumeration of equally likely cases. Doing this was more calculational expediency than metaphysical assumption. The calculus of probabilities was only able to deal efficiently with equally likely cases (prior to Laplace's later work). In this context an assumption of equally likely causes would have been understood by Laplace to mean that the problem is specified—the "causes" enumerated and defined—in such a way that they are equally likely, not that any specification or list of causes may be taken a priori equally likely. The principle of insufficient reason (to use a later name for the assumption that causes not known to have different a priori probabilities should be assumed a priori equally likely) would therefore be less a metaphysical axiom than a simplifying and approximative assumption invoked to permit calculation. Because Laplace considered a priori probabilities to be statements relative to our a priori knowledge of the causes of events, he supposed that if the causes under one specification were known not to be equally likely, we would respecify them in such a way that they were equally likely, say, by subdividing the more likely causes.[1] There is, however, no indication that he was aware of the difficulties inherent in this approach, difficulties that would be pointed out later.

Where did Laplace get his principle? The statement and approach were so different from those of Bayes that we can discount the possibility that Laplace had seen Bayes's essay.[2] Bayes's development had been an orderly

1. In 1781 and 1786 Laplace dealt explicitly with nonuniform a priori probabilities in the context of inference for the binomial distribution; I shall comment on this work later.

2. In addition to the internal dissimilarity of Bayes's and Laplace's statements, there is other evidence to suggest that Bayes's essay was not noticed on the continent prior to about 1780. In particular, it was apparently not known to Condorcet in 1774 when he wrote an introduction to the volume containing Laplace's memoir, nor was it referred to in the report of the examiners of the first memoir Laplace read to the Academy announcing his approach (read 10 March 1773; see Stigler, 1978b). Apparently Lagrange was also ignorant of Bayes's work because on 13 January 1775 he wrote to Laplace, after receiving a copy of the 1774 memoir, praising "the novelty of this material." He also wrote, "it is a very important new branch that you have added to the theory of chances . . . this theory has taken a new face and become a new science" (Lagrange, 1892, pp. 58–60). For other comments on the lack of knowledge of Bayes before 1780, see Stigler (1975a, 1978b).

(albeit obscure) progression, starting with a definition of probability and a derivation of the relation $P(A_i|E) = P(A_i \cap E)/P(E)$. In Bayes's work the relationship is derived explicitly and rigorously from first principles, including a rather explicit argument for the assumption of equally likely causes. In Laplace's work the relationship emerges full grown, without proof, and with only a tacit reliance on equally likely causes. Laplace gave no indication how he had arrived at the principle other than the earlier cited hint that he was led in this direction by consideration of means. When this principle is presented later in the *Théorie analytique des probabilités* (Laplace, 1812, p. 182), an orderly proof is given, with the assumption of equally likely causes made explicit. The closest he comes to a derivation in his early work, however, is a curious one presented in his 1781 memoir. Rephrased in modern notation, it can be described as follows (Laplace, 1781, pp. 415–417): Consider the occurrence of the event E *twice*. Let E_1 denote its occurrence on the first trial, E_2 its occurrence on the second trial, and E its generic occurrence on any single given trial. Now, $P(E_2|E_1)$ may be computed in two different ways:

$$\text{(i)}\quad P(E_2|E_1) = P(E_1 \cap E_2)/P(E_1),$$

$$\text{(ii)}\quad P(E_2|E_1) = \sum_i P(E|A_i)P(A_i|E).$$

The first of these is the definition of conditional probability; the second follows from what we now call the law of total probability, with the tacit assumption that E_1 and E_2 are conditionally independent given the cause A_i:

$$\begin{aligned} P(E_2|E_1) &= \sum_i P(E_2 \cap A_i|E_1) \\ &= \sum_i P(E_2|A_i \cap E_1)P(A_i|E_1) \\ &= \sum_i P(E_2|A_i)P(A_i|E_1) \\ &= \sum_i P(E|A_i)P(A_i|E). \end{aligned}$$

Laplace then used the supposition of equally likely causes A_i to give

$$\begin{aligned} P(E_1) &= \frac{1}{n}\sum_i P(E_1|A_i) \\ &= \frac{1}{n}\sum_i P(E|A_i) \end{aligned}$$

and

$$\begin{aligned}
P(E_1 \cap E_2) &= \frac{1}{n} \sum_i P(E_1 \cap E_2 | A_i) \\
&= \frac{1}{n} \sum_i P(E_1 | A_i)\, P(E_2 | A_i) \\
&= \frac{1}{n} \sum_i P(E | A_i)\, P(E | A_i) \\
&= \frac{1}{n} \sum_i [P(E | A_i)]^2 .
\end{aligned}$$

Substituting these into (i) and equating (i) and (ii) gave

$$\sum_i P(E|A_i)\, P(A_i|E) = \frac{\sum_i [P(E|A_i)]^2}{\sum_i P(E|A_i)} .$$

Now in like manner, by considering the chance that E occurs $k + 1$ times given that it occurs once, he had

$$\sum_i [P(E|A_i)]^k\, P(A_i|E) = \frac{\sum_i [P(E|A_i)]^{k+1}}{\sum_i P(E|A_i)} .$$

For $k = 1, 2, \ldots, n - 1$, these are $n - 1$ equations linear in the $P(A_i|E)$'s, to which we can add $\Sigma_i P(A_i|E) = 1$. Thus they can be solved; clearly the solution is

$$P(A_i|E) = \frac{P(E|A_i)}{\sum_j P(E|A_j)} .$$

A close consideration of the 1774 memoir, however, will suggest that this argument may have been merely an afterthought. In fact the principle probably had sprung most naturally from Laplace's peculiarly limited notion of a conditional probability distribution, inspired by the "fiducial" manner in which he had expressed the problem of determining a best mean.

The Choice of Means

Laplace's first treatment of this problem dates from early 1772 and was appended as Problem 3 to the 1774 memoir. His treatment was extremely limited—it only extended to the case of three observations! This case was already enough to enmesh him in a fifteenth degree equation; and there can be no doubt that it was this analytic inelegancy in the simplest nontrivial case that convinced him that the approach did not merit publication because of its limited usefulness. Laplace's treatment did, however, contain important new ideas that were to characterize much of his and others' later work. First, he introduced a new criterion for the choice of a "best mean"; and second, he showed how, at least in principle, the new inverse probability could be brought to bear on nontrivial practical problems.

Laplace's discussion was conducted in reference to three figures (see Figure 3.1). Laplace considered the problem to be as follows: Given three observed times of a phenomenon (a, b, and c) along the time axis AB, find the point V that we should take as the true time of the phenomenon. In determining this point he supposed that if V is the true instant of the phenomenon, then the probability of an observation differing from the truth by an amount x was given by a curve $y = \phi(x)$, shown as ORM in the middle diagram. Two points remained to be settled. What curve should be taken as the error curve $\phi(x)$? And given $\phi(x)$, how should the mean be determined?

Laplace listed three conditions that $\phi(x)$ should satisfy: First, the curve should be symmetrical about V "because it is just as probable that the observation deviates from the truth to the right as to the left." Second, the curve must decrease toward the axis KP "because the probability that the observation differs from the truth by an infinite distance is evidently zero." Third, the area under the curve must be one "because it is certain that the observation will fall on one of the points of the line KP." These conditions did not determine $\phi(x)$, but they did permit some general reflections on the problem.

If a modern statistician were to approach the problem in the context of Laplace's formulation (as I have thus far described it), he or she might proceed as follows: Given the true instant V, the probability of three observations assuming values a, b, and c is

$$\phi(a, b, c|V) = \phi(a - V)\,\phi(b - V)\,\phi(c - V),$$

where the ϕ on the left is a modern generic expression for "probability density of" and the ϕ's on the right are Laplace's error curve. Then Laplace's principle, or Bayes's theorem with a uniform prior distribution on V, would give the distribution of V given a, b, c as

$$\phi(V|a, b, c) \propto \phi(a, b, c|V),$$

or,

$$\phi(V|a, b, c) = \frac{\phi(a - V) \cdot \phi(b - V) \cdot \phi(c - V)}{\int \phi(a - U) \cdot \phi(b - U) \cdot \phi(c - U)dU}.$$

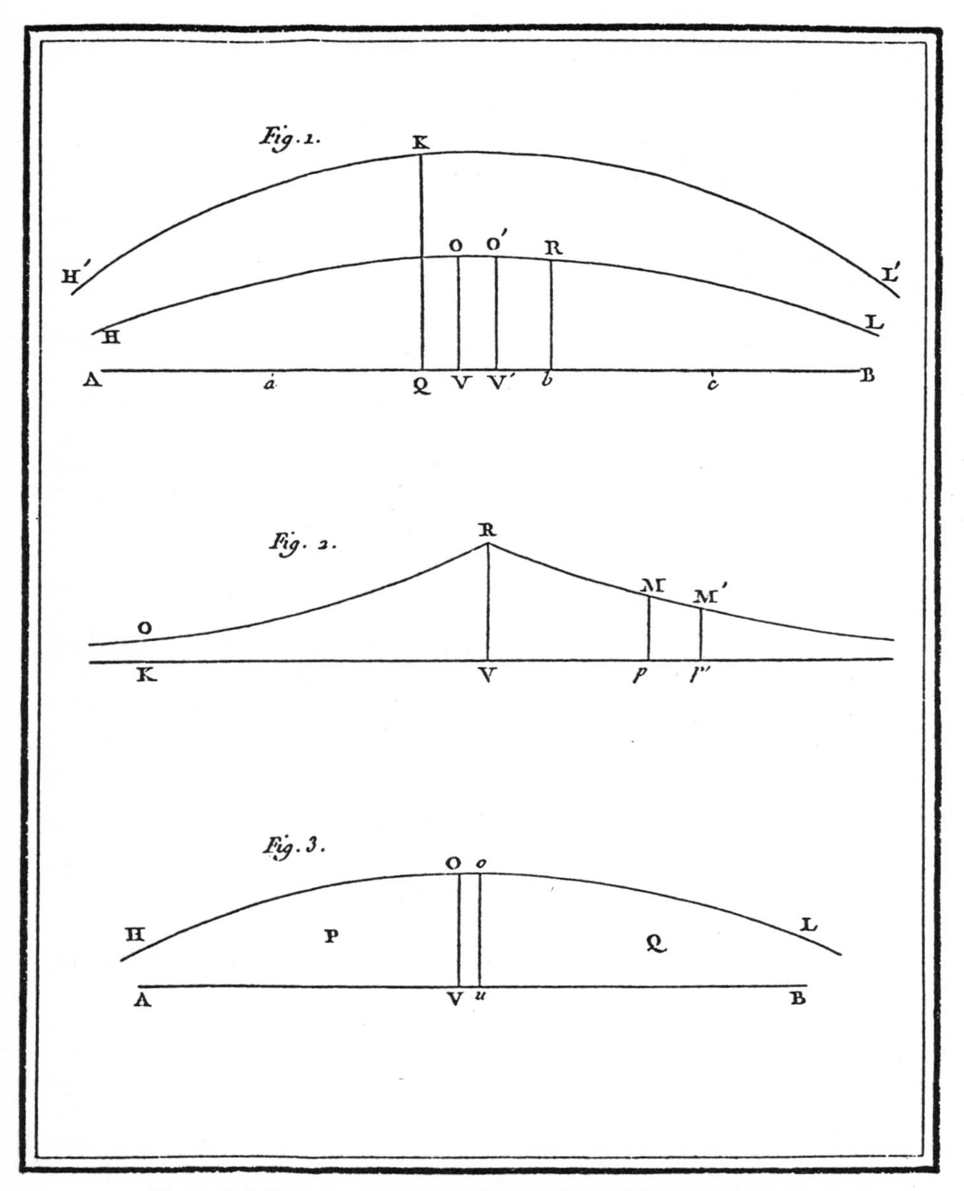

Figure 3.1. Diagrams accompanying Laplace's 1774 memoir on inverse probability. His Figure 2 shows the double exponential density, and his Figures 1 and 3 show posterior distributions, the latter illustrating his argument that the posterior median minimizes the posterior expected error. (From Laplace, 1774, facing p. 656.)

This is essentially how Laplace *did* proceed—but with one subtle difference that may have helped him grope toward his inverse principle in his original 1772 attack on this problem.

Instead of proceeding as above, Laplace reexpressed the problem in terms of differences in a way that helped emphasize the fiducial nature of the problem. Laplace let $p = b - a$ and $q = c - b$ be the differences between the observations; and he let $x = V - a$, the correction that should be added to a in order to find V. Then he gave the probability of the differences x, p, and q as $\phi(x)\phi(p - x)\phi(p + q - x)$. Laplace then invoked his principle. If the "true instant" were at V' instead of V, and $x' = V' - a$, "then, this probability would be $= \phi(x')\phi(p - x')\phi(p + q - x')$. It follows then from our fundamental principle of article II that the probabilities that the true instant of the phenomenon is at the points V or V', are to each other as $\phi(x)\phi(p - x)\phi(p + q - x):\phi(x')\phi(p - x')\phi(p + q - x')$." That is, his principle gave the probability distribution of the correction x given p and q to be proportional to $y = \phi(x)\phi(p - x)\phi(p + q - x)$, the curve HOL in the top diagram of Figure 3.1. By this reexpression—by his submersion of the fixed, true instant V in an expression involving only differences—Laplace was able to liberate his view from one of cause-to-effect in the same way Simpson had. It was this liberation, I conjecture, that led Laplace (as it had led Bayes) to the general treatment of inverse probability, after he had seen how it might proceed in one intuitively well understood example.

In terms of our modern generic expression for "probability density of," Laplace had found $\phi(x, p, q) = \phi(x)\phi(p - x)\phi(p + q - x)$. The point V had disappeared from view, its presence only felt through the first error x, a random and unobservable quantity. It was x that then became the object of the investigation, the "cause" to be found from the observable "events" p and q.[3] The distribution of these events given the cause x, $\phi(p, q|x)$, was then proportional to $\phi(x, p, q)$. Therefore, by the principle, the distribution of the cause was also $\phi(x|p, q) \propto \phi(x, p, q)$. This formalization of Laplace's reasoning can be deceptive, for it suggests a deeper understanding of conditional probability than was in fact available to Laplace. Nonetheless, it provides an accurate and concise statement of the end result of his analysis thus far.

The problem then became how to use this curve of the probabilities of the correction x to choose a best single value for x. For the modern statistician, the question would be how, given the posterior distribution of the true instant V, to decide on a single best value. Laplace described two criteria for choosing this value. One might, he said, choose that value of x

3. We now recognize x as a pivotal quantity and p and q as ancillary statistics; hence Laplace's treatment is an early form of a type of conditional analysis later discussed by R. A. Fisher.

such that it is equally probable that the true x falls before it or after it. He called this the *mean of probability;* we call it the posterior median. Alternatively, one might choose that x "that *minimizes* the sum of the errors to be feared multiplied by their probabilities." Laplace called this the *mean of error,* or *astronomical mean,* because, he said, "it is that which astronomers should give preference to." We now recognize this as a description of the value that minimized the posterior expected loss, loss being measured as the absolute value of the difference between the chosen x and the true x. Laplace then showed in perhaps the most elegant simple proof of early mathematical statistics that these two means were identical! The astronomical mean and the mean of probability were one and the same and could be found by bisecting the area under the curve HOL.

Laplace's proof was mechanical in nature and made reference to the third diagram in Figure 3.1. He noted that if P was the center of gravity of AVOH, and z' was the distance from P to OV, and if Q was the center of gravity of BuoL, and z the distance from uo to Q, and if the distance from V to u was an infinitesimal dx, then the posterior expected loss if V is the selected value is $Mz + Nz' + \frac{1}{2}y(dx)^2$, where M is the area of BuoL, N is the area of AVOH, and y is the height OV. Similarly, the expected loss if u is selected is $M(z - dx) + N(z' + dx) + \frac{1}{2}y(dx)^2$. If V is to be the minimizing value, these must be equal. Their difference is $Ndx - Mdx$. Thus we must have $N = M$.

The Deduction of a Curve of Errors in 1772–1774

At a theoretical level Laplace had solved the problem of choosing the best mean among discordant measurements. To minimize the expected error, one simply had to bisect the probability distribution of the true value V (or, equivalently, of the correction x to be made to the first measurement). Up to this point, Laplace's treatment was valid for any number of observations, notwithstanding the fact that the exposition was in terms of only three. The restriction to three observations did serve notice, though, of the difficulties Laplace had encountered in trying to implement his theoretical solution, because, to actually choose a mean in practice using Laplace's criterion, a specific form for the curve of errors $\phi(x)$ was needed.

Of all aspects of the problem of choosing a mean, the determination of the curve of errors proved to be the thorniest. Laplace made two major efforts in this direction, and he considered neither to be totally successful. The first attempt was presented in his 1772 work, and there his lack of success is evident both in his need to restrict himself to three observations and in his initial suppression of the work. His second attempt, even more formidable, was presented in a memoir read on 8 March 1777; again publication was deferred, this time to 1781, at which time the material appeared near the end of the memoir. Both of Laplace's attempts were remarkable technical achievements for their time, however; and both are worth considering for what they tell us about his early views of this problem and the conceptual difficulties in its solution.

It is, of course, extremely easy to specify an error curve, and it was this extreme simplicity that bedeviled early efforts by Laplace and others. All early workers, from Simpson on, accepted it as given that the curve should be symmetrical and that the chance of an error should decrease toward zero as the magnitude of the error increased. Laplace himself had repeated these conditions in his 1774 memoir, as I have noted. The problem was that there were *too many* possibilities, and the choice of just one was critical to the mean to be obtained. As Laplace wrote, "But of an infinite number of possible functions, which choice is to be preferred?"

Earlier writers had confronted the choice of an error curve by making an arbitrary selection. Simpson, for example, had based his choice of the uniform and triangular distributions upon mathematical expediency. For his limited aim (showing that an arithmetic mean was better than a single observation) this arbitrariness was not a major drawback—his audience only needed to accept the distribution as qualitatively correct in order to believe his qualitative conclusion—that averaging increased accuracy. Laplace, however, was attempting an exact computation, and he needed both a convincing case for any choice of $\phi(x)$ he might make[4] *and* a mathematical analysis equal to the task of calculating the best mean for that curve.

Within the context of post-Newtonian scientific thought, the only acceptable grounds for the choice of an error distribution were to show that the curve could be mathematically derived from an acceptable set of first principles. As the inverse square law was the touchstone of mathematical astronomy, so the principle of equally likely cases was that of mathematical probability. If a choice of a curve of errors was to be found acceptable, it must be reducible in some sense to a description in terms of cases supposed equally likely, or indifferently indistinguishable. Both of Laplace's early derivations fall within this paradigm.

4. The case of Daniel Bernoulli illustrates the difficult reception an inadequately justified solution to the problem could expect. His memoir (Bernoulli, 1778) provided such a scant case for either his error curve (essentially $\phi(x) = a^2 - x^2$, for $|x| < a$) or his mode of choosing a mean (essentially an adumbration of maximum likelihood) that Euler appended a critical commentary to the published version and the work was largely ignored by both contemporaries and posterity. See Pearson and Kendall (1970) for a translation of the memoir and commentary and Sheynin (1972) for relevant discussion. An early draft of Bernoulli's memoir, without any explicit error curve and no "maximum likelihood," was circulated in Berlin in 1769 and is today at the university library in Basel. For an outline of the early draft, see Bernoulli (1785).

In his earliest work, Laplace actually entertained and dismissed two different possible curves before settling on one that today bears his name. He first noted that, if (referring to the middle diagram in Figure 3.1) the points P and P′ were supposed equally probable, the curve $\phi(x)$ would have to be a constant, a possibility he rejected as contrary to the assumption that large errors were less likely than were small ones. He then briefly considered (and also rejected) the possibility that the *decrease* in $\phi(x)$ (as x increases) was constant; that is, he rejected the triangular density.[5] As he had rejected a priori the assumptions that we have no reason to suppose values of $\phi(x)$ different and that we have no reason to suppose values of $d\phi(x)/dx$ different, he now pushed this principle of insufficient reason one step further and found a hypothesis he *could* accept. Since (for positive x) we could suppose both $\phi(x)$ and $d\phi(x)/dx$ decreased with x, what can we say about the relationship between their two *rates* of decrease?

> Now, as we have no reason to suppose a different law for the ordinates than for their differences, it follows that we must, subject to the rules of probabilities, suppose the ratio of two infinitely small consecutive differences to be equal to that of the corresponding ordinates. We thus will have $\frac{d\phi(x+dx)}{d\phi(x)} = \frac{\phi(x+dx)}{\phi(x)}$. Therefore $\frac{d\phi(x)}{dx} = -m\phi(x)$, which gives $\phi(x) = \mathscr{C}e^{-mx}$. Thus this is the value that we should choose for $\phi(x)$. The constant $\mathscr{C}$ should be determined from the supposition that the area of the curve ORM equals unity, which represents certainty, which gives $\mathscr{C} = \frac{1}{2}m$. Therefore $\phi(x) = \frac{m}{2}e^{-mx}$, e being the number whose hyperbolic logarithm is unity. (Laplace, 1774). [Of course it was understood here that x is taken as positive. We would now write $\phi(x) = (m/2)e^{-m|x|}$, but Laplace lacked a notation for absolute value.]

Pushing this principle of indifference, or of insufficient reason, to apply to *rates* of decrease, Laplace was thus led to an elegant expression for an error curve, what we now call the double-exponential or Laplace distribution. But Laplace soon learned, as have generations of graduate students in this present century, that this elegance of expression does not bring with it a corresponding ease of analysis for statistical problems. The difficulty of evaluating the best mean by bisecting the curve HOL forced Laplace to limit his discussion to only three observations, and even then he encountered such severe difficulties that his great analytic skills barely sufficed. And in the process he made a subtle and revealing error.

Laplace began simply enough. By supposing (with no loss of generality)

5. The reasoning was vintage Laplace: "We can also easily see that this diminution cannot be constant, that it must become less as the observations deviate more from the truth."

$a < b < c$, he evaluated the integral of $\phi(x)\phi(p - x)\phi(p + q - x)$ by considering its different forms over each of the four regions determined by the observations. He then proceeded to find the "best" value of x according to his criterion. For the case where $p = b - a > q = c - b$, it turned out to be

$$(1) \qquad x = p + \frac{1}{m} \ln(1 + \frac{1}{3} e^{-mp} - \frac{1}{3} e^{-mq}).$$

Now, this result permitted one important conclusion to be drawn, but it also presented Laplace with a new and very hard problem. The conclusion he drew was that the arithmetic mean was not the best method of combining observations, for that mean would correspond to the correction $x = (2p + q)/3$, a result that only follows from equation (1) "when $m = 0$ or is infinitely small." Laplace rejected such a value of m, saying (in reference to the middle diagram of Figure 3.1) that it "makes all points on the line KP equally probable, at least up to an extremely large distance, which is very unlikely both by the nature of the problem and by the result of calculation, as we shall see in a moment." The "calculation" he referred to was needed to resolve the new problem that (1) presented: The best value of x depends on the unspecified value of m.

The dependence of statistical analyses upon the relationship between the scale of measurement and the likely size of measurement errors had arisen before Laplace's time. Many astronomers had informally discarded extremely discordant observations. Thomas Simpson had on external grounds arbitrarily specified an upper limit for his error distribution in his 1755 paper on the arithmetic mean. Simpson (1755, pp. 90–91) described his limits as "the limits of the errors to which any observation is subject. These limits, indeed, depend on the goodness of the instrument, and the skill of the observer; but I shall suppose here, that every observation may be relied on to 5 seconds." But prior to Laplace, no analytic treatment of this problem based internally on the observations themselves had been attempted. Allowance for a scale factor such as m is not an altogether simple matter. There is no unanimous agreement today on how to best handle the problem. Parameters such as m are often referred to as "nuisance parameters," as they are not the immediate object of attention; but they can have a big effect upon the determination of more important quantities. The fact that Laplace even attempted to come to grips with this problem showed a bold imagination.

Laplace approached the problem of unknown scale m as he had the earlier one, by applying his principle of inverse probability. He had earlier noted that the probability distribution of x, p, and q given m was given by $\phi(x)\phi(p - x)\phi(p + q - x)$. Then the distribution of just p and q given m was the integral of this function with respect to x; and by his principle the distribution of m given p and q was proportional to this integral. He had earlier found this integral (the area in Figure 3.1 under HOL) to be

$$\frac{m^2}{4} e^{-m(p+q)} \left(1 - \frac{1}{3} e^{-mp} - \frac{1}{3} e^{-mq}\right).$$

Thus, he observed, "the probability of m is proportional to

$$m^2 dm\, e^{-m(p+q)} \left(1 - \frac{1}{3} e^{-mp} - \frac{1}{3} e^{-mq}\right)."$$

In modern notation we could describe his chain of reasoning as follows:

$$\phi(x, p, q|m) = \phi(x)\phi(p - x)\phi(p + q - x),$$

so

$$\phi(p, q|m) = \int_{-\infty}^{\infty} \phi(x)\phi(p - x)\phi(p + q - x)dx,$$

and so, by the principle,

$$\phi(m|p, q) \propto \int_{-\infty}^{\infty} \phi(x)\phi(p - x)\phi(p + q - x)dx$$

$$\propto m^2 e^{-m(p+q)} \left(1 - \frac{1}{3} e^{-mp} - \frac{1}{3} e^{-mq}\right).$$

He went on to conclude that since this distribution put negligible probability near zero, "the probability that $m = 0$ or is infinitely small (the supposition that leads to the method of arithmetic means) is infinitely less than that m equal any finite [that is, nonzero] quantity whatever." This was the calculation he had referred to earlier.

The Genesis of Inverse Probability

Laplace then tackled the problem of determining the best correction x, when m is unknown; and here he made a subtle error. His analysis was cryptic. With reference to the top diagram in Figure 3.1, he wrote:

> Next, if we denote by y the probability, corresponding to m, that the true instant of the phenomenon falls at a distance x from the point a, the whole probability that this instant falls at this distance will be proportional to
>
> $$\int ym^2\, dm \cdot e^{-m(p+q)} \cdot \left(1 - \frac{1}{3} e^{-mp} - \frac{1}{3} e^{-mq}\right),$$
>
> the integral being taken from $m = 0$ to $m = \infty$. If we then construct a new curve H′K L′ on the axis AB whose ordinates are proportional to this quantity, the ordinate KQ which divides the area of this curve in two equal parts

cuts the axis at the point we should take as the mean between the three observations.

The area of this new curve will evidently be proportional to the integral of the product of the area of the curve HOL by

$$m^2 dm\, e^{-m(p+q)} \left(1 - \frac{1}{3} e^{-mp} - \frac{1}{3} e^{-mq}\right).$$

Then since in order to determine x under a particular supposition for m we have

$$m^2 e^{-m(2p+q-x)} = m^2 e^{-m(p+q)} \left(1 + \frac{1}{3} e^{-mp} - \frac{1}{3} e^{-mq}\right),$$

we will have

$$\int m^4\, dm\, e^{-m(3p+2q-x)} \left(1 - \frac{1}{3} e^{-mp} - \frac{1}{3} e^{-mq}\right)$$
$$= \int m^4\, dm\, e^{-m(2p+2q)} \left(1 + \frac{1}{3} e^{-mp} - \frac{1}{3} e^{-mq}\right)\left(1 - \frac{1}{3} e^{-mp} - \frac{1}{3} e^{-mq}\right),$$

where the integrals go from $m = 0$ to $m = \infty$. (Laplace, 1774, pp. 640–641)

In modern notation, we may interpret this as follows: Laplace's y seems to be described as being $\phi(x|p, q, m)$, and the integral given in the first of these paragraphs is then

$$\int_0^\infty \phi(x|p, q, m)\phi(m|p, q)dm.$$

Laplace, then, is telling us that the posterior median (say, x_0) may be found by solving the equation

$$(2) \qquad \int_{-\infty}^{x_0} \int_0^\infty \phi(x|p, q, m)\phi(m|p, q)dmdx$$
$$= \frac{1}{2} \int_{-\infty}^{\infty} \int_0^\infty \phi(x|p, q, m)\phi(m|p, q)dmdx.$$

Since

$$\int_0^\infty \phi(x|p, q, m)\phi(m|p, q)dm = \int_0^\infty \phi(x, m|p, q)dm$$
$$= \phi(x|p, q),$$

equation (2) is equivalent to

$$\int_{-\infty}^{x_0} \phi(x|p, q)dx = \frac{1}{2} \int_{-\infty}^{\infty} \phi(x|p, q)dx,$$

and Laplace's assertion, as interpreted here, is quite correct. Had he actually found and solved equation (2), he would have arrived at the right answer to his problem. But the second paragraph in the preceding quotation shows that he did something rather different and that his understanding of conditional probability had not yet fully crystalized. There he appears to claim, erroneously, that "evidently" the area of this new curve, namely (as interpreted above),

$$\int_{-\infty}^{\infty}\int_{0}^{\infty} \phi(x|p, q, m)\phi(m|p, q)dmdx,$$

will be proportional to the integral of the area of HOL $[\int\phi(x, p, q|m)dx]$ times $\phi(m|p, q)$, that is, proportional to

$$\int_{0}^{\infty}\int_{-\infty}^{\infty} \phi(x, p, q|m)\phi(m|p, q)dxdm.$$

This would certainly be true if

$$\phi(x|p, q, m) \propto \phi(x, p, q|m),$$

and indeed this is correct, but only if the "constant" of proportionality is permitted to depend on m as well as on p and q [because $\phi(x|p, q, m) = \phi(x, p, q|m)/\phi(p, q|m)$]. But this is not a permissible substitution in an integral with respect to m; the "evident" claim is false and, consequently, so is the statement that follows it, which may be paraphrased in modern notation as "since when m is known, x_0 is determined by the equation

$$\int_{-\infty}^{x_0} \phi(x, p, q|m)dx = \frac{1}{2}\int_{-\infty}^{\infty} \phi(x, p, q|m)dx,$$

it is here found from

$$\int_{-\infty}^{x_0}\int_{0}^{\infty} \phi(x, p, q|m)\phi(m|p, q)dmdx$$
$$= \frac{1}{2}\int_{-\infty}^{\infty} \phi(x, p, q|m)\phi(m|p, q)dmdx."$$

The point is that throughout all his analyses Laplace had been rather cavalier in his substitution of a joint density for a conditional density, and vice versa, claiming that the result is equivalent up to proportionality. He clearly knew that (i) $\phi(u, v) \propto \phi(u|v)$ and that (ii) $\phi(u, v) \propto \phi(v|u)$. But his repeated use of proportionality statements allowed the fact that the "constants" of proportionality were functions of one of the variables to disappear from view. The full statements (i) $\phi(u, v) = \phi(u|v)\phi(v)$ and (ii) $\phi(u, v) = \phi(v|u)\phi(u)$ would have kept this dependence in full view and prevented that error. As long as u remained fixed within an analysis, it mattered not whether $\phi(u, v) = \phi(v|u)\phi(u)$ or $\phi(u, v) \propto \phi(v|u)$ was stated. But when, as in the investigation with unknown m, the expression was substituted in an integral with respect to the "fixed" variable, it mattered a great deal.

It is not simply that Laplace lacked the explicit notation I have employed in this discussion, he lacked the clarified notion of conditional probability that we possess today and that *requires* this notation. To Laplace in 1774 a conditional probability distribution *was* the joint distribution, up to a constant of proportionality, and the need to be more explicit about this constant had not occurred to him. Hence he was able to produce his principle out of thin air instead of deriving it as Bayes had. If $\phi(u, v) \propto \phi(u|v)$ and $\phi(u, v) \propto \phi(v|u)$, then *of course* $\phi(u|v) \propto \phi(v|u)$. This principle was fully evident from Laplace's vague notion of conditional probability distribution, and no demonstration was required. Laplace's unclarified notion of conditional probability had actually eased the way toward his principle once his development of a fiducial statement of the inference problem had shown that such a principle was needed.

The vague notion of conditional probability Laplace did possess in 1774 had served him well until he boldly pursued his approach into the realm of unknown m. Even there, he never seems to have realized his error, although he did pay a heavy price nonetheless. With some effort he managed to integrate the aforementioned erroneous equation for the posterior median x_0, only to find that it yielded an equation of the fifteenth degree in x_0. Incredibly, he was able to prove that only one of the roots of this equation fell in the interval $0 < x_0 < p$; and he showed how, with great labor, this root could be found iteratively. He even evaluated the root for $q/p = 0, 0.1, 0.2, \ldots, 1.0$ and gave the results in a table. It is easy to see (to borrow one of Laplace's favorite phrases) why he had limited attention to only three observations, although in the now time-honored tradition of mathematical statisticians, he added the unconvincing claim that "it is clear that the solution is entirely similar for any number whatever."

It is ironic that, had Laplace proceeded correctly, he would have had a slightly simpler analytic problem. For his prior distributions, $\phi(x|p, q, m) \propto \phi(x, p, q|m)/\phi(m|p, q)$ (where the "constant" of proportionality depends only on p and q), and the problem reduces to solving

$$\int_{-\infty}^{x_0} \int_0^{\infty} \phi(x, p, q|m)dmdx = \frac{1}{2} \int_{-\infty}^{\infty} \int_0^{\infty} \phi(x, p, q|m)dmdx,$$

or

$$\int_0^{\infty} m^2 e^{-m(2p+q-x_0)}dm = \int_0^{\infty} m^2 e^{-m(p+q)} \left(1 + \frac{1}{3} e^{-mp} - \frac{1}{3} e^{-mq}\right)dm,$$

a simpler pair of integrals than that Laplace actually dealt with.

The correct equation reduces to a cubic in x_0, namely, $(2p + q - x_0)^3 = [(p + q)^{-3} + (2p + q)^{-3}/3 - (p + 2q)^{-3}/3]^{-1}$. By my calculation the solu-

Table 3.1. Roots of Laplace's equation.

	q/p										
x_0/p	0	0.1	0.2	0.3	0.4	0.5	0.6	0.7	0.8	0.9	1.0
Laplace	0.860	0.894	0.916	0.932	0.944	0.955	0.965	0.975	0.984	0.992	1.000
Correct	0.878	0.911	0.930	0.944	0.955	0.964	0.972	0.980	0.987	0.994	1.000
Arithmetic mean	0.667	0.700	0.733	0.767	0.800	0.833	0.867	0.900	0.933	0.967	1.000

tions for different q/p are even further from the corrections that give the arithmetic mean than those given by Laplace (Table 3.1).

Laplace's Memoirs of 1777–1781

Laplace did not remain satisfied with his first error curve; I expect that he blamed it for the analytic difficulties that had held him to a consideration of only three observations. It is tempting to imagine that he could, at this point, have gone on to consider the curve $\propto e^{-mx^2}$ and could have found that the solution for any number of observations follows with relative ease. There are two reasons why this was not to be. First, having found that the curve e^{-mx} (Laplace's notation, x taken as positive) caused incredible difficulties, it is hard to imagine that he would have guessed that the apparently more complex curve e^{-mx^2} would be easier to handle. In fact, no recorded discoverer of the normal curve has come upon it by this route, neither De Moivre, nor Laplace, nor Gauss, nor Adrain. The second and overriding reason is that even if Laplace had come upon e^{-mx^2} by mathematical guesswork, such a guess would have been statistically meaningless within the context of eighteenth century (or even early nineteenth-century) thought. Two centuries later, with our present wealth of empirical experience on the shapes, forms, and properties of probability distributions, a speculation of this type would be imaginable; but in the 1770s it was not. To the Newtonians of that period—and Laplace was the supreme post-Newton Newtonian—a theoretical error curve could only have meaning if it was derived from first principles; and the first principle of probability theory was that all problems could be solved by reducing consideration to the study of equally likely cases. If e^{-mx} would not do, Laplace would look for a different way to apply the principle of insufficient reason. In 1777 he found such a way.

Lagrange's memoir "On the usefulness of the method of taking the mean among the results of many observations, in which we examine the advantages of this method by the calculus of probabilities, and where we resolve different problems relative to this matter" was finally published in 1776; and it appears to have been the publication of that memoir that inspired Laplace to reconsider the problem. From a statistical point of view Lagrange's memoir was of rather limited scope; but it contained one mathematical development that was to prove to be of key importance in later statistical investigations. Lagrange had focused on the arithmetic mean and its probability distribution. And the primary point of novelty in this long and systematically developed memoir was the technique Lagrange employed for determining that distribution, namely, an extension of De Moivre's use of generating functions. Lagrange developed it far beyond Simpson's earlier work.[6] Starting with the simplest error distribution, where the only possible errors were -1, 0, and 1, Lagrange proceeded to successively more general discrete cases and finally to the consideration of a few continuous cases. The memoir is effectively a demonstration of how what we now call the Laplace transform can be used to determine the distribution of the mean error. Lagrange's application of the technique was limited to a few cases [such as $\phi(x) \propto p^2 - x^2$, and $\phi(x) \propto \cos x$] where he could recognize the nth power of the transform as itself a transform; he lacked an inversion formula. Later (in 1785 and 1810) Laplace was to seize on this technique, develop an inversion formula, and proceed to use it to prove the central limit theorem. But in 1777 Lagrange's memoir provided a rather different spur.

Lagrange's study of means had been exclusively based upon the arithmetic mean. He had taken the point of view that the error distribution and the quantity to be determined are fixed and that the value of that mean was to be judged in terms of probabilistic statements about its possible values. He determined, for example, the correction for bias needed when the center of gravity of the error distribution was not zero. He did deal briefly with the question of how one might estimate an unknown discrete error distribution. In the process he derived a multivariate normal distribution as an approximation to the multinomial distribution in much the same way that De Moivre had approximated the binomial distribution. But even there, although his description of interval estimates of the unknown probabilities was ambiguous, I think it is fair to say that his work was untouched by any real sense of inverse probability.

On 8 March 1777 Laplace read a memoir to the Academy of Sciences, "Researches on the mean to be chosen among the results of many observations." It remained unpublished until 1979 (when it was discovered in the Academy's archives by Charles C. Gillispie) although its main results were

6. There is no indication in the memoir (Lagrange, 1776) that Lagrange had read Simpson. Possibly he had not, as there is little similarity between their analyses and Lagrange's notation is different from that of Simpson. Citation was an inexact art in the eighteenth century, however; and it is possible that Lagrange had scanned Simpson's *Tracts* at one time or another and subsequently redeveloped and greatly extended the subject from his own point of view.

incorporated in the 1781 "Memoir on probabilities." Laplace began by citing Lagrange's memoir and summarizing the problem it treated and the results obtained. "But," Laplace wrote, "it appears to me that this inquiry requires principles other than those which have been used by that learned author. The importance of this problem in physics has led me to return to it here, and I shall endeavor to treat it with all the precision and clarity that one may wish in a matter which is touched by such subtle metaphysical aspects and which has, until now, been little known."

The issue of principle that Laplace was raising with Lagrange is still a central one in debates on the foundations of statistics. The question is, Should one choose and evaluate a statistical procedure (in this case, a mean) before the observations are made, on the basis of the probability distribution of the procedure in repeated samples as Simpson and Lagrange had done, or should one make this choice a posteriori after (and conditional on) observation, as Laplace had done in his 1772–1774 work? In his earlier work Laplace had not even acknowledged the existence of this question. In 1777 he at least made a clear statement of the choice involved, even if the argument he gave for his own preference—self-evidence—would not satisfy a modern skeptic. He wrote:

> The problem we are concerned with may be regarded from two different points of view, depending on whether we consider the observations before or after they are made. In the first case, the search for the mean we should take among the observations consists in determining *a priori* which function of the results of the observations is most advantageously taken for the mean result. In the second case, the search for this mean consists in determining a similar function *a posteriori,* that is, paying regard to the distances between the respective observations. We see easily that these two manners of regarding the problem must lead to different results, but it is at the same time clear that the second is the only one that should be used. Nevertheless, it is from the first of these viewpoints that the question has, prior to now, been treated by Geometers. Thus however ingenious their researches have been, they can only be of very little use to observers. (Laplace, 1777, p. 229)

After repeating the statement of the problem as he had given it in 1774, including the proof that the posterior median minimized the posterior expected error, Laplace restated this distinction more explicitly:

> Suppose that the number of observations is n, and let $p, p^{(1)}, p^{(2)}, \ldots, p^{(n-1)}$ be the different times for the phenomenon given by the observations. Let $\psi\{p, p^{(1)}, p^{(2)}, \ldots, p^{(n-1)}\}$ be the function we seek. Then if the observations are supposed not yet made, the quantities p, $p^{(1)}$, $p^{(2)}$, etc. could receive an infinity of different values, and the function $\psi\{p, p^{(1)}, p^{(2)}, \text{etc.}\}$ would give an infinity of errors corresponding to these values; the sum of these errors multiplied by the [probabilities of] the corresponding values of $p, p^{(1)}, p^{(2)}$, etc. should be a *minimum*. If the observations are supposed to have been made, the

quantities $p, p^{(1)}$, etc. are constant, and the function $\psi\{p, p^{(1)}, p^{(2)}, \text{etc.}\}$ should be such that for these given values the error will be a *minimum*. We thus see even more clearly than above the difference between these functions, depending on whether we consider the observations before or after they are made. (Laplace, 1777, p. 231)

Laplace's statements are not free from ambiguity when viewed through modern eyes, and different modern schools of thought could quote extracts from these quotations to serve different ends. Laplace's first characterization of the a posteriori mean as "paying regard to the distances between the respective observations" smacks of a modern "conditioning on an ancilliary" and might have appealed to Ronald A. Fisher. The latter quotation is more explicitly Bayesian, but with no tinge of subjective probability. In truth, the attribution of either of these views to Laplace would be unwarranted. He had (as none before him had) seized on and articulated the distinction between a priori and a posteriori criteria, but he had not here (nor elsewhere) fully developed the foundations of an a posteriori view. His application of his mathematical development was usually to be consistent with a Bayesian view when considering problems of estimation or the determination of unknown quantities.

The Error Curve of 1777

The question of the specification of an error curve remained, and a major portion of Laplace's memoirs of 1777 and 1781 were given over to an extremely complex argument as to why the curve

$$y = \frac{1}{2a} \log\left(\frac{a}{|x|}\right), \qquad |x| \leq a,$$

should be taken as an error distribution. Here a represented the upper limit of the possible errors. The argument leading to this curve was as difficult as any mathematical argument Laplace attempted in this period, and it seems reasonable to speculate that it was his success in the mathematics of this intricate investigation that led him to emphasize the result. The curve itself was more difficult to work with than his earlier exponential distribution had been.

Laplace's argument was intended to arrive at an expected random, symmetric, unimodal density. In modern terminology, we can outline his procedure as follows: Suppose that the unit interval is broken into $n + 1$ pieces (or spacings) by choosing n points at random in the interval. Suppose these $n + 1$ spacings are ordered in decreasing order: $d_1 > d_2 > \ldots > d_{n+1}$, where $d_1 + d_2 + \ldots + d_{n+1} = 1$. Let n be very large, and erect ordinates of a curve at heights $d_1, d_2, \ldots, d_{n+1}$, equally spaced along the abscissa between 0 and a. Calculate the expected values of these ordinates and let n pass to the limit. The result, Laplace found, was a curve proportional to $\log(a/x)$ on the interval 0 to a. By symmetry, this gave the curve as proportional to $\log(a/|x|)$ on $(-a, a)$. The condition that the total area be 1 led to the result

$$\frac{1}{2a} \log\left(\frac{a}{|x|}\right).$$

By basing his argument upon a random partition of the unit interval (that is, upon a random probability distribution), Laplace had maintained the spirit of the principle of insufficient reason. No distribution was favored a priori. Only the suppositions of symmetry and unimodality, and the need to deal with the expectation of the random distribution rather than the distribution itself, led from the chaos of the random partition to the smooth (if intractable) curve $(1/2a)\log(a/|x|)$ (Figure 3.2).

Laplace's labor in producing this result was immense, however. In 1777 he started with a discrete partition and passed to the limit. The 1979 printing of this argument consumes the better part of twenty pages. In 1781 the argument started with a continuous partition and attempted to present a more orderly flow; it still required eighteen pages (Laplace, 1781, pp. 396–413). Even in the 1812 *Théorie analytique des probabilités,* where portions of the development were presented separately, it occupied thirteen pages (Laplace, 1812, pp. 262–274). And the end result of this

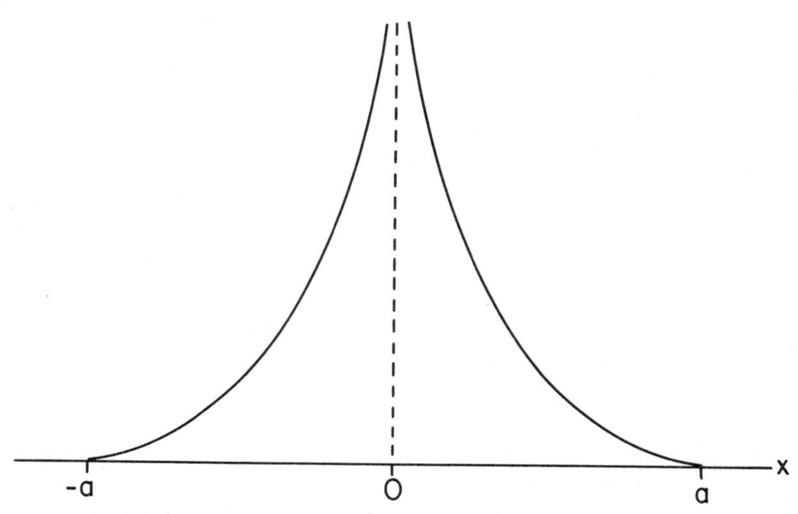

Figure 3.2. Laplace's 1777 error curve, $y = (2a)^{-1} \log(a/|x|)$. (Reconstructed from Laplace, 1777.)

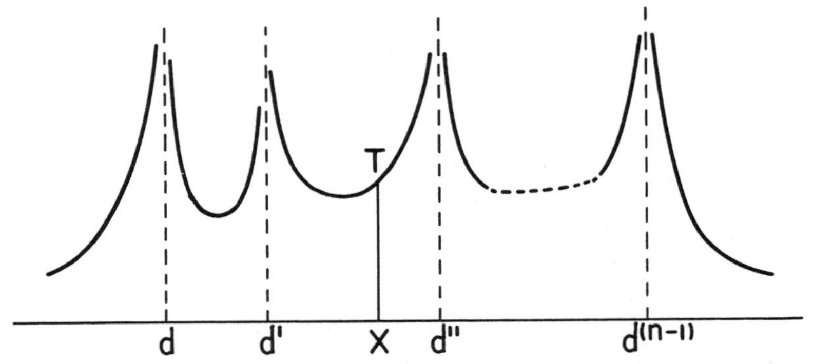

Figure 3.3. A portion of the posterior density of an astronomical quantity based upon observations at d, d', d'', . . . $d^{(n-1)}$, with errors having Laplace's 1777 distribution (Figure 3.2). The line XT represents a bisection of the area, X being the posterior median. (Reconstructed from Laplace, 1777.)

hard-won battle was unworkable in practical terms. Only in the long-dormant 1777 manuscript did he attempt to work with it. Figure 3.3, reconstructed from Laplace's description, shows the posterior distribution of an astronomical quantity based on several observations subject to this error distribution; it is as unattractive a curve for analysis as statistics has produced in over two centuries. Surely it was with understatement that Laplace admitted in his 1781 memoir (p. 479) that only in "very delicate" investigations, such as of the transit of Venus across the sun, would one actually attempt an analysis with this curve. Clearly more progress toward a usable statistical analysis of observations would have to await a more congenial error curve.

Bayes and the Binomial

I have argued that it was the consideration of astronomical observations that first led Laplace to inverse probability, and I have suggested that the same motivation lay behind Bayes's earlier consideration of this question. But regardless of whether or not this thesis is accepted, there is no doubt that both men wasted no time in bringing this, a truly Copernican revolution in statistical concept, to bear on Jacob Bernoulli's original problem. Bernoulli's clear wish had been to show how one could learn from experience about nature, learn of the proportion of "fertile" tickets in an urn from repeated draws. His conceptual stance, his mathematics, his discrete urn model, and his lack of a yardstick for the measurement of uncertainty all had conspired to deny him a satisfactory solution to his problem. In the last half of the eighteenth century the problem was finally to yield.

As mentioned earlier, the first limited success was reached by Thomas Bayes, probably in the period shortly after Simpson's 1755 article (Bayes died in 1761.) Bayes's essay had no discernible impact on early work, despite prominent publication in the *Philosophical Transactions* of the Royal Society of London in 1764. It sank from view and reference almost from its first appearance until the time Laplace's 1781 memoir was published. At that time Condorcet seems to have rediscovered it, as mentioned in his preface to the volume. Even then it remained obscure; and if it exerted any important influence in its first century, it was only on Laplace's later reformulation of principles he had already come upon independently. But it is worth asking why such a work could fail so completely —indeed, why Bayes himself did not publish it in his lifetime instead of leaving it (and £100) for Richard Price, a man whose whereabouts he seemed somewhat unsure of when he wrote his will in December 1760. Barnard (1958) quotes Bayes's will, executed 12 December 1760, as referring to "Richard Price, now I suppose preacher at Newington Green." That Bayes died four months after making his will suggests that he was in declining health and that he would have had the opportunity to communicate his work to the Royal Society had he wished to do so.

Bayes's essay "Towards solving a problem in the doctrine of chances" is extremely difficult to read today—even when we know what to look for. Part of the difficulty comes from the adoption of the geometric mode of description used by Newton, a mode that would have been less perplexing to Bayes's contemporaries. Perhaps of more serious consequence is the fact that Bayes found it necessary to wrestle simultaneously with the fundamental nature of probability while attempting its inversion. Still, although Bayes's solution was difficult to comprehend, the *problem* was stated clearly. We may wonder why no early reader persevered; or if one did, why they found the work not worthy of public notice or application.

Price's introduction and Bayes's succinct statement left no doubt about the concern of the essay:

PROBLEM

> *Given* the number of times in which an unknown event has happened and failed: *Required* the chance that the probability of its happening in a single trial lies somewhere between any two degrees of probability that can be named.

In modern notation we would write the problem as follows: Let X be the number of times the event happens in n trials and θ the probability it happens in a single trial. Then Bayes required $\mathrm{P}(a < \theta < b|X)$. His solution proceeded as an essentially axiomatic derivation, starting from a fundamental notion of probability as the subjective value of a future contract. Bayes went on to prove his Proposition 3, that (in modern notation) if E_1 and E_2 are events ordered in time, $P(E_2|E_1) = P(E_1 \cap E_2)/P(E_1)$. This was followed by a different proof of his Proposition 5: $P(E_1|E_2) = P(E_1 \cap E_2)/P(E_2)$. Bayes's Proposition 5, where E_1 is an event determined antecedent to E_2, is Bayes's theorem in its first incarnation. The fact that Bayes regarded these propositions as fundamentally different and as needing proof shows him to have been a subtle thinker; indeed his philosophical and formal attitude toward the foundation of probability stands as an interesting contrast to the more intuitive and less-examined view in Laplace's early works. (See Shafer, 1982, for a careful analysis of Bayes's reasoning.)

When Bayes finally came to treat the problem of the binomial directly, he did so in terms of an ingeniously contrived physical analogue that differed dramatically from Bernoulli's urn. In fact, the difference between the physical structures Bernoulli and Bayes employed was absolutely crucial to Bayes's success. Bernoulli's urn had had two obvious drawbacks as a vehicle for inverting probability: It had an intrinsic discrete structure (it would have been too cumbersome even to attempt a specification of a distribution over all possible urns at that time), and it was inherently asymmetric with respect to the constitution of the urn (the tickets) and the chance mechanism (the draws). Bayes's mechanism had neither defect.

Bayes imagined a square table or plane (ABCD in Figure 3.4) and two balls W and O. Later writers, including Karl Pearson (1920b) and R. A. Fisher (1959, p. 128), have promoted it to a billiard table, but the Reverend Bayes was neither so specific nor so frivolous. The plan was this: First the ball W was to be thrown across the table in such a manner that it was equally likely to come to rest at any point on the table. (Actually, Bayes's description was more felicitous, hinting at an awareness of the need to describe continuous random quantities by areas—"there shall be the same probability that it rests upon any one equal part of the plane as another.") Second, the ball O was to be rolled in the same manner as W n times and the number of times it came to rest to the right of W counted. The position of W corresponds to our θ; the number of times O is to the right of W is our X.

In terms of Bayes's diagram, W determines a line *os* through W and parallel to sides BC and AD; θ is the ratio of A*o* to AB. Our X is then the number of times O comes to rest in the rectangle *os*DA; Bayes called this latter event M.

Given Bayes's specification that the final resting place of each ball is uniformly distributed over the square, it is clear that θ is uniformly distributed on the unit interval; and once θ is determined X has a binomial distribution where θ is the probability of a "success" or fertile case on a single trial. The distribution of θ, which has now lost the discrete character

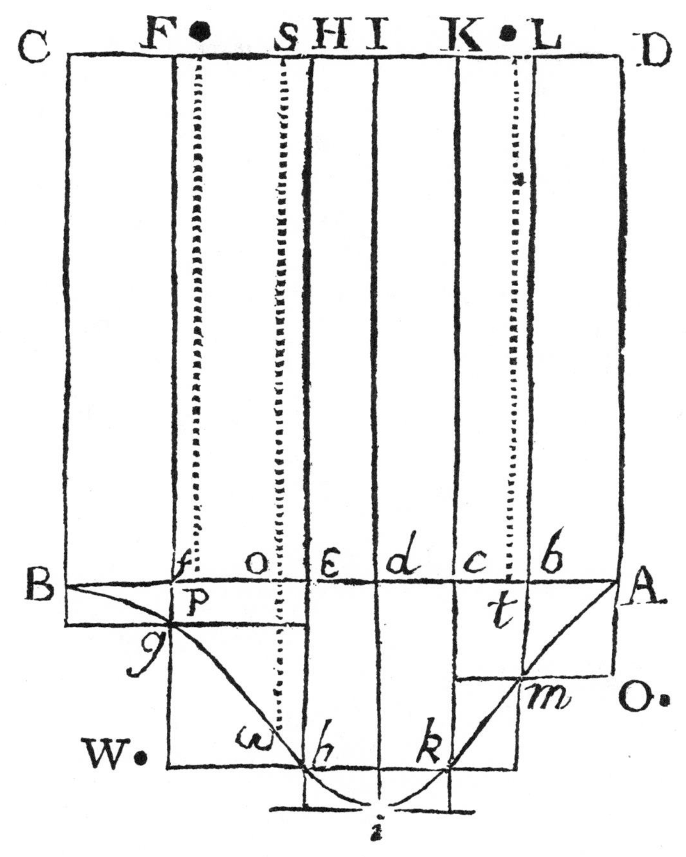

Figure 3.4. Bayes's "Billiard Table": a square table ABCD and two balls W and O. The ball W is rolled and comes to a rest on the line os, then the ball O is rolled n times and the number of times it comes to a rest in the rectangle osDA is recorded. The lower curve is meant to show the posterior density of the fraction Ao / AB, given the recorded count. Bayes's upper and lower approximations to this density are also shown. (From Bayes, 1764, p. 385.)

it would have had in Bernoulli's urn, is unambiguously specified, and the problem has acquired a newly symmetric character: Both draws and urn are represented by identical balls. The way is open for an unusual appearance of a fiducial reversal of conceptual positions, a reversal that would not have been possible with Bernoulli's urn.

With the physical mechanism thus clearly described, Bayes went on to solve the problem he had posed. Treating the roll of W as the first of the events (E_1 in our notation) and the n rolls of O as the second event (our E_2), he wished to apply his Proposition 5. From the uniform distribution of W and the fact that, given the position (θ) of W, the distribution of the number of occurrences of M (X) is binomial, Bayes was able to use Proposition 3 to derive the joint probability that θ falls between b and f *and* M occurs p times. In our notation, Bayes found

$$\text{(3)} \qquad \text{P}(b < \theta < f \cap X = p) = \int_b^f \binom{n}{p} \theta^p (1-\theta)^{n-p}\, d\theta$$

(Actually, Bayes described this probability as the area fghikmb in Figure 3.4, where the curve drawn was $y = Ex^p r^q$ and his x corresponds to our θ, r to $1 - \theta$, $E = \binom{n}{p}$, and $q = n - p$.) Then, setting $b = 0$ and $f = 1$ gave

$$\text{P}(X = p) = \int_0^1 \binom{n}{p} \theta^p (1-\theta)^{n-p}\, d\theta,$$

an integral he was later to evaluate as equal to $1/(n + 1)$, and Proposition 5 led to

$$\text{P}(b < \theta < f | X = p) = \frac{\text{P}(b < \theta < f \cap X = p)}{\text{P}(X = p)} = \frac{\displaystyle\int_b^f \binom{n}{p} \theta^p (1-\theta)^{n-p}\, d\theta}{\displaystyle\int_0^1 \binom{n}{p} \theta^p (1-\theta)^{n-p}\, d\theta}.$$

The proof was remarkable for its careful treatment of probability densities, treating all probabilities as areas and not relying (in the later manner of both French and English mathematicians) on the interpretation of densities as attaching infinitesimal probability to points. The basic relation (3) was derived by approximating the curve above and below by rectangles (the approximating piecewise linear functions are shown in Bayes's figure) and by showing that for any such approximating lines the probability (3) must lie between the bounding areas.

Formally and abstractly the problem was solved, yet there remained two major obstacles to overcome before the solution could be a practical one. First, the appropriateness of Bayes's physical mechanism had to be established, as an analogue to the problems of civil life that Bernoulli had envisaged. Second, the mathematical problem of evaluating the integral (3) had to be met. It is a tribute to Bayes's insight that he sensed both needs and made explicit attempts to deal with them.

The justification of his clever physical mechanism, Bayes's billiard table, was the central philosophical question that was to be at the root of all later controversy over the application of Bayes's result. Could nature's selection of an unknown chance θ be likened to the rolling of a ball across a leveled table? Bayes met this point in a scholium that immediately followed the result. His rule, which he sought to justify, was that in situations where we know nothing a priori about the probability θ that an event will occur—or as he put it, "in the case concerning the probability of which we absolutely know nothing antecedently to any trials made concerning it"—then that probability could be taken, like the ball on the table, to be uniformly distributed and the derived result (3) put to use. Many commentators on the work, including Karl Pearson (1920b), Harold Jeffreys (1939, p. 34), R. A. Fisher (1959, chap. 2), and Ian Hacking (1965, pp. 199–200), have interpreted his argument for this rule as an appeal to what is now, after Laplace, termed the principle of insufficient reason: Because we have no reason to favor one value for θ over another, equal intervals of values may be described as equally likely. The immediate criticism of this argument (already dealt with in Edgeworth, 1885a, and echoed in Fisher, 1959, p. 16) is that we are equally ignorant about other monotone functions of θ, say $\frac{1}{2}\cos^{-1}(1 - 2\theta)$, which might just as well then be taken as uniformly distributed, leading to different results. But this is not the way Bayes actually reasoned.

Price made clear in the introduction that Bayes had originally proceeded in essentially just this manner, taking the uniform distribution of θ as an axiomatic "postulate," but had then decided on a different approach. Karl Pearson (1978, p. 364) has justly reproached nineteenth-century interpretations of Bayes, and his comments could apply to others as well: "His [Bayes's] actual method of approaching the problem is somewhat obscure and it may lead to the equal distribution of chances, but that is not a ground for criticizing him on the basis of a postulate which he distinctly avoids." Pearson's own explanation of Bayes's scholium, however, is no more accurate than those he criticized; and I believe the other commentators cited (Jeffreys, Fisher, Hacking) misinterpreted Bayes.

What, then, did Bayes actually intend? All readers agree that Bayes's clear aim was to draw an analogy between the billiard table and other applications where the probability θ was unknown. The question is, At what logical point was the analogy to be verified? Most have supposed that it was at the first stage, at the throw of the first ball W: Because we are ignorant of the outcome θ for the ball W and because θ is clearly uniformly distributed, then in other situations where we are ignorant of θ, it must also be uniformly distributed. This is the principle of insufficient reason, but it is not Bayes's argument. This inferred argument ascribes a distribution to "ignorance" in a unique way that is properly vulnerable to criticism in a way that Bayes's actual argument is not. To see what Bayes actually meant, we must first see what he actually said. The key passage from the scholium is this:

> From the preceding proposition it is plain, that in the case of such an event as I there call *M*, from the number of times it happens and fails in a certain number of trials, without knowing anything more concerning it, one may give a guess whereabouts it's probability is, and, by the usual methods computing the magnitudes of the areas there mentioned, see the chance that the guess is right. And that the same rule is the proper one to be used in the case of an event concerning the probability of which we absolutely know nothing antecedently to any trials made concerning it, seems to appear from the following consideration; viz. that concerning such an event I have no reason to think that, in a certain number of trials, it should rather happen any one possible number of times than another. For, on this account, I may justly reason concerning it as if its probability had been at first unfixed, and then determined in such a manner as to give me no reason to think that, in a certain number of trials, it should rather happen any one possible number of times than another. But this is exactly the case of the event *M*. For before the ball *W* is thrown, which determines it's probability in a single trial, (by cor. prop. 8) the probability it has to happen p times and fail q in $p + q$ or n trials is the ratio of *AiB* to *CA*, which ratio is the same when $p + q$ or n is given, whatever number p is; as will appear by computing the magnitude of *AiB* by the method of fluxions . . . And consequently before the place of the point *o* is discovered or the number of times the event *M* has happened in n trials, I can have no reason to think it should rather happen one possible number of times than another.
>
> In what follows therefore I shall take for granted that the rule given concerning the event *M* in prop. 9 is also the rule to be used in relation to any event concerning the probability of which nothing at all is known antecedently to any trials made or observed concerning it. And such an event I shall call an unknown event.

It is not θ that Bayes stated is uniformly distributed, but X! He reasons thusly: for the billiard table it is clear from our description that θ is uniformly distributed. A mathematical consequence of this is that the marginal distribution of X is uniform:

$$\begin{aligned} P(X = p) &= P(0 < \theta < 1 \cap X = p) \\ &= \int_0^1 \binom{n}{p} \theta^p (1 - \theta)^{n-p}\, d\theta \\ &= \frac{1}{n+1}, \quad \text{regardless of } p. \end{aligned}$$

Now let us consider another possible application of inverse probability, say, following Bernoulli, a tennis match of n games, where we "absolutely know nothing" of the probability that player 1 will win a single match. This, Bayes said, would mean that we would have "no reason to think that" in the n trials, there "should rather happen any one possible" value of X than another. That is, "knowing nothing" of the value of θ takes on the operational meaning that $P(X = p) = 1/(n + 1)$ for all p. A possible, even likely rationale that Bayes may have had in mind for this is that if (to the contrary) for some p^* and p we have $P(X = p^*) > P(X = p)$, then $X = p^*$ is more likely than $X = p$. But then our "expectation" (Bayes's definition of probability) that X will be p^* is greater than for p. And our "expectation" for X/n, a natural estimate of θ even in Bernoulli's time, is greater for one value than another—surely that would contradict a belief we "absolutely know nothing" of θ. A future contract or bet on p^*/n would be worth more than one on p/n. Now, of course, this argument is a cousin to the principle of insufficient reason, but it is not the same because it is applied to the empirically observable X rather than to the unobservable θ. "Knowing nothing" is plausibly *defined* to mean $P(X = p) = 1/(n + 1)$; a tennis match where we know nothing is one where a priori all outcomes, $X = 0$ or 1 or 2 or . . . or n, (and thus all ratios X/n) are equally likely. If we can recognize a situation as fitting this description, then, Bayes tells us, it is a situation analogous to the billiard table and the reasoning that he has derived applies.

The argument is ingenious. It is free of the objection Fisher and others raised to what they, I think mistakenly, took to be his argument. For if $P(X = p) = 1/(n + 1)$, then $P[f(X) = f(p)] = 1/(n + 1)$ for any strictly monotone function f; knowing nothing about X is knowing nothing about $f(X)$, since X (unlike θ) is discrete. What Bayes did not know is that part of the argument hinged on a subtle mathematical point. The condition that $P(X = p) = 1/(n + 1)$ for all p is a *necessary* condition that θ be a priori uniformly distributed, but it is not a *sufficient* condition unless we insist it hold for all p *and* all n. The distribution of X for a single value of n does not determine the distribution of θ; prior distributions other than the uniform can also lead to $P(X = p) = 1/(n + 1)$. If $P(X = p) = 1/(n + 1)$ for all p *and* all $n \geq 1$, then the objection disappears because all moments of the prior distribution of θ are then determined, and hence (since that prior distribution is supported by the unit interval and the distribution is thus determined uniquely by its moment-generating function) θ must be a priori uniformly distributed. In any event, the argument is strictly limited to the binomial situation, and it puts a rather severe restriction on "knowing nothing," showing that Bayes did not insert the word "absolutely" carelessly. This interpretation of the scholium is discussed more fully in Stigler (1982b); see also Edwards (1978) and Dale (1982).

Some readers of Bayes, notably Fisher (1959, p. 17), have expressed the view that it was uncertainty over the philosophical soundness of his rule that made Bayes hesitant to publish. I see no such uncertainty in the essay; on points of philosophy he was sure and sound. His acceptance of subjectively determined probabilities seems complete and unambiguous; and his early doubt as to how to operationally characterize a prior distribution on an unobservable θ vanished with the clever argument that threw the force of the characterization on the observable X. Rather, it would appear that if Bayes was actually reluctant to publish, it was over a point in mathematics. It involved Bayes's attempt to deal with the second remaining point necessary for the practical application of his rule, the evaluation of the integral (3).

The evaluation of the integral $\int_0^{f} \theta^p (1 - \theta)^q \, d\theta$, Bayes noted (although he used x where we use θ), would complete the solution. This integral is now known as the incomplete beta function, and its difficulty depends upon the values of p and q: When they are small, nothing could be simpler; when they are large, the problem is a formidable one. The first extensive tables of this function were not compiled until this century, when the students in Karl Pearson's laboratory were pressed into reluctant service as "computers." A story, possibly apocryphal, still circulates in University College London of a student who resigned in disgust after a week, telling Pearson of his plans for a different career and announcing, "As far as I am concerned, the Table of the Incomplete Beta Function may *stay* incomplete."

Bayes noted that the integral was susceptible to relatively easy analysis when either p or q is small: for q small he would expand $(1 - \theta)^q$ by the binomial series, multiply by θ^p, and integrate term by term. But he recognized that this was impractical for the most important case in which p and q are both large, because the series that resulted converged too slowly.

At the point where Bayes turned to large p and q, the character of the essay changes as Richard Price substituted his own abridged rendition of Bayes for the original and added an appendix. This much is clear, however: Through an exceedingly painstaking and tortured analysis, Bayes sought to bound the incomplete beta above and below. His solution was no more than a Pyrrhic victory because his bounds were far too complex for practical evaluation and were not even very close.[7] In one of his applications of Bayes's results in the appendix, Price tried to evaluate, for $p = 1{,}000$ and $q = 100$, the chance that θ was within $1/110$ of $p/(p + q) = 10/11$. His best effort (which required much calculation) was successful in showing only that this chance was between 0.7953 and 0.9405. Although we must sympathize with Bayes and Price for the difficulties involved, we can also see why Bayes might well have felt his work incomplete in an important way and why contemporary readers would not have seen it as providing useful solutions to practical problems. If the acceptance of bounds this crude were ever to come, it would only be through a convincing demonstration by example that they were adequate

7. Dale (1982) shows that one of Price's developments of Bayes's analysis can be pushed a bit further to produce a useful normal approximation. Neither Bayes nor Price did this, however.

for practical problems. Such a demonstration would be hard today, and Price's own application to lotteries may have even worked to the detriment of the point: The difference 0.9405 − 0.7953 = 0.1452 would not have seemed small to anyone familiar with games of chance.

Laplace the Analyst

Bayes's treatment of the binomial may be regarded as mathematically incomplete. Whether or not that is accepted as the reason for his reluctance to publish, it is a sufficient explanation for the lack of attention his work received after it was published. A decade later Laplace was led to the same problem; but he was armed with a far greater analytic skill and consequently his solution was richer and more complete.

Laplace's analysis was spread out over four decades, from his first approach in the seminal paper of 1774 on inverse probability to the more polished presentation of the *Théorie analytique* of 1812. Indeed, it may be seen as the touchstone of all his work on probability, from the use of generating functions for the solution of difference equations to the expansion of integrals to the central limit theorem itself. He began with the problem Bayes had considered, the approximation of the posterior distribution for a binomial probability, and he progressed by degrees of refinement and generalization until he possessed a mathematical apparatus for an extraordinary array of inference problems, from ratio estimators based on partial censuses to the linear observational equations that Mayer had studied. I shall not attempt a full investigation of this work (although if our focus were more upon the mathematics of probability, the needed effort would be amply repaid). Rather, I shall survey the initial stages and later look closely at an application of the apparatus in its fully developed form.

Laplace's first serious memoir on probability contained the seeds of what was to follow. Shortly after introducing his principle of inverse probability, Laplace went on to state and prove a theorem. He supposed that in drawing $p + q$ times from an urn containing white and black tickets in an unknown ratio x, p white tickets and q black tickets were obtained. The problem was to determine the unknown ratio x. Laplace's theorem was this:

> *One can suppose that the numbers* p *and* q *are so large that it becomes as close to certainty as one wishes that the ratio of the number of white tickets to the total number of tickets contained in the urn is included between the two limits* $\frac{p}{p+q} - w$ *and* $\frac{p}{p+q} + w$. w *can be supposed less than any given quantity.* (Laplace, 1774, pp. 626–627; italics in original)

In modern terminology, we say that Laplace was asserting the posterior consistency of the empirical ratio $p/(p+q)$. By making the number of trials larger, one could guarantee the a posteriori closeness of $p/(p+q)$ to x, to whatever degree was required, to whatever fraction of certainty was desired. That is, $P\{|x - p/(p+q)| \leq w | p, q\} \to 1$ as $p + q \to \infty$, for any $w > 0$, where the probability in question is with respect to the a posteriori distribution of x, conditional on p and q.

There is an almost uncanny sense in which Laplace's theorem seemed directed toward a mysterious passage the English psychologist David Hartley had inserted in his 1749 *Observations on Man.* After a clear statement of De Moivre's limit theorem, Hartley had written:

> An ingenious Friend has communicated to me a Solution to the inverse Problem, in which he has shewn what the Expectation [that is, probability] is, when an Event has happened p times, and failed q times, that the original Ratio of the Causes for the Happening or Failing of an Event should deviate in any given Degree from that of p to q. And it appears from this Solution, that where the Number of Trials is very great, the Deviation must be inconsiderable. (Hartley, 1749, vol. 1, p. 339; Stigler, 1983)

Not only is the notation (p and q) that of Laplace, so is the aim. Laplace, like Hartley's Friend (and unlike Bayes), focused here upon posterior consistency rather than upon approximating the posterior probability of intervals. Although we may never know whether Hartley's Friend could prove the claim, Laplace not only could prove the theorem as stated, his proof carried the seeds of an extensive theory.

Laplace had begun by directly applying his principle to the binomial situation to conclude that the posterior distribution of the unknown ratio x was given by

$$\frac{x^p (1-x)^q \, dx}{\int_0^1 x^p (1-x)^q \, dx} = \frac{(p+1)(p+2)\cdots(p+q+1)}{1 \cdot 2 \cdot 3 \cdot \ \ldots \ \cdot q} \cdot x^p (1-x)^q \, dx.$$

The problem then became essentially that which had bedeviled Bayes: Evaluate the integral of this expression from $p/(p+q) - w$ to $p/(p+q) + w$. To this end Laplace took the first step toward what was to become known as Laplace's method of the asymptotic evaluation of definite integrals. The integrand $x^p(1-x)^q$ is at a maximum at $p/(p+q)$; Laplace effectively expanded $x^p(1-x)^q$ around that point, writing

$x = p/(p+q) + z$ and

$$\int_{\frac{p}{p+q}-w}^{\frac{p}{p+q}+w} x^p(1-x)^q \, dx$$

$$= \frac{p^p q^q}{(p+q)^{p+q}} \int \left(1 + \frac{p+q}{p} z\right)^p \left(1 - \frac{p+q}{q} z\right)^q dz$$

(and after some analysis)

$$\cong \frac{p^p q^q}{(p+q)^{p+q}} \int_0^w 2 \exp\left[-\frac{(p+q)^3}{2pq} z^2\right] dz,$$

Multiplying both sides by $[(p+1)(p+2) \ . \ . \ . \ (p+q+1)]/(1 \cdot 2 \cdot 3 \cdot \ . \ . \ . \ \cdot q)$ and applying Stirling's formula gave the result, in modern notation, that

$$P(|\frac{p}{p+q} - x| \leq w|p, q) \cong \frac{2}{\sqrt{2\pi\sigma^2}} \int_0^w \exp[-z^2/2\sigma^2] \, dz$$

$$= \frac{2}{\sqrt{2\pi}} \int_0^{w/\sigma} \exp[-u^2/2] \, du$$

where $\sigma^2 = pq/(p+q)^3$. We now recognize this as the normal approximation to the beta posterior distribution, but in 1774 this was not Laplace's goal. Instead he evaluated[8] the integral

$$\int_0^\infty \exp[-u^2/2] \, du = \sqrt{\frac{\pi}{2}}$$

and concluded that as long as $p + q \to \infty$ (and hence our $\sigma^2 \to 0$, so $w/\sigma \to \infty$) $P(|p/(p+q) - x| \leq w|p, q) \to 1$. He did not stop there; he went on to refine the analysis to obtain an approximation for the difference $1 - P(|p/(p+q) - x| \leq w|p, q)$.

Laplace's 1774 analysis had been directed toward a demonstration of posterior consistency. It contained the essence of what was needed to overcome Bayes's analytic stumbling block, but it was not to be until the 1780s that Laplace distilled that essence. In 1774 he was content to conclude consistency and reanalyze a problem of classical probability theory from his newfound Bayesian perspective. That problem was "the problem of points," or, What is the conditional expected gain of each of two players when a sequence of games of chance for a fixed total stake is interrupted?

In 1781 Laplace expanded upon his 1774 analysis to derive arbitrarily

8. This may have been the first time this integral was explicitly evaluated, although it followed rather directly from an integral of Euler's. Gauss (1809, p. 212) referred to this as "an elegant theorem of Laplace." (See Figure 4.1.)

accurate series approximations to the incomplete beta integral. He was able to manipulate the series into forms that permitted calculation, but the effort involved was Herculean and the results far from elegant. By 1785, though, the technique had undergone further development. In 1781 he had focused upon the incomplete beta, upon the integral $\int x^p (1-x)^q\, dx$ for p, q large. In 1785 he published a long memoir, one dealing more generally with integrals of products of functions raised to high powers and focusing clearly upon normal approximations to such integrals. To facilitate the evaluation of these approximations Laplace further derived a series for the normal integral (1785, p. 230).[9] For small T,

$$\int_0^T e^{-t^2}\, dt = T - \frac{1}{3} T^3 + \frac{1}{1 \cdot 2} \cdot \frac{T^5}{5} - \frac{1}{1 \cdot 2 \cdot 3} \cdot \frac{T^7}{7} + \frac{1}{1 \cdot 2 \cdot 3 \cdot 4} \cdot \frac{T^9}{9} - \cdots,$$

For large T,

$$\int_T^\infty e^{-t^2}\, dt = \frac{e^{-T^2}}{2T} \left(1 - \frac{1}{2T^2} + \frac{1 \cdot 3}{2^2 T^4} - \frac{1 \cdot 3 \cdot 5}{2^3 T^6} + \cdots \right).$$

Laplace had succeeded in providing the tools needed for inference about a binomial probability, in a sense completing a line of work begun by Jacob Bernoulli and capping eighty years of fitful, difficult theoretical and conceptual development. And the first application of these ideas he attempted was not to games of chance, but rather to the analysis of statistics on the sex ratio at birth. From the Academy's *Mémoires* for 1771 he found that over the twenty-six-year span 1745–1770, 251,527 boys and 241,945 girls had been born in Paris. Was this an indication of a greater propensity for male births? Letting x represent the probability that a given birth is male, Laplace found from a straightforward application of his analysis that the posterior probability (in modern notation) was

$$P(x \leq \frac{1}{2} \mid p = 251{,}527,\ q = 241{,}945) = 1.1521 \times 10^{-42}.$$

Thus Laplace regarded it as morally certain that $x > \frac{1}{2}$ (1781, pp. 429–431).[10] The preponderance of males was even greater in London, where data from 1664 to 1757[11] gave 737,629 boys and 698,958 girls. Laplace

9. Later, in the *Mécanique céleste,* he added a continued fraction expansion, which he also gave in the *Théorie analytique* (Laplace, 1799–1805, vol. 4, p. 255; 1812, p. 104).

10. This calculation was repeated in Laplace (1786b, pp. 317–318), and, with data from fourteen more years, in Laplace (1812, pp. 377–380).

11. These are the years given in Laplace (1781, p. 432). Two later republications of these data gave the span as 1664–1758 (Laplace, 1786b, p. 317; 1812, p. 377).

went on to ask whether the effect was constant: Was the probability of a male birth in Paris the same as that in London? The data gave the relative frequencies of 0.50971 in Paris and 0.51346 in London. Letting $u =$ probability of a male birth in Paris and $s =$ probability of a male birth in London, Laplace extended his analysis to calculate the posterior probability,

$$P(u > s|\text{data}) = \frac{1}{410{,}458}.$$

He concluded that it was very probable that there was in London some cause that, more so than in Paris, "facilitates the birth of boys; it may depend upon climate, food, or customs" (1781, pp. 458–466).[12] When a similar calculation comparing the probability of a male birth in Paris with that in the Kingdom of Naples (0.51162) gave a posterior probability of about 1/100, however, Laplace demurred: "This probability . . . is not sufficiently extreme for an irrevocable pronouncement" (1786b, p. 325).

Nonuniform Prior Distributions

As I have already remarked, Laplace's frequent assumption of uniform prior probabilities—of equally likely "causes"—was not a blind metaphysical assumption that whatever was unknown was necessarily equally likely to be any of its possible values. Rather, it was an implicit assumption that for ease of analysis the problem had been specified in such a manner that this principle of insufficient reason was reasonable and that, if such were not the case, other assumptions or other prior specifications would be called for. One example in support of this claim comes at the end of his 1774 memoir; there he analyzes a problem involving the tossing of a coin, where the probability of heads is unknown but is uniformly distributed over a small neighborhood of $\frac{1}{2}$ rather than over the whole range of possible values.[13]

More explicit general statements can be found in two later memoirs. The clearest statement came in 1786 in the context of the problem of making inference about a binomial distribution. With x as the unknown probability and $y = x^p (1 - x)^q$ as before (so that $ydx/\int ydx$ gave the posterior distribution in the case where x was a priori uniform), Laplace wrote:

12. Later Laplace (1786b, p. 323) gave a slightly different approximation for the probability, namely, 1/410,158. In both cases he was content to describe the posterior odds as "better than 400,000 to 1."

13. He also investigated multinomial versions of this problem (one where the true probabilities of the faces of a die are uniform over $1/6 \pm 1/q$ each) and took a more detailed look at this same problem for the simpler case of a "three-sided" die (Laplace, 1774).

> If the values of x, considered independently of the observed results, are not all equally possible (but their probabilities can be expressed as a function z of x), it suffices to change y to yz in the preceding expressions. This amounts to supposing all values of x are equally possible and considering the observed results as consisting of two independent results whose probabilities are y and z. We can in this manner reduce all cases to those where we suppose the different values of x to be equally possible, and for this reason we adopt this hypothesis in the following research. (Laplace, 1786b, p. 303)

Thus nonuniform prior distributions were allowed but unnecessary: The analysis for uniform prior distributions was already sufficiently general to encompass all cases, at least for the large sample problems Laplace had in mind.

An earlier attempt to make a similarly general statement had misfired in an interesting way. In the 1781 memoir Laplace had said he would permit the values of x to have a priori probabilities $u(x)$ and those of $1 - x$ to have a priori probabilities $s(x)$. Apparently treating x and $1 - x$ as a priori independent, he gave the posterior distribution as $usydx/\int usydx$, (1781, p. 469). He did not separate u and s in the analysis, however, being content to note that the maximum of the posterior corresponded to the root of $0 = d(us)/us + dy/y$. By 1786 he must have realized the nature of his error.

The Central Limit Theorem

After 1786 Laplace seems to have set probability to the side for a while. It was at just this time that he was succeeding in accounting for the anomalies in the orbits of Jupiter and Saturn, and it is natural that astronomical matters would have increasingly captured his attention. He had tried out his "calculus of probabilities" upon one real example—the sex ratio data—and he even took one step further toward applying the work in the social sciences by showing how the techniques he had devised could be used to estimate a population's size based upon a census of a small portion of the population and upon birth and death statistics for the whole population (Laplace, 1786c). This latter application is now called ratio estimation (see Chapter 5). But these were only indications of the work's potential for applicability, and they lacked the immediate and dramatic impact of the work on Jupiter and Saturn. Only after the completion of the fourth volume of the *Mécanique céleste* in 1805 was Laplace free to reexamine this work.

Laplace's major result in probability theory is now called the central limit theorem, where central is to be understood as meaning fundamental. It was read to the Academy on 9 April 1810 (Laplace, 1810) and can be most simply described as a major generalization of De Moivre's limit theorem: Any sum or mean (not merely the total number of successes in n trials) will, if the number of terms is large, be approximately normally distributed. Regularity conditions and exceptional cases would come later, but even without such refinements the achievement was major and the analysis that produced it a triumph.

It is not entirely clear when Laplace did this work. The basis of the analysis was the use of what we now call Fourier transforms or characteristic functions. They were a direct outgrowth of a technique Lagrange had employed in 1776 (now called the Laplace transform), itself a development of De Moivre's and Simpson's generating functions. But what made it work was an inversion formula of Laplace's own devising. De Moivre, Simpson, and Lagrange had known that if, for example (and with modern notation), $X_1, X_2, \ldots, X_n$ were independent random variables and $P(X_i = k) = p_k$, then the probability that the sum $S_n = X_1 + X_2 + \ldots + X_n$ took on the value k would be the coefficient of s^k in the expansion of

$$(p_0 + sp_1 + s^2p_2 + \ldots)^n.$$

The question was, How does one recover that coefficient in a form amenable to analysis? Lagrange (1776), working with the continuous version $[\int s^a\phi(a)da]^n$ had effectively provided a small dictionary of transforms that would permit the analyst to recognize the answer in a few simple cases, but no general method was known.

Laplace had written an extensive memoir on generating functions in 1782, emphasizing their use in solving difference and differential equations. By 1785 he had a further, remarkably fruitful idea. Consider (again, in modern notation) the simple generating function

$$(s^{-1}p_{-1} + sp_1)^n = s^{-n}a_{-n} + s^{-n+1}a_{-n+1} + \ldots + a_0 + \ldots + a_ns^n$$

where $p_{-1} = p_1 = 1$, and consider the problem of extracting the coefficient a_k of s^k in the expansion. Laplace's idea was to set $s = e^{t\sqrt{-1}}$, getting

$$(e^{-t\sqrt{-1}} + e^{t\sqrt{-1}})^n = e^{-nt\sqrt{-1}}\,a_{-n} + \ldots + a_0 + \ldots + a_ne^{nt\sqrt{-1}}.$$

The problem of recovering the middle term (supposing n even) was then one of finding the term that was constant [and thus free of s (or t)]. Since $e^{t\sqrt{-1}} = \cos t + \sqrt{-1}\sin t$, it followed that the nonconstant terms could be annihilated by integration: $s^k = e^{kt\sqrt{-1}} = \cos(kt) + \sqrt{-1}\sin(kt)$, and so $\int_0^{2\pi} \cos(kt)dt = \int_0^{2\pi} \sin(kt)dt = 0$ unless $k \neq 0$. If $k = 0$, the first integral equaled 2π, and the second still vanished. Thus the middle term could be recovered by integration; it equaled

$$\frac{1}{2\pi}\int_0^{2\pi} (e^{-t\sqrt{-1}} + e^{t\sqrt{-1}})^n\,dt = \frac{2^n}{2\pi}\int_0^{2\pi} \cos^n(t)dt,$$

and this last integral was one whose asymptotic analysis was Laplace's specialty. Other terms (say, the kth) could be recovered by multiplying the generating function by $s^{-k} = e^{-kt\sqrt{-1}}$ before integrating. Actually, in 1785 Laplace only gave this argument for a few simple symmetric cases (pp. 270–278), but he must have realized that it worked quite generally. By the time the *Théorie analytique* was published, it had become a fully developed tool (Laplace, 1812, pp. 83–84). By that time, combined with his unrivaled ability to derive asymptotic approximations to integrals, it had enabled him to show that quite general sums or averages had distributions well approximated by the normal curve.

Yet, despite this triumph, something was missing. Laplace had an extensive probability apparatus developed for the analysis of binomial distributions. He also had developed principles for choosing the mean among several observations, although here his unfortunate choice of error curves had severely limited the practical use of the principles. Meanwhile Laplace had played a role in producing usable and sensible ways of combining observations in complex situations, a line of work that had culminated in 1805 with Legendre's publication of the method of least squares. What was missing was any connection between these two lines of work. In 1809 Gauss provided the key. His work on this was short, but the catalytic effect was immense. Within two years a remarkable synthesis was to be achieved.

4. The Gauss–Laplace Synthesis

Carl Friedrich Gauss (1777–1855)

PUBLICATION of the method of least squares in 1805 marked only the end of the beginning, to use Winston Churchill's phrase, of the development of mathematical statistics. Legendre had proposed a method of combining observations that was immediately seen as applicable in a variety of astronomical and geodetic problems, that was relatively simple and straightforward in its application, and that was based on an easily understood and intuitively reasonable criterion.[1] But seen through modern eyes, and even through at least one pair of contemporary eyes, something was missing: a formal consideration of probability and its relationship to least

1. That the method was easily understood by Legendre's contemporaries is evident from the correct accounts given by other writers in both France and Germany before the end of 1806, as noted earlier. There is, however, one amusing bit of evidence that even excellent

squares. Such a consideration was crucial to the method's usefulness and further development because, without it, no assessment of the accuracy of the method's results was possible. The method of least squares produced results that could be called "best," as they minimized the sum of squared errors and produced an appealing mechanical equilibrium; but as long as the stochastic nature of the observations was unmentioned, the quantification of uncertainty, the answer to the question How good is "best"? was not possible.

Gauss in 1809

The next stage of development was triggered by the publication in 1809 of *Theoria Motus Corporum Coelestium in Sectionibus Conicis Solum Ambientium* ("The Theory of the Motion of Heavenly Bodies Moving about the Sun in Conic Sections")[2] by Carl Friedrich Gauss (1777 – 1855). The main topic of Gauss's book was a masterly investigation of the mathematics of planetary orbits. At the end, however, he added a section on the combination of observations. In that section he addressed essentially the same problem Mayer, Euler, Boscovich, Laplace, and Legendre had addressed, but with one significant difference: Gauss couched the problem in explicitly probabilistic terms. Gauss supposed that there were μ unknown linear functions

$$V = ap + bq + cr + ds + \text{etc.}$$
$$V' = a'p + b'q + c'r + d's + \text{etc.}$$
$$V'' = a''p + b''q + c''r + d''s + \text{etc.}$$
$$\text{etc.}^{3}$$

of observables $a, b, c, \ldots$ with ν unknown coefficients $p, q, r \ldots$, and that these are found by direct observation to be M, M', M'', etc. Gauss supposed that the possible values of errors $\Delta = V - M$, $\Delta' = V' - M'$,

astronomers were capable of misconstruing the method's capabilities. In February 1808 the first class (physics and mathematics) of the Institut (the old Academy of Sciences) presented a report to the emperor on the progress of mathematics since the revolution. The report was edited by Delambre, and he included an account of Legendre's method, writing that "it consists of setting the sum of the squares of all the errors equal to zero" (Delambre, 1810, p. 182.) Whether this overoptimistic account of the method's potential was put in to impress Napoleon, or was merely a slip, is not known.

2. This book was translated into English in 1857 by Charles Henry Davis (an admiral in the U.S. Navy who was influential in the founding of the U.S. Naval Observatory; Admiral Davis was also the brother-in-law to Professor Benjamin Peirce of Harvard).

3. The use in this manner of *etc.* by Gauss and others, instead of using double subscripts with explicitly given ranges of values for the subscripts as is now common, may be due to (1) the fact that for Gauss's purposes his notation is easier to understand than ours would be to those unfamiliar with it, (2) difficulties with printing subscripts, (3) etc.

$\Delta'' = V'' - M''$, etc. had probabilities given by a curve $\varphi(\Delta)$, $\varphi(\Delta')$, $\varphi(\Delta'')$, etc. The ingredients of this formulation were all drawn from earlier works, in particular from various writings of Laplace. And so was the approach Gauss chose to resolve this dilemma of having more linear equations than unknowns: He used Laplace's form of Bayes's theorem. He supposed all values of the unknowns equally likely a priori; this assumption led to the choice of those values for p, q, etc. that maximized

$$\Omega = \varphi(\Delta)\varphi(\Delta')\varphi(\Delta'') \text{ etc.}$$

as the "most probable" system of values. Now, as Gauss noted, these most probable values could be found by setting the derivatives of Ω with respect to $p, q, r, \ldots,$ equal to zero and solving the resulting simultaneous equations. But these equations involved the error curve $\varphi(\Delta)$; before Gauss could proceed further he needed a formula for the error curve.

Thus far Gauss had gone little, if any, beyond what Laplace had done for simple measurements. But he struck out in a new direction with his approach to the error curve. Rather than taking Laplace's tack of starting with the principle of insufficient reason in some guise and reasoning to a curve and then to a method of combining observations, Gauss reversed the process. He noted that a priori he could only make general statements about $\varphi(\Delta)$: It would be at a maximum at $\Delta = 0$; it would be symmetric; and it would be zero outside the range of possible errors. Instead of imposing further conditions directly, he assumed the conclusion! He adopted as an axiom the principle that the most probable value of a single unknown observed with equal care several times under the same circumstances is the arithmetic mean of the observations, and he proved in this case (that is, where $V = V' = V'' = \ldots = p$) that taking $p = (1/\mu)(M + M' + M'' + \text{etc.})$ maximizes Ω only when

$$\varphi(\Delta) = \frac{h}{\sqrt{\pi}} e^{-h^2\Delta^2},$$

for some positive constant h, where h could be viewed as a measure of precision of observation. He then showed how in the more general situation this error distribution led to the method of least squares as providing values for p, q, r, etc. that maximize Ω (Figure 4.1).

Viewed from the perspective of more than a century and a half later, we can see much to criticize in Gauss's argument. First and foremost, the argument was a logical aberration—it was essentially both circular and non sequitur. In outline its three steps ran as follows: The arithmetic mean (a special, but major, case of the method of least squares) is only "most probable" if the errors are normally distributed; the arithmetic mean is "generally acknowledged" as an excellent way of combining observations so that errors may be taken as normally distributed (as if the "general"

212 LIBR. II. SECT. III.

$$\frac{dv}{dr}\varphi' v + \frac{dv'}{dr}\varphi' v' + \frac{dv''}{dr}\varphi' v'' + \text{etc.} = 0$$

$$\frac{dv}{ds}\varphi' v + \frac{dv'}{ds}\varphi' v' + \frac{dv''}{ds}\varphi' v'' + \text{etc.} = 0$$

Hinc itaque per eliminationem problematis solutio plene determinata deriuari poterit, quamprimum functionis φ' indoles innotuit. Quae quoniam a priori definiri nequit, rem ab altera parte aggredientes inquiremus, cuinam functioni, tacite quasi pro basi acceptae, proprie innixum sit principium triuium, cuius praestantia generaliter agnoscitur. Axiomatis scilicet loco haberi solet hypothesis, si quae quantitas per plures obseruationes immediatas, sub aequalibus circumstantiis aequalique cura institutas, determinata fuerit, medium arithmeticum inter omnes valores obseruatos exhibere valorem maxime probabilem, si non absoluto rigore, tamen proxime saltem, ita vt semper tutissimum sit illi inhaerere. Statuendo itaque $V = V' = V''$ etc. $= p$, generaliter esse debebit $\varphi'(M - p) + \varphi'(M' - p) + \varphi'(M'' - p) +$ etc. $= 0$, si pro p substituitur valor $\frac{1}{\mu}(M + M' + M'' + \text{etc.})$, quemcunque integrum positiuum exprimat μ. Supponendo itaque $M' = M'' =$ etc. $= M - \mu N$, erit generaliter, i. e. pro quouis valore integro positiuo ipsius μ, $\varphi'(\mu - 1)N = (1 - \mu)\varphi'(-N)$, vnde facile colligitur, generaliter esse debere $\frac{\varphi'\Delta}{\Delta}$ quantitatem constantem, quam per k designabimus. Hinc fit $\log \varphi\Delta = \frac{1}{2}k\Delta\Delta + \text{Const.}$, siue designando basin logarithmorum hyperbolicorum per e, supponendoque Const. $= \log \varkappa$,

$$\Delta\varphi = \varkappa e^{\frac{1}{2}k\Delta\Delta}$$

Porro facile perspicitur, k necessario negatiuam esse debere, quo Ω reuera fieri possit maximum, quamobrem statuemus $\frac{1}{2}k = -hh$; et quum per theorema elegans primo ab ill. Laplace inuentum, integrale $\int e^{-hh\Delta\Delta}d\Delta$, a $\Delta = -\infty$ vsque ad $\Delta = +\infty$, fiat $= \frac{\sqrt{\pi}}{h}$, (denotando per π semicircumferentiam circuli cuius radius 1), functio nostra fiet

$$\varphi\Delta = \frac{h}{\sqrt{\pi}}e^{-hh\Delta\Delta}$$

178.

Functio modo eruta omni quidem rigore errorum probabilitates exprimere certo non potest: quum enim errores possibiles semper limitibus certis coërceantur,

Figure 4.1. Gauss's 1809 derivation of the normal density. (From Gauss, 1809, p. 212.)

scientific mind had already read Gauss!); finally, the supposition that errors are normally distributed leads back to least squares. Even Gauss himself, returning to this subject twelve and thirty years later was to find this chain of reasoning unpalatable. Except for one circumstance, Gauss's argument might have passed relatively unnoticed, to join an accumulating pile of essentially ad hoc constructions, a bit neater than some but less compelling than most.[4] That one circumstance was the reaction it elicited from Laplace.

Gauss's book reached Paris as early as May 1809 (Plackett, 1972), and the exact sequence of events can only be surmised. Laplace did not read his memoir presenting the central limit theorem to the Academy until April 1810, but internally it shows no indication that he had seen Gauss's work when it was prepared (Laplace, 1810). The emphasis is almost wholly upon analysis; there is no hint that Laplace was at the time aware of the possibility of thinking of e^{-t^2} as an error curve. A powerful tool is produced, but it is a tool without an apparent higher purpose, to be admired for its beauty alone. The only applications are to simple means, such as the mean inclination of comets' orbits. By the end of 1810 all that was changed—dramatically and irrevocably.

Reenter Laplace

Laplace must have encountered Gauss's work soon after April 1810, and it struck him like a bolt. Of course, Laplace may have said, Gauss's derivation was nonsense, but he, Laplace, already had an alternative in hand that was not—the central limit theorem. Before seeing Gauss's book Laplace had not seen any connection between the limit theorem and linear estimation, but almost immediately afterward he could see how it all fit together. Laplace rushed a short sequel to his memoir to press in time for it to appear at the end of the same volume (Figure 4.2). In the sequel he cited Gauss, restated the limit theorem in a more usable form, and showed how it could provide a better rationale for Gauss's choice of $\varphi(\Delta)$ as an error curve: If the errors of Gauss's formulation were themselves aggregates, then the limit theorem implied they should be approximately distributed as what would later be called the normal, or Gaussian,[5] curve $\varphi(\Delta)$. (This argument of Laplace was later to become known as the hypothesis of elementary errors; see Chapter 5). And once Gauss's choice of curve was given a rational basis, the entire development of least squares fell into place, just as Gauss had showed. In fact, Laplace improved upon Gauss by taking advantage of his own 1774 result and concluding that the least squares

4. Indeed, a contemporary work by the American Robert Adrain (1808) did go essentially unnoticed; see Stigler (1978c).

5. The name of this curve is discussed at some length in Stigler (1980b).

SUPPLÉMENT AU MÉMOIRE

Sur les approximations des formules qui sont fonctions de très-grands nombres.

Par M. LAPLACE.

J'AI fait voir dans l'article VI de ce Mémoire, que si l'on suppose dans chaque observation, les erreurs positives et négatives également faciles; la probabilité que l'erreur moyenne d'un nombre n d'observations sera comprise dans les limites $\pm \frac{rh}{n}$, est égale à

$$\frac{2}{\sqrt{\pi}} \cdot \sqrt{\frac{k}{2k'}} \cdot \int dr . c^{-\frac{k}{2k'} \cdot r^2}$$

h est l'intervalle dans lequel les erreurs de chaque observation peuvent s'étendre. Si l'on désigne ensuite par $\varphi\left(\frac{x}{h}\right)$ la probabilité de l'erreur $\pm x$, k est l'intégrale $\int dx . \varphi\left(\frac{x}{h}\right)$ étendue depuis $x = -\frac{1}{2}h$, jusqu'à $x = \frac{1}{2}h$; k' est l'intégrale $\int \frac{x^2}{h^2} . dx . \varphi\left(\frac{x}{h}\right)$, prise dans le même intervalle: π est la demi-circonférence dont le rayon est l'unité, et c est le nombre dont le logarythme hyperbolique est l'unité.

Supposons maintenant qu'un même élément soit donné par n observations d'une première espèce, dans laquelle

Figure 4.2. Laplace's 1810 statement of the central limit theorem. (From Laplace, 1810, p. 559.)

estimates, because they bisected the posterior distribution, minimized the expected posterior error.

In retrospect the choice of an error curve may seem like a minor portion of the problem, but the slowness of progress before 1800 shows that it was in fact central. The remarkable circumstance that the curve that led to the simplest analysis also had such an attractive rationale was conceptually liberating. Even in 1809 Gauss had gone beyond simply observing that this curve led to least squares, because he explored further remarkable consequences and developed ways of simplifying the method's use. Gauss noted that if the errors were distributed according to the curve $\varphi(\Delta)$, then the unknowns p, q, r, etc., were a posteriori distributed according to a distribution proportional to e^{-h^2W}, where $W = \Delta^2 + \Delta'^2 + \Delta''^2 +$ etc. He proceeded to show how this fact could be used to assess the method's precision: by determining the marginal a posteriori distributions of the unknowns separately and their attendant "precision" [which we would now describe as $1/(\sqrt{2}$ · standard deviation)]. He also showed how the procedures described could be generalized to measurements M, M', M'', etc. that were made with unequal (but known) precisions and to unknown functions V, V', V'', etc. that were not linear in p, q, r, etc. (only locally linear). And he discussed an elimination algorithm to speed computation and simultaneously solve for the unknown p, q, r, etc.

In addition to the specious case he gave for the choice of error curve, Gauss's development had other limitations. In the discussion of the precision of his estimates of the unknowns, Gauss only mentioned relative precision. An estimate of p was described as having precision 4.96 times that of a single measurement (h being put equal to unity). No attempt was made to translate this to a statement expressing the uncertainty of the estimate of p in its own scale of measurement, and no scheme was presented to determine the unknown precision h of a single measurement. Such criticism is gratuitous, however, for it overlooks the fundamental synthesis that resulted from this—merging a century's work along two distinct lines into one. Legendre's principle of least squares had nearly come of age as the basic tool of mathematical statistics.

Or was it Legendre's principle? Gauss deeply affronted Legendre by referring to the method of least squares as "our principle" (*principium nostrum* in Latin) and by claiming that he, Gauss, had been using the method since 1795. The ensuing priority dispute, and another one involving the law of quadratic reciprocity of number theory, exacerbated the relationship between the two men. The heat of the dispute never reached that of the Newton–Leibniz controversy, but it reached dramatic levels nonetheless. Legendre appended a semianonymous attack on Gauss[6] to the 1820 version of his *Nouvelles méthodes pour la détermination des orbites des comètes*, and Gauss solicited reluctant testimony from friends that he had told them of the method before 1805. Plackett (1972) reviews most of the evidence. A recent study of this and further evidence (Stigler, 1981)

6. The attack, which is translated in Stigler (1977a), carried the by-line "par M * * *" (by Monsieur * * *).

suggests that, although Gauss may well have been telling the truth about his prior use of the method, he was unsuccessful in whatever attempts he made to communicate it before 1805. In addition, there is no indication that he saw its great general potential before he learned of Legendre's work. Legendre's 1805 appendix, on the other hand, although it fell far short of Gauss's work in development, was a dramatic and clear proclamation of a general method by a man who had no doubt about its importance.

Laplace's 1810 supplement was only the first sign of the great burst of intellectual energy Gauss's book had unleashed. In that supplement Laplace had confined his view to that of inverse probability, content to show how his limit theorem could improve upon Gauss's argument for taking the normal curve as an error distribution and to use his own 1774 theorem to strengthen the conclusion from one stating that least squares provided "most probable" estimates to the claim that least squares provided the most accurate estimates (in the sense of smallest posterior expected error). By the next year, however, Laplace had had time to develop a whole new line of attack, one that did not involve inverse probability.

Laplace's new work, which used the central limit theorem in a quite different manner, was presented in a long memoir that was published in 1811. It was largely repeated in the following year's book, the *Théorie analytique des probabilités* (Laplace, 1812, pp. 312–329). Laplace started with the simplest situation, where, to a close approximation, the errors were related to the observations by $\epsilon^{(i)} = p^{(i)}z - \alpha^{(i)}$. In his notation (which was closer to that of Legendre than to that of Gauss), the $p^{(i)}$ and $\alpha^{(i)}$ were given by observation, the unknown coefficient z was to be determined, and the $\epsilon^{(i)}$ were errors. In this 1811 effort Laplace departed from Gauss by avoiding any assumptions about the distribution of errors (other than that the errors were symmetrically distributed about zero). Instead, he noted that if a linear function of the errors $\Sigma m^{(i)}\epsilon^{(i)}$ was equal to zero, then since

$$\sum m^{(i)}\epsilon^{(i)} = z\sum m^{(i)}p^{(i)} - \sum m^{(i)}\alpha^{(i)},$$

he could, once the weights $m^{(i)}$ were specified, solve this equation to find

$$z = \frac{\sum m^{(i)}\alpha^{(i)}}{\sum m^{(i)}p^{(i)}}.$$

He supposed that the weights were positive or negative integers. This approach gave him a whole class of estimates to consider, one estimate for each specification of the weights. At first glance this is a curious way of treating the problem; but of course it was essentially an algebraic formulation of the way Laplace had generalized Mayer's method: Given a system of more inconsistent equations than unknowns, take as many linear combinations of the equations as there are unknowns and solve for the unknowns. Here he was proceeding with but a single unknown. He even assumed (although this was not essential to the analysis) that the weights $m^{(i)}$ were all whole numbers; thus the sum was really just a simple aggregate of the original observational equations (as in the case of Mayer's method), but with the ith equation counted $m^{(i)}$ times in the aggregation (in a manner of weighting reminiscent of his treatment of arc lengths in the *Mécanique céleste*). Only the weights remained to be specified, and here the Gaussian catalyst led him in a new direction. For now Laplace was thinking probabilistically; errors were to be taken as random, symmetrically distributed about zero. He would not have $\Sigma m^{(i)}\epsilon^{(i)} = 0$ exactly, for it would differ by a random amount. Thus the above ratio would not *equal* z exactly, but it would err by an amount u:

$$z = \frac{\sum m^{(i)}\alpha^{(i)}}{\sum m^{(i)}p^{(i)}} + u,$$

where

$$u = \frac{\sum m^{(i)}\epsilon^{(i)}}{\sum m^{(i)}p^{(i)}}.$$

Once he had expressed the problem in these terms, Laplace was ready to address the choice of weights. By a slight extension of his central limit theorem he could show that the error u was (if the number of equations was large) approximately normal and that, for that limiting distribution, not only was the expected error *(la valeur moyenne de l'erreur à craindre)* proportional to

$$\frac{\sqrt{\sum m^{(i)2}}}{\sum m^{(i)}p^{(i)}}$$

but also the probability of any interval symmetric about zero varied inversely with this quantity. It was then a simple exercise in calculus to conclude that this quantity was smallest if the $m^{(i)}$'s were proportional to the $p^{(i)}$'s. But that was just the least squares estimate,

$$\frac{\sum p^{(i)}\alpha^{(i)}}{\sum p^{(i)2}}.$$

Laplace's conclusion can be summarized in modern terms: All estimates of z that are linear functions of the observed $\alpha^{(i)}$ (where the coefficients of the $\alpha^{(i)}$ may depend upon the $p^{(i)}$) are approximately normally distributed, and within this class the least squares estimate has the smallest expected error. He went on to generalize this conclusion to linear estimation problems with several unknowns (Laplace, 1811; 1812, pp. 322–329) in an argument that effectively derived the multivariate normal limiting distribution of two or more least squares estimates (Figure 4.3). These conclusions, although based on asymptotic arguments, did not involve stringent hypotheses about the error curve. Laplace (1811; 1812, p. 321) was even bold enough to suggest again that the curve $(1/2a)\log(a/|x|)$ he had derived in 1777 could be considered as a possible error curve.

Laplace had been led to least squares as being most accurate for large numbers of equations in two senses: smallest expected error and most likely to be near the quantity being estimated. He would later come to refer to it as "the most advantageous method." Twelve years later Gauss published an extension to this argument; he noted that the analysis really only involved second moments and that if one was content to measure accuracy by expected squared error then the conclusion held without regard to the number of equations. The argument was no longer asymptotic in nature (Gauss, 1823, 1855). This extension has come to be called the Gauss–Markov theorem.

A remarkable synthesis had been achieved, but the development was far from finished. Methods of determining the accuracy of the estimates were yet to be devised, and algorithms to simplify computation were yet to be invented. Many hands and minds would be involved in the process—Laplace and Gauss, Poisson and Bienaymé and Cauchy, and a host of other astronomers, mathematicians, and geodesists. The technical complexity was such that it is hard to appreciate what was accomplished and impossible to gauge accurately the depth of contemporary understanding by looking further at the works of the period describing purely mathematical development. Accordingly, I shall now turn to an application of the technology in Laplace's last work, an application published after he died in 1827.

A Relative Maturity: Laplace and the Tides of the Atmosphere

The theoretical discussions we have encountered thus far leave open many questions. How would Laplace have estimated residual variation? Compared means? Dealt with dependent errors? Assessed the significance of the value of an estimated coefficient? The *Théorie analytique* contains some examples, but they are stripped of their context and incomplete in matters of detail. To understand how the techniques were used we need to look at a fully developed application. In the *Mécanique céleste,* Laplace had successfully treated the problem of how to apply gravitational theory to explain the effect of the moon upon the tides of the sea. In 1823 he attempted to apply the same theory to the effect of the moon upon the tides of the atmosphere. Specifically, he wished to measure the moon's effect on the les intégrales étant prises depuis ϖ et ϖ' égaux à $-\pi$, jusqu'à ϖ et ϖ' égaux à π. Cela posé;

En suivant exactement l'analyse du numéro précédent, on trouve que la fonction précédente se réduit à très-peu près à

$$\frac{1}{4\pi^2}.\iint d\varpi.d\varpi'.c^{-l\varpi\sqrt{-1}-l'\varpi'\sqrt{-1}-\frac{k''}{k}.a^2.[\varpi^2.S.m^{(i)2}+2\varpi\varpi'.S.m^{(i)}n^{(i)}+\varpi'^2.S.n^{(i)2}]},$$

k et k'' ayant ici la même signification que dans le numéro cité. On voit encore, par le même numéro, que les intégrales peuvent s'étendre depuis $a\varpi = -\infty$, $a\varpi' = -\infty$, jusqu'à $a\varpi = \infty$ et $a\varpi' = \infty$. Si l'on fait

$$t = a\varpi + \frac{a\varpi'.S.m^{(i)}n^{(i)}}{S.m^{(i)2}} + \frac{kl.\sqrt{-1}}{2k''a.S.m^{(i)2}},$$

$$t' = a\varpi' - \frac{k}{2k''a}.\frac{(l.S.m^{(i)}n^{(i)} - l'.S.m^{(i)2}).\sqrt{-1}}{S.m^{(i)2}.S.n^{(i)2} - (S.m^{(i)}n^{(i)})^2};$$

si l'on fait ensuite

$$E = S.m^{(i)2}.S.n^{(i)2} - (S.m^{(i)}n^{(i)})^2;$$

la double intégrale précédente devient

$$c^{-\frac{k}{4k''a^2.E}.[l^2.S.n^{(i)2}-2ll'.S.m^{(i)}n^{(i)}+l'^2.S.m^{(i)2}]}$$

$$\times\iint\frac{dt.dt'}{4\pi^2.a^2}.c^{-\frac{k''t^2}{k}.S.m^{(i)2}-\frac{k''t'^2.E}{k.S.m^{(i)2}}}.$$

En prenant les intégrales dans les limites infinies positives et négatives, comme celles relatives à $a\varpi$ et $a\varpi'$, on aura

$$\frac{1}{\frac{4k''\pi}{k}.a^2\sqrt{E}}.c^{-\frac{k}{4k''a^2}.\frac{l^2.S.n^{(i)2}-2ll'.S.m^{(i)}n^{(i)}+l'^2.S.m^{(i)2}}{E}}. \quad (o)$$

Il faut maintenant, pour avoir la probabilité que les valeurs de l et de l' seront comprises dans des limites données, multiplier cette quantité par $dl.dl'$, et l'intégrer ensuite dans ces limites. En nommant X cette quantité, la probabilité dont il s'agit sera donc

Figure 4.3. Laplace's 1812 derivation of the bivariate normal limiting distribution of two linear functions of observational errors. (From Laplace, 1812, p. 324.)

barometric pressure at Paris by comparing the daily variations in pressure at each of the phases of the moon.

Because the moon exerts such a strong influence upon the sea, Newton and later scientists believed that a similar effect upon the atmosphere must exist. Laplace sought to prove the existence of such an effect, which he thought could be due either directly to gravitational attraction or indirectly to the rising and falling of the sea. His tidal theory indicated that the magnitude of the effect could be gauged by comparing the changes in barometric pressures between 9:00 A.M. and 3:00 P.M. on the four days surrounding the syzygies (those two days in each month when the moon, earth, and sun are aligned) with the daily change in pressure on the four days surrounding the quadratures (those two days in each month when the moon, earth, and sun form a right angle).

To take advantage of the available data, which was an eight-year series of barometric measurements taken three times a day at the Paris Observatory, at 9:00 A.M., noon, and 3:00 P.M., Laplace reduced his theory to a system of linear equations:

$$x \cos(2iq) + y \sin(2iq) = E_i,$$

$$y \cos(2iq) - x \sin(2iq) = F_i,$$

for $i = -1, 0, +1, +2$. Here x and y are unknown quantities, to be estimated from the data; q is a known quantity, the synodic movement of the moon; and E_i and F_i are calculated from the data, the index i representing the day of the phase of the moon. The manner in which E_i and F_i are calculated from the data is important to the analysis:

$$(1) \qquad E_i = A_i'' - A_i + B_i - B_i'',$$

$$(2) \qquad F_i = \{2A_i' - (A_i + A_i'') - 2B_i' + (B_i + B_i'')\}\left(1 + \frac{1}{19}\right),$$

where A_i is the mean, taken over eight years, of the 9:00-A.M. measurements of barometric pressure for the ith day after a syzygy, A_i' the same for the noon measurements, A_i'' the same for the 3:00-P.M. measurements. The B_i, B_i', and B_i'' represent the corresponding mean measurements for the ith day after quadrature. That is, E_i is the mean daily (9:00 A.M. to 3:00 P.M.) barometric change for the ith day after syzygy minus the mean change for the ith day after quadrature, and F_i is proportional to the difference between the mean rates of change, or second differences, for the same two days.

Laplace then assumed that the deviations, due to "irregular causes," of the daily measurements from their mean heights all followed the same law, not necessarily normal. Treating the different measurements as having been made independently, he applied a method of analysis we might now call weighted least squares. He had presented this method in the third supplement (1820) of the *Théorie analytique,* where he called it "the most advantageous method." As explained by Laplace in the third supplement, the method differed from the least squares of Legendre, which Laplace emphasized was only a special case of his method, in that the observations (in the present case, the E_i's and F_i's) might be calculated as linear combinations of other independent observations, each possibly having a different distribution. This structure would then be exploited to find the best method of combining the observations, in the manner of his 1811 memoir, through an appeal to his limit theorem.

Having applied this method to calculate x and y, Laplace then addressed the question of assessing "the probability with which these observations indicate a lunar tide." Earlier in the paper, Laplace (1823) had noted that it was not enough to combine a large number of observations in the most advantageous manner; rather one must have "a method for determining the probability that the error in the obtained results is contained in narrow limits, a method without which one risks presenting the effects of irregular causes as laws of nature; this has happened often in meteorology."

To test the hypothesis that barometric changes were not influenced by the phase of the moon, he compared the mean change on 792 days near syzygies with the mean change on 792 days near quadratures and found that, based upon the approximation given by the central limit theorem, when there was no actual regular difference in barometric changes, chance alone would produce a difference in means no larger than that actually observed with probability 0.843. Laplace stated that this number was not large enough to confirm the existence of an actual difference. He calculated that if the lunar effect were actually of the estimated size it would require nine times as much data to confirm its existence. In modern terminology Laplace stated that the observed difference between means would be significant at the 0.01 level only if it were based on seventy-two years of data.

In addition to the interest this paper has as an example of the application of statistics to a delicate scientific problem a century and a half ago, a more revealing aspect of it is that in his investigation Laplace had made three subtle but important errors.

First, Laplace's implicit assumption that different barometric readings were independent was severely violated, for within-day readings were quite highly correlated. One facet of this correlation did not escape Laplace's attention: It was precisely his observation that day-to-day variations were much larger than within-day variations that had led him to formulate his theory in terms of within-day changes rather than in terms of the readings alone. In fact, basing his analysis upon differences thus avoided a pitfall that was to trouble meteorologists for the better part of the next century—for day-to-day variations are so great that many later workers thought the moon's effect far less significant than Laplace had. But even though he was aware of one implication of this correlation, he ignored it in his application of the "most advantageous method," an omission leading to an incorrect weighting of the observations.

Had Laplace taken this correlation into account, he might have treated the variances of E_i and F_i as being approximately in the ratio 3 : 1 rather than in the ratio 1 : 3 that he actually used. Laplace's use of 1 : 3 for the ratio of the variance of E_i to that of F_i neglected the factor $1 + (1/19)$ and was based upon the definitions [equations (1) and (2)] and the assumptions that all A's and B's were independent with the same variance. That is, assuming that measurements taken at different times of the day are independent and equally variable, the variances are

$$\begin{aligned}\operatorname{Var}(E_i) &= \operatorname{Var}(A_i'' - A_i + B_i - B_i'')\\ &= \operatorname{Var}(A_i'') + \operatorname{Var}(A_i) + \operatorname{Var}(B_i) + \operatorname{Var}(B_i'')\\ &= 4\sigma^2, \text{ say; \quad and}\end{aligned}$$

$$\begin{aligned}\operatorname{Var}(F_i) &= \operatorname{Var}\left\{\left(1 + \frac{1}{19}\right)[2A_i' - (A_i + A_i'') - 2B_i' + (B_i + B_i'')]\right\}\\ &= \left(1 + \frac{1}{19}\right)^2 [4 \operatorname{Var}(A_i') + \operatorname{Var}(A_i) + \operatorname{Var}(A_i'')\\ &\quad + 4 \operatorname{Var}(B_i') + \operatorname{Var}(B_i) + \operatorname{Var}(B_i'')]\\ &= \left(1 + \frac{1}{19}\right)^2 12\sigma^2.\end{aligned}$$

Except for the factor $(1 + 1/19)^2$, these are in the ratio 1 : 3. On the other hand, if we let $U_i = A_i'' - A_i'$ and $V_i = A_i' - A_i$, then $U_i + V_i = A_i'' - A_i$ and $V_i - U_i = 2A_i' - (A_i + A_i'')$. An analysis of limited meteorological data from Paris for 1823 suggests that the correlation of U_i and V_i was not far from $\frac{1}{2}$, a value that would give the ratio 3 : 1, again neglecting the factor $1 + 1/19$. Let $\operatorname{Var}(U_i) = \operatorname{Var}(V_i) = \tau^2$; then

$$\begin{aligned}\operatorname{Var}(U_i + V_i) &= \operatorname{Var}(U_i) + \operatorname{Var}(V_i) + 2 \operatorname{corr}(U_i, V_i)\sqrt{\operatorname{Var}(U_i)\operatorname{Var}(V_i)}\\ &= \tau^2 + \tau^2 + 2 \cdot \frac{1}{2} \cdot \tau^2\\ &= 3\tau^2; \quad \text{and similarly}\end{aligned}$$

$$\begin{aligned}\operatorname{Var}(U_i - V_i) &= \tau^2 + \tau^2 - 2 \cdot \frac{1}{2} \cdot \tau^2\\ &= \tau^2,\end{aligned}$$

giving the 3 : 1 ratio.

Laplace's second error was in his test of significance, where he compared the two mean changes. Even though his use of the central limit theorem was correct (at least if one assumes that the readings on different days are independent with equal variances), he had estimated the variance by pooling the two samples as one rather than by pooling two estimates of the variance, one based on changes near syzygies and the other on changes near quadratures. This error is particularly interesting because it is clear from his other work, in particular from the third supplement, that Laplace knew how to estimate a variance by using a residual sum of squares rather than a total sum of squares; it was only in this situation of testing a null hypothesis that he felt compelled to estimate the variance under the assumption that the null hypothesis was true. Laplace's use of the total sum of squares would tend to inflate greatly his estimate of the variance if the null hypothesis were false.

Laplace's third error is just as interesting. It lies in his implicit assumption that the eight equations are independent. This assumption is questionable because both E_i and F_i involve the same measurements. Now, the actual effect of this error is small: Because E_i involves differences of changes and F_i, second differences, E_i and F_i would be uncorrelated if the variance of the change from 9:00 A.M. to noon equaled that of the change from noon to 3:00 P.M.—and this seems to be not far from the truth. But I think this was a lucky accident, and one that Laplace did not fully understand. It is remarkable, however, that he did sense the possible difficulty a bit later. In his very last paper, which appeared after his death, he returned to this problem and repeated his 1823 analysis—but with one major difference.

In the third supplement to the *Théorie analytique,* Laplace had presented the equation

$$l^{(s)}x + p^{(s)}y + q^{(s)}z + \ldots = a^{(s)} + m^{(s)}\gamma^{(s)} + n^{(s)}\lambda^{(s)} + r^{(s)}\delta^{(s)} + \ldots$$

as a stereotype for one of the "equations of condition" in situations in which his most advantageous method would be applicable. Here the $l, p, q, a, m, n,$ and r represented given coefficients; $x, y,$ and z were quantities to be estimated; and γ, λ, and δ were independent observations with possibly different distributions. Laplace viewed the right-hand side as a derived observation, one calculated from others according to known rules. Except for the previously mentioned lack of independence, the equations of condition [(1) and (2)] for the lunar tides are of this form. One can easily imagine how pleased Laplace must have been to apply his theory in its most general form to a problem that had not been formulated when the theory was developed. But in 1827, in his "Mémoire sur le flux et reflux lunaire atmosphérique," Laplace returned to this problem with an additional three years of data and handled it differently, announcing, "I have determined, *with special care,* the factors by which one must multiply the different equations of condition in order to obtain the most advantageous results" [emphasis added].

In 1827 Laplace first solved the equations of condition in pairs for x and y, to obtain eight new equations, four involving each of x and y. Then he collected together those terms involving the same measurements and applied his most advantageous method, using only the special case relative to a single unknown. I believe that he would not have changed his analysis in this way if he had not been sensitive to the fact that, because E_i and F_i are both calculated from the same measurements, they may not be independent. Thus this change in his analysis shows that in 1827 Laplace was able to correctly deal with multiple linear regression problems with correlated errors and a known covariance structure.

Laplace did repeat his other, earlier errors: assuming that different measurements were independent and assessing whether or not his estimate of x differed significantly from zero by using a pooled sample to estimate the variance of daily pressure changes. Here he estimated x to be 0.031758 and calculated the probability that chance alone would produce an estimate within the limits ± 0.031758 to be 0.3617, stating,

> If this probability had closely approached unity, it would indicate with great likelihood that the value of x was not due solely to irregularities of chance, and that it is in part the effect of a constant cause which can only be the action of the moon on the atmosphere. But the considerable difference between this probability and the certainty represented by unity shows that, despite the very large number of observations employed, this action is only indicated with a weak likelihood; so that one can regard its perceptible existence at Paris as uncertain. (Laplace, 1827)

Ironically, even though Laplace was able to recognize and take account of the possible dependence between observations in this one instance where his notation made the form of the dependence clear (for example, with A_i appearing in both E_i and F_i), the final section of this 1827 paper indicates that this understanding was very limited and perhaps tied to the situation where the dependence was notationally explicit. In the last section of this, his last paper, Laplace undertook to investigate the homogeneity of the data over the course of the year. Specifically, he sought to determine whether an apparent difference in the mean changes in barometric pressure over four quarters of the year was in fact significant; the variation is shown in Table 4.1.

Laplace chose to handle this problem (now called an analysis of variance problem) by performing a sequence of four separate tests. First he considered the difference between the February to April mean change and the overall mean change, $0.940 - 0.763 = 0.177$, and evaluated the probabil-

Table 4.1. Mean diurnal variation in barometric pressure at Paris, 1816–1826.

Period	Mean change (mm) 9:00 A.M. to 3:00 P.M.
Nov. to Jan.	0.557
Feb. to Apr.	0.940
May to July	0.752
Aug. to Oct.	0.802
Nov. to Oct.	0.763

Source: Laplace (1827), as given in his *Oeuvres complètes.*

ity that a difference of this size or larger would be due solely to chance. He found the value 0.0000015815 for this probability by means of a continued fraction expansion for the normal integral and judged that it was extremely likely that the discrepancy was indicative of some "constant" cause. He repeated his analysis for the other three quarters, finding that the November to January mean change also differed significantly from the overall mean but that the discrepancies between the other quarterly means and the overall mean could "without improbability be attributed solely to the irregularities of chance."

Although Laplace's analysis was correct in many respects, including his recognition of the effect that the differing sample sizes would have upon the relative variances of a quarterly mean and the overall mean, his derivation of the distribution of the difference between these means was based implicitly upon the assumption that they were independent. Earlier, where his notation had made it explicitly clear that the same measurements were entering into the calculation of two different quantities, for example, E_i and F_i, Laplace seems to have noticed this difficulty and allowed for it in his analysis. Here in his final paper, where he only presented the means numerically and used no notation for them that made explicit the fact that the quarterly means determined the overall mean, Laplace missed noticing the correlation between these means. Not surprisingly, he did not allow for the dependence among his four significance tests.

Also, as he had before, Laplace based his estimate of the variance of daily barometric changes upon the pooled eleven-year sample, grouped into 132 months, that is, upon the total sum of squares (Table 4.2).

Interestingly, Laplace's general conclusions in all these investigations seem to have been correct, despite these errors in his analyses. According to Chapman (1951), Laplace could scarcely have chosen a worse location than Paris to attempt to measure the lunar atmospheric tide. Although the tide does exist, its effect is extremely small; and it was not until 1945 that its magnitude at Paris was successfully determined! When contemporary (1823) data are used to estimate within-quarter variability of changes in

Table 4.2. Mean change in barometric pressure, 9:00 A.M. to 3:00 P.M., 1816–1826.

Month	1816	1817	1818	1819	1820	1821	1822	1823	1824	1825	1826	Mean
Jan.	0.513	1.234	0.840	0.751	0.310	0.288	0.521	0.762	0.882	0.756	0.599	0.677
Feb.	0.846	0.685	1.306	0.802	0.912	1.077	1.081	1.104	0.662	0.886	0.863	0.929
Mar.	0.836	0.568	1.085	0.861	0.750	0.576	0.425	0.514	1.042	1.141	0.882	0.797
Apr.	0.894	1.118	1.040	1.071	1.256	0.956	0.880	0.914	0.873	1.222	0.887	1.010
May	0.613	0.840	1.045	1.163	0.975	0.642	0.692	0.646	0.440	0.887	1.002	0.813
June	0.596	0.820	0.805	0.821	0.396	0.579	0.797	0.660	0.704	0.849	0.744	0.707
July	0.537	0.686	1.083	0.720	0.376	0.584	0.918	0.521	0.714	1.084	0.888	0.737
Aug.	0.951	0.702	0.989	0.961	0.763	0.929	0.950	0.812	0.632	0.687	1.024	0.854
Sept.	0.534	0.719	0.828	0.808	0.712	0.579	0.958	1.047	0.863	0.889	0.871	0.801
Oct.	1.043	0.903	0.750	0.374	1.271	0.793	0.448	0.564	0.679	0.535	0.908	0.751
Nov.	0.014	0.624	0.438	0.331	0.249	0.964	0.598	0.665	0.474	0.734	0.906	0.545
Dec.	0.730	0.246	0.696	0.476	0.600	0.173	0.331	0.243	0.614	0.459	0.371	0.449
Mean	0.676	0.762	0.909	0.762	0.714	0.678	0.717	0.704	0.715	0.844	0.828	0.756

Source: Bouvard (1827).

Note: The marginal means are Bouvard's; the March mean should evidently read 0.789.

barometric pressure, Laplace's statements relative to the simultaneous comparison of the four quarterly means can be confirmed at the 0.05 level, using Newman's (1939) multiple range test. This conclusion would remain true if the data upon which both analyses are based are corrected; the figure 0.940 should apparently be 0.910, to be consistent with the data in Table 4.2. It seems reasonable to surmise that Laplace, or Bouvard, misread a handwritten 1 as 4.

The Situation in 1827

Laplace died on the fifth of March in 1827, just before his seventy-eighth birthday, and his death marked the end of an era. By 1827 the theoretical development of the Gauss–Laplace synthesis had reached a level of relative maturity. Laplace's *Théorie analytique des probabilités* had been through three editions and acquired four supplements. Gauss had published his extension of Laplace's theory in 1823, effectively abandoning his circular 1809 argument for the normal as an error curve. Ironically, by virtue of its extreme simplicity Gauss's 1809 derivation of least squares took on a life of its own. It became a staple of textbooks even into the twentieth century and thus was the most widely circulated of Gauss's statistical works. Together with his elimination algorithms for computing least squares estimates, Gauss's 1809 derivation was thus the most influential of his statistical works, probably because the 1823 argument was not generally understood and did not play an important role until much later.[7]

The methods related to the Gauss–Laplace synthesis became widely known as well. In 1814 Legendre republished his appendix on least squares in the *Mémoires de l'Institut de France* (the journal of the successor to the Academy of Sciences) to remind the world of his priority; and it was soon translated into English (Harvey, 1822). Laplace's developments of this theory were trumpeted on the front page of the official French newspaper, the *Moniteur Universel* for 11 January 1812, and portions of them were incorporated in widely circulated texts. Silvestre Lacroix's 1816 *Traité élémentaire du calcul des probabilités* contained a nice exposition of both the theory and methods; and Louis Puissant, who had already included Legendre's treatment of least squares in his 1805 *Traité de géodésie,* expanded upon that to include Laplace's probabilistic methods in a later supplement (Puissant, 1805, pp. 137–141; 1827, pp. 39–42).

This dissemination continued into the middle of the century. Laplace's *Théorie analytique* may have been rough going for all but the most mathematical minds, but Augustus De Morgan digested and re-presented almost the whole of it in his massive article "Theory of Probabilities" in the

7. There were few who showed a clear understanding of this work of Gauss during his lifetime, Ellis (1844) being one notable exception. Even Poincaré (1896, p. 168; 1912, p. 188) misconstrued its nature.

Encyclopaedia Metropolitana (1845). Other less extensive presentations were given in the *Penny Cyclopaedia* (1833–1843) (the articles "Probability," "Mean," "Least squares," and "Weight of observations" were all by De Morgan) and the *Encyclopaedia Britannica* (seventh edition, 1839, article "Probability" by Galloway); and there were numerous discussions in handbooks and encyclopedias in other languages. In 1877 Mansfield Merriman compiled a list of "writings relating to the method of least squares." It contained 70 titles published between 1805 and 1834 and 179 titles published between 1835 and 1864—and it was far from complete at that.

This burgeoning literature included theoretical refinements of all types, such as attempts to "prove" the method without recourse to probability (for example, Donkin, 1844) and criteria for rejecting doubtful observations (for example, Peirce, 1852). Applications were even more numerous and in some cases astoundingly ambitious. The 1858 Ordnance Survey of the British Isles required the reduction of an immense mass of data through the use of least squares. The main triangulation was cast as a system of 1554 equations involving 920 unknowns. Even though they broke the system into 21 pieces of no more than 77 unknowns each before attempting a solution, the calculations took two teams of human "computers," working independently and in duplicate, two and a half years to complete (Clarke, 1858; 1880, p. 243; Ordnance Survey, 1967).

The Gauss–Laplace synthesis brought together two well-developed lines—one the combination of observations through the aggregation of linearized equations of condition, the other the use of mathematical probability to assess uncertainty and make inferences—into a coherent whole. In many respects it was one of the major success stories in the history of science. Yet it also poses a puzzle, for the applications of this marvelous statistical technology were widespread only geographically; to an amazing degree they remained confined to the narrow set of disciplines that spawned them. They became commonplace in astronomical and geodetic work while remaining almost unknown in the social sciences, even though their potential usefulness in the social sciences was known from the time of Bernoulli and even though some early social scientists (for example, Adolphe Quetelet) were also astronomers.

This puzzle suggests that the situation was not so simple as it might appear at first glance, that major conceptual barriers remained to be overcome before the new technology could spread to the social sciences and fulfill the more-than-a-century-old promise. The difficulties were all the greater because the conceptual barriers were not well understood, much less well articulated. The story of the struggle to surmount these problems and the slow development and intellectual diffusion of statistical ideas between 1827 and 1900 will be the topic of the remainder of this work.

PART TWO

The Struggle to Extend a Calculus of Probabilities to the Social Sciences

5. *Quetelet's Two Attempts*

Adolphe Quetelet (1796–1874), as portrayed in 1822

ALTHOUGH the works of Bernoulli and Laplace foreshadowed the application of probability to the measurement of uncertainty in the social sciences, the works of Quetelet represent the first steps toward making this wish a practical reality. Quetelet did not accomplish a great deal toward this end—in some respects he failed totally—but his groping and limited successes help us understand both the problems faced and the way they were viewed by those who might have triumphed over them. A close look at Quetelet's work shows that the lapse in the social application of probability, a lapse that to a naive modern glance might appear to be an oversight, is really a sign of severe conceptual difficulties.

Adolphe Quetelet[1] was born on 22 February 1796 in Ghent, Belgium, and in 1819, for a dissertation on the theory of conic sections, he received the first doctorate of science degree awarded by the new University of Ghent. Quetelet's early bent was toward pure mathematics, with a strong secondary interest in literature.[2] From 1819 on he taught mathematics at the Athenaeum in Brussels. With his election in 1820 to the Académie Royale des Sciences et Belles-Lettres de Bruxelles, he began a half-century of domination of Belgian science. Quetelet wrote a dozen books, founded a journal (*Correspondance mathématique et physique,* eleven volumes, 1825–1839) to which he contributed the lion's share of the articles, and still found time to fill the pages of the Academy's *Mémoires, Bulletin,* and several other journals. A bibliography of his published work contains more than 300 works (*Bibliographie nationale,* 1897, pp. 216–228).

Soon after 1820 Quetelet became active in a movement to found a new observatory. He appears to have entered into this project with all the extraordinary zeal and energy that marked his later career, although he was innocent of any knowledge of practical astronomy. Now, Quetelet was an entrepreneur of science as well as a scientist (he was active in the founding of more statistical organizations[3] than any other individual in the nineteenth century). He had first obtained government support for the project and only then sought to remedy the minor deficiency in his knowledge. His remedy was to have a profound effect upon the direction and scope of his life's work. At the expense of the state Quetelet set out for Paris in December 1823 to learn from the master scientists of the age. From Arago and Bouvard he learned of the practice of astronomy and meteorology and the business of running an observatory. But he did not stop there—he became widely acquainted with the circle of Bouvard's friends; and most important, he learned of probability and its application from Joseph Fourier and perhaps from the aging Laplace himself.

Fourier probably had a greater direct influence on Quetelet than did Laplace, who by 1824 was nearly seventy-five years old and was concentrating his energies on a final, supplementary volume to the *Mécanique céleste.*[4] But regardless of the source, Quetelet acquired a disposition

1. His full name was Lambert Adolphe Jacques Quetelet. The last name is sometimes accented (Quételet), a practice Hankins (1908, p. 9) believed originated with his Paris publishers.

2. Prior to 1823 he wrote the libretto to an opera, a historical survey of romance, and much poetry. See Hankins (1908, p. 11).

3. These organizations included the (Royal) Statistical Society of London, the Statistical Section of the British Association, the International Statistical Congresses, to name only those outside of Belgium.

4. Reichesberg (1893, p. 450, cited by Hankins, 1908, p. 20) says Quetelet received instruction from Laplace himself on the theory of probability; no source is given, however. Quetelet's own recollections (1869, vol. 2, p. 446) mention only lessons from Bouvard and Fourier, so this claim may be an exaggeration.

toward the application of probability that was to color all of his later writing. He did not truly master the subject—the whole of his visit to Paris lasted but three months, and, as Quetelet was fond of writing in other contexts, "the effect is proportional to the cause." Yet even this brief exposure was sufficient to introduce him to the principles to a degree that would eventually enable him to write three introductory books on probability (in 1828, 1846, and 1853).

The de Keverberg Dilemma

Quetelet's principal career within Belgium was as an astronomer and a meteorologist at the Royal Observatory in Brussels, but his international reputation was as a statistician and a sociologist. His curiosity seems to have been as unlimited as his energy, and by 1826 he had become a regional correspondent for his country's statistical bureau. Prior to 1830 Belgium and Holland were united as the Kingdom of the Low Countries (Pays-Bas), and Quetelet's early statistical work involved research on population: the analysis of past population data, and the planning of a census for 1829. His initial encounter with problems of a census had an immense influence upon his later handling of statistical problems. Quetelet had apparently returned from France with the idea that Laplace's method of ratio estimation could be usefully employed in his country. In Quetelet's second statistical memoir, "Researches on population, births, deaths, prisons, poor houses, etc. in the Kingdom of the Low Countries" (1827), he criticized past estimates of the Kingdom's population as being based upon inexact partial censuses taken under difficult political conditions. He argued that attempts to update those figures with data solely on population changes due to births and deaths did not allow for migration. Quetelet initially proposed adopting Laplace's method: "It is thus desirable that the government prepare a new census, after the method proposed by M. De Laplace; the data that we have at present can only be considered as provisional and are in need of correction."

"Laplace's method" had been proposed in the 1780s (Laplace, 1786c) and employed in 1802, and Laplace had subsequently described it in his *Essai philosophique sur les probabilités.*[5] In Laplace's own application, it consisted of determining the number of births in France in the past year from the birth registers (which were considered to be quite accurate) and

5. The mathematics underlying this method, which may be viewed as a forerunner of modern "ratio estimation" methods, had been developed by Laplace in the 1780s. It has been discussed by Karl Pearson (1928; 1978, pp. 462–465), by Sheynin (1976), and most recently by Cochran (1978). In the English translation of the *Essai* (*A Philosophical Essay on Probabilities,* Dover, 1951) the discussion occupies pp. 66–67. A fuller mathematical description can be found in his *Théorie analytique des probabilités* [pp. 391–401 of the first (1812) edition].

multiplying this number by the ratio of population to births. This latter ratio was estimated, not by a complete census of the entire country, but by taking a census only in a few carefully selected communities. As Laplace described the selection, "The most precise means of obtaining [the ratio of population to births] consists, (1.) in choosing departments distributed in an almost equal manner over the whole surface of the country, so as to render the general result independent of local circumstances; (2.) in carefully enumerating at a given time, the inhabitants of several communities in each of these departments; (3.) by determining the mean number of the annual births for each community from the registers of births during several years which precede and follow this period. This number, divided by that of the inhabitants, will give the ratio of the annual births to the population in a manner that is the more accurate as the enumeration is more extensive . . . In thirty departments spread out equally over the whole of France, communities have been chosen which would be able to furnish the most exact information" (Laplace, 1814, pp. 100–101). Laplace lacked any concept approaching that of a random sample, although his purposive sample of communities had clearly been designed to achieve some of the goals we now ascribe to random sampling ("to render the general result independent of local circumstances").[6] Laplace also furnished a mathematical analysis of the accuracy of his ratio estimate (Cochran, 1978).

Quetelet clearly was attracted by the possibility of employing this method with its potential for great saving of time and effort at little or no cost in accuracy. He even went so far as to advance some tentative calculations of the 1824 population of the Low Countries, calculations based upon the ratios of population to births and to deaths found for France. But he did not put the plan into operation. Instead he helped launch plans for a complete census in 1829, and he seems never again to have entertained the possibility of basing population estimates upon incomplete samples. This abrupt turnabout[7] seems to have been due primarily to Quetelet's acceptance of an argument put forward by the Baron de Keverberg in notes that Quetelet appended to his published 1827 memoir. Keverberg, who apparently was serving as an official advisor on state matters, began his note with the type of disclaimer that has become a cliché, "No one is more convinced

6. In fact, given the political organization of the time it is quite conceivable that Laplace's sample would have given more accurate results than a random sample—if the major sources of regional bias were avoided. The selected sample communities had been those with the most energetic and intelligent mayors.

7. Lazarsfeld (1961, p. 309), citing Stephan (1948), speaks of the lack of continued use of sampling methods in social science research as a "discontinuity" in methodological development. As I hope to make clear, I feel the lack of follow-up to Laplace's pioneering effort has a deeper conceptual explanation than simply that later scientists were unaware of or ignored this work.

than I of the utility of statistical research." Then he went on to make a cogent attack on the practical usefulness of Laplace's method, an attack that must have articulated the doubts then prevalent among those who might otherwise have employed statistical methods on social data. Keverberg expressed great respect for Quetelet but differed with him on the best means of attaining the desired goals.

> I only know of Laplace's method through the explanation which M. Quetelet has been kind enough to give me. If I have correctly understood the method, it consists of taking a precise census of population, but only at a few given places in a country, and of then comparing the results so obtained with the mean of the numbers of births and deaths for those places. These latter can be easily obtained for each place and for the totality of the country. The population of each section of the country would then be determined without being verified by a direct census, by employing the ratios derived rigorously for each given place as applying to each section which obeys the same laws of reproduction and mortality.
>
> If it were easy to divide any arbitrary country according to the differences among these laws, the procedure as described would lead without doubt to the proposed end. But it is here that difficulties appear which, it seems to me, are nearly impossible to overcome.
>
> The law regulating mortality is composed of a large number of elements: it is different for towns and for the flatlands, for large opulent cities and for smaller and less rich villages, and depending on whether the locality is dense or sparsely populated. This law depends on the terrain (raised or depressed), on the soil (dry or marshy), on the distance to the sea (near or far), on the comfort or distress of the people, on their diet, dress, and general manner of life, and on a multitude of local circumstances that would elude any a priori enumeration.
>
> It is nearly the same as regards the laws which regulate births.
>
> It must therefore be extremely difficult, not to say impossible, to determine in advance with any precision, based on incomplete and speculative knowledge, the combination of all of these elements that in fact exists. There would seem to be infinite variety in the nature, the number, the degree of intensity, and the relative proportion of these elements. It is then doubtful that we will often find populous regions which, in this regard, can be assimilated the one with the other, and combined in the same category. If such a division of the kingdom could be accomplished on an approximately exact basis, it is likely that it would consist of such a large number of parts that there would be little advantage in terms of work saved.
>
> In my opinion there is only one way of attaining exact knowledge of the population and the elements of which it is composed, and that is an actual and complete census, the formation of a register of the names of all inhabitants, together with their ages and professions. (Keverberg, 1827, pp. 176–177)

Thus Keverberg argued that the need to guard against lack of homogeneity in birth and death rates would require that the country be subdivided into nearly as many sampling units as there were people and thus destroy any potential savings.[8] Even if Quetelet had been disposed to reject this argument, his tentative calculations employing French birth and death rates to estimate the Belgian population would have given him pause. Separate use of birth and death rates gave wildly different answers, and both were quite different from the results based on the past partial censuses.[9] In addition, Quetelet presented estimates of both ratios for nineteen districts in the Low Countries: Both tabular and graphical presentations showed marked inhomogeneity, even at this level of subdivision (Figures 5.1 and 5.2). Keverberg's warning could have caused Quetelet to suspect that further subdivision would only increase the inequality, and he had no guidance on how fine a subdivision to employ.

Whether it was Keverberg's argument or Quetelet's own growing experience with social data, the result was the same. Quetelet discarded any notion he had had of using either Laplace's nonrandom sampling technique or its accompanying probabilistic analysis. Henceforth Quetelet was a convert to complete censuses and insisted that firm inferences could only be based on large amounts of data. He frequently cited Keverberg's notes as an authoritative critique of problems of determining population size.[10] From his first brush with social data Quetelet emerged with a view that tended to make statistical analysis of the type we are discussing impossible. He was acutely aware of the infinite number of factors that could affect the quantities he wished to measure, and he lacked the information that could tell him which were indeed important. He, like the generation of social scientists to follow, was reluctant to group together as homogeneous, data that he had reason to believe was not. A major direction for his life's work was to emerge from this, namely, the examination and sifting of social data to reveal those factors that did have an important influence.

In some respects Quetelet's view of his social data was like Euler's view of astronomical data. To be aware of a myriad of potentially important factors, without knowing which are truly important and how their effect may be felt, is often to fear the worst. After Quetelet had learned for himself (in a way Laplace never did) the extent to which birth and death rates could vary, he could not bring himself to treat large regions as homogeneous, he could not think of a single rate as applying to a large area, and a fortiori he could not conceive of measuring the uncertainty of an estimate where the very thing being estimated was uncertain and ill-defined.

8. He was also aware of and discussed intelligently the difficulties that would have to be overcome in a complete census.

9. The French birth ratio gave an estimate of the combined population of the Low Countries and Luxembourg of 6,924,424, the death ratio gave 5,351,628, and the censuses gave 5,992,666. Quetelet ascribed the discrepancy to the fact that his country's population was increasing much more rapidly than that of France.

10. See, for example, Quetelet and Smits (1832, p. 3) and Quetelet (1835, vol. 1, p. 304; 1842, p. 53).

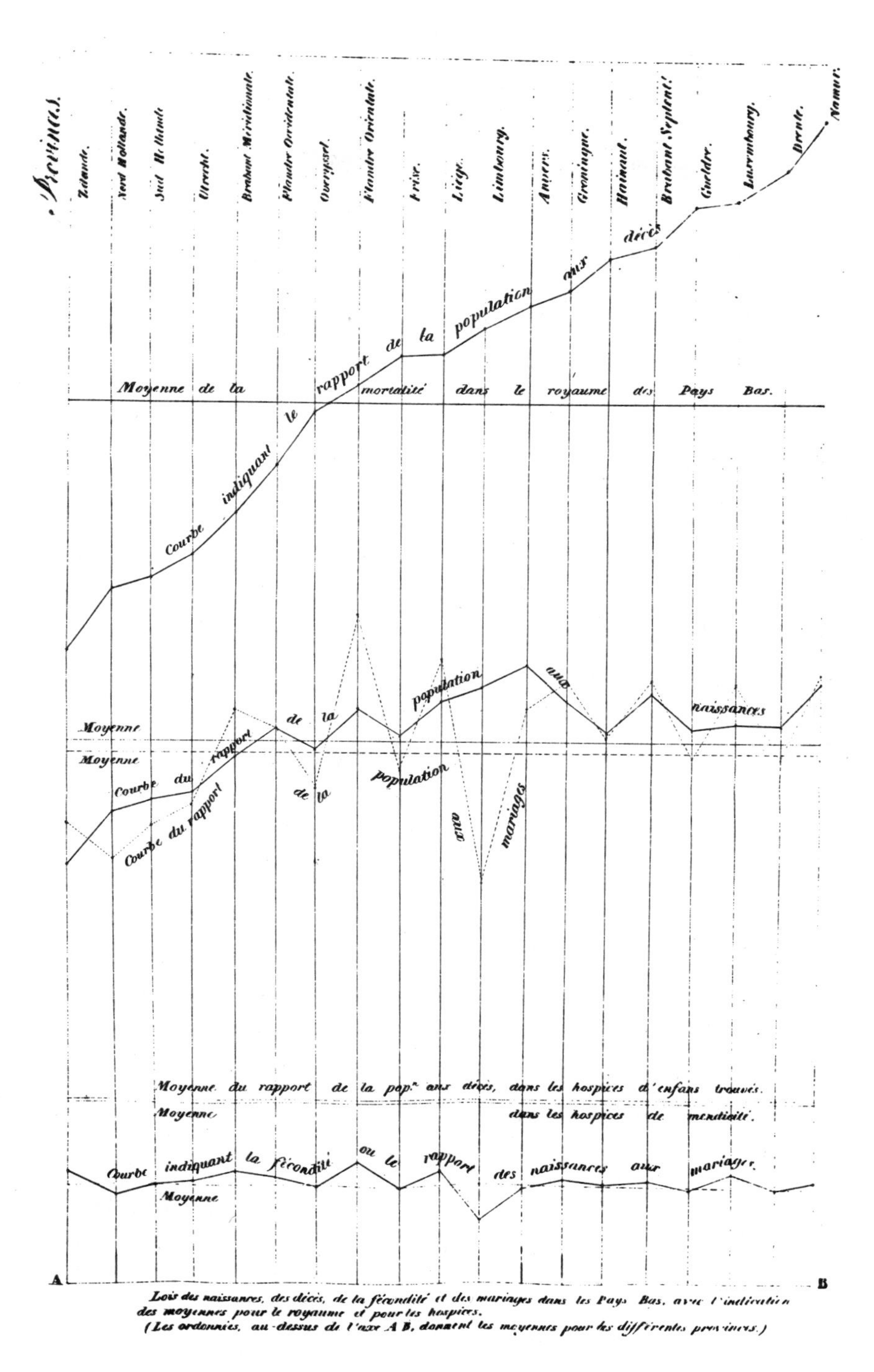

Figure 5.1. Quetelet's 1827 diagram showing the variations over different regions in Belgium in rates of birth, death, and marriage. (From Quetelet, 1827.)

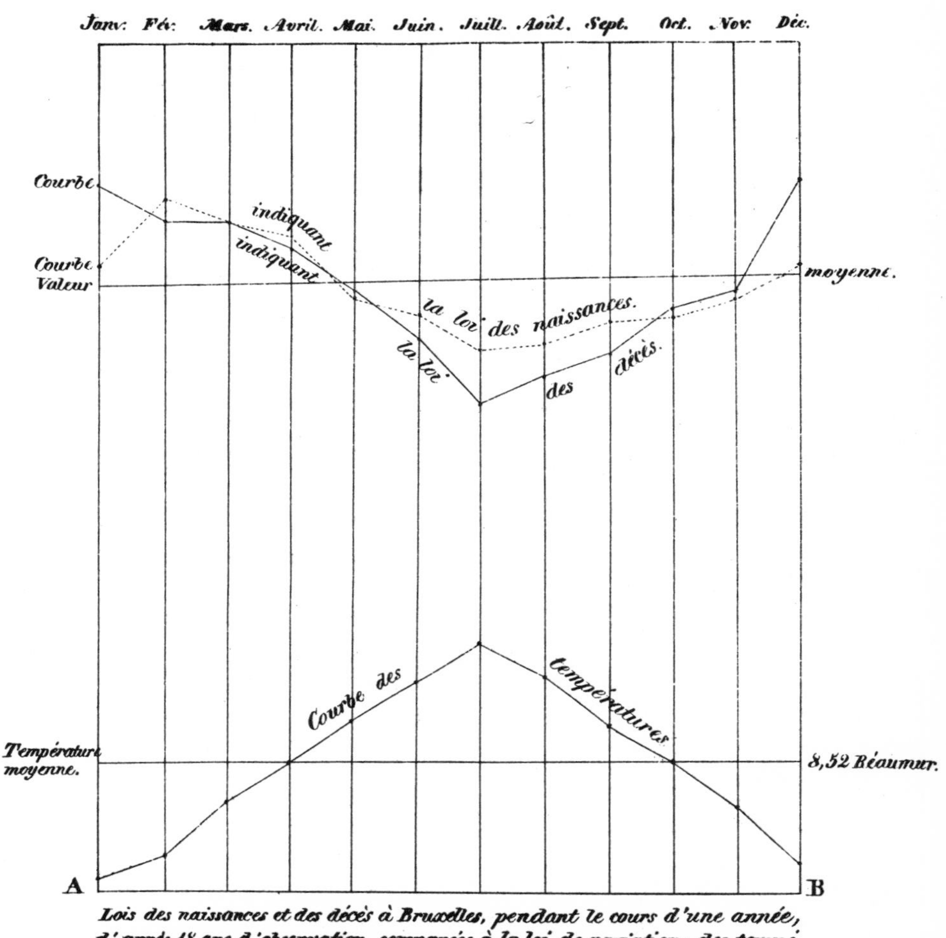

Figure 5.2. Quetelet's 1827 diagram showing the variations over different months in temperature and the rates of birth and death. (From Quetelet, 1827.)

Euler, and those before him, had been reluctant to combine observations made under different circumstances because experience did not permit them to group these observations in ways that could be treated homogeneously. Similarly Quetelet, and those who followed, became increasingly aware of differences that required measurement and had to be allowed for. Quetelet's first instinct had been to group observations crudely and treat ratios as constant in large, superficially similar districts. But Keverberg's comments and his own empirical evidence had robbed him of his boldness. Astronomers had overcome a similar case of intellectual cold feet in the previous century by comparing observation with fact, by comparing prediction with realization. Success had bolstered confidence in the combination of observations. Social scientists were to require massive empirical data gathered under a wide variety of circumstances before they gained the astronomers' confidence that the quantities they measured were of sufficient stability that the uncertainty of the estimates was itself susceptible to measurement.

The Average Man

Quetelet made two important advances toward the statistical analysis of social data: the first of these was formulating the concept of the average man, the second the fitting of distributions. Quetelet's first awakening to the variety of relationships latent in society may have come with his investigation of population data, but his interests soon spread. From 1827 through 1835 he examined scores of potentially meaningful relationships through the compilation of tables and the preparation of graphical displays. With few exceptions he compared only two characteristics at a time, but within this constraint his curiosity was seemingly boundless. He examined birth and death rates by month and city, by temperature, and by time of day. He calculated the month of conception from the birth month and tried to relate it to marriage statistics. He investigated mortality by age, by profession, by locality, by season, in prisons, and in hospitals. He considered other human attributes: height, weight, growth rate, and strength. Quetelet's interests also extended to moral qualities: statistics on drunkenness, insanity, suicides, and crime. In 1835 he collected a number of earlier memoirs and added to them to form the two-volume book that was to gain him an international reputation as a social scientist: *Sur l'homme et le développement de ses facultés, ou essai de physique sociale.* It was translated into English in 1842 as *A Treatise on Man and the Development of His Faculties.*[11]

11. When Quetelet himself referred to this book, as he did in Quetelet (1842, p. vii) and Quetelet (1848, p. vii), he called it simply *Physique sociale.* When he published an expanded version in 1869, he reversed the order of the title.

There was no mistaking Quetelet's aim in this book: to lay the groundwork for a social physics, to conduct a rigorous, quantified investigation of the laws of society that might some day stand with astronomers' achievements of the previous century. His beginning was only tentative, and he was careful (sometimes to the point of being apologetic) not to claim more success than he could defend; but he was eloquent and his zeal caught the public eye. A rare, three-part review in the *Athenaeum* concluded by remarking, "We consider the appearance of these volumes as forming an epoch in the literary history of civilization."[12]

The focal point of much of this work, and the source of much of the attention that it received, was Quetelet's introduction of the average man *(l'homme moyen)*. The average man was a fictional being in his creator's eye, but such was his appeal that he underwent a transformation, like Pinocchio or Pygmalion's statue, so that he still lives in headlines of our daily papers. Actually, despite his exclusive use of the singular masculine noun, there was in Quetelet's work not just one average man, but a whole species of average men and average women. The concept initially came about as Quetelet sought to summarize anthropometric data for the purpose of investigating the relationships between and differences among groups of people. From the data on a large number of French conscripts, for example, an average height and an average weight could be calculated. These values would then be taken as the height and weight of the "average" French conscript. A comparison could then be made between this, perhaps nonexistent, French conscript and his counterpart in Belgium, or Scotland, or wherever comparable data was available. From height and weight Quetelet went on to consider other characteristics of such an "average" individual: The propensity of the average man to commit a crime or become drunk could be calculated as the arrest or drunkenness rate for the group in question, and these rates or propensities could then be compared among different populations or even between the two sexes. Given enough time and ingenuity, an entire statistical profile could be constructed for an average man. Any aspect of a population that was susceptible to measurement would produce a measurement for that population's average man, either a physical measurement such as a height or a "propensity" such as a murder rate.

The idea of the average man caught the imagination in 1835 as it does now. As a psychological ploy it was a brilliant device. It captured the

12. The review is in the issues of the *Athenaeum* for 8 August 1835 (pp. 593–595), 15 August 1835 (pp. 611–613), and 29 August 1835 (pp. 658–661). The review was anonymous. Incidentally, recent Darwinian scholarship argued that Darwin's reading of this review and subsequently of Quetelet and Malthus played an important role in his development of the theory of natural selection; see Schweber (1977, p. 232).

egalitarian idea of a common man in a precise and apparently scientific way that was quite congenial to nineteenth-century political thought, and it served a valid and useful statistical purpose. It also attracted much criticism. The anonymous reviewer in the *Athenaeum* appreciated the usefulness of the concept for physical and medical measurements but demurred on moral measurements. In several expansive passages Quetelet had elevated the average man in terms such as these: "If an individual at any given epoch of society possessed all the qualities of the average man, he would represent all that is great, good, or beautiful."[13] The reviewer felt that in moral spheres such mediocrity would be unappealing. Cournot (1843, pp. 213–214) was even harsher, arguing that a totally average man, if forced to exist, would be an unviable monstrosity, just as the average of several different right triangles (the triangle formed from the averages of their respective sides) will not be a right triangle.

The average man, as perceived by the public and as found in some of Quetelet's later, more rhetorical descriptions, was an important contribution to sociology and political science—a vivid and concrete symbol of a society, an embodiment of the target and the ultimate benefactor of nineteenth-century social reform. But Quetelet's average man was in fact a more complex creature than naive accounts would indicate. He was Quetelet's device for allowing a beginning of a "social physics," the gatekeeper to a mathematical social science.

To Quetelet there was not one, but many average individuals, and his view of them underwent a slight evolution in his writings. As I have indicated, there was, at least potentially, a different average man for each different way of categorizing people. There would be an average man at each age, of each race, in each country, and at each combination of these. As Quetelet put it, the average man is always "conformable to and necessitated by time and place" (Quetelet, 1835, vol. 2, p. 274; 1842, p. 100; see also Hankins, 1908, chap. 3). It was the relationship between these average men that was the focus of Quetelet's attention, their rates of development and their differences and similarities. The average man was a device for smoothing away the random variations of society and revealing the regularities that were to be the laws of his "social physics."

Quetelet's use of the average man was founded upon the belief that if there is no change in any underlying causal relationship—if there is a "persistence of causes"—then there will be a tendency for the average of large aggregates of even unhomogeneous data to be stable. Thus if there is instability or differences between averages, this is evidence of a lack of persistence or difference of causes. As Quetelet expressed it, "effects are

13. Quoted from the *Athenaeum* review, 29 August 1835, p. 661. The corresponding passage is at p. 276 of Quetelet (1835, vol. 2) and p. 100 of Quetelet (1842).

proportional to causes."[14] This belief was borrowed from observational astronomy; but more important, it was founded on a growing empirical support that Quetelet was collecting. Crime rates, birth ratios, suicide rates: All these and more showed such an appearance of stability in large numbers over time that even the existence of free will seemed to be called into question. Quetelet italicized this as a fundamental principle: *"The greater the number of individuals observed, the more do individual peculiarities, whether physical or moral, become effaced, and allow the general facts to predominate, by which society exists and is preserved"* (Quetelet, 1835, vol. 1, p. 12; 1842, p. 6).

Quetelet's interpretation of the average man underwent a slight evolution from 1835 to 1848. This individual was first introduced as a counterpart of a physical center of gravity, with attention being given primarily to the limits around this central value within which others of his class might vary. By 1846, when Quetelet published his *Lettres à S.A.R. Le Duc Régnant de Saxe-Cobourg et Gotha sur la Théorie des Probabilités appliquée aux sciences morales et politiques,* attention had shifted to the center itself. The average man, the center of gravity in both physical and moral senses, had been singled out as the "type" for the race and occasionally elevated to a standard of beauty at which nature aims, as a marksman aims at the center of a target.

We can, I think, best appreciate the significance of Quetelet's average man, both as he and as his contemporaries saw it, if we liken it to a modern statistical construction designed to serve the same end. We might, for purposes of illustration, consider a single characteristic of a single individual k of age i and sex j. The characteristic might be denoted Y_{ijk} (Quetelet used no such notation) and could be a continuously measured characteristic such as the individual's height, or it could be an indicator,

$$Y_{ijk} = \begin{cases} 1 & \text{if individual convicted of crime} \\ 0 & \text{otherwise.} \end{cases}$$

Quetelet knew that individuals' characteristics could not be represented by a deterministic law, but he believed that averages over groups could be so represented. In effect, he would have

$$Y_{ijk} = \mu_{ij} + \epsilon_{ijk},$$

where μ_{ij} would represent the "type" for a large group of individuals, homogeneous with respect to age (i) and sex (j), and ϵ_{ijk} would represent

14. This recurs in Quetelet's writings; see Quetelet (1835, vol. 1, p. 33; 1842, p. 10; 1848, p. 75) and Hankins (1908, p. 103) for examples. Lazarsfeld (1961, p. 306) takes it to mean a literally linear relationship, but few instances can be found to support such a precise interpretation. In most instances Quetelet clearly intended that the reader interpret the statements in a vague, qualitative way.

the single individual's deviation from that type. By averaging measurements for large numbers of individuals Quetelet hoped to recover the "type" μ_{ij}; by considering different categories he could determine (approximately) deterministic laws satisfied by these "types." For example, he fit the simple growth curve

$$y + \frac{y}{1000(T-y)} = ax + \frac{t+x}{1+\frac{4}{3}x},$$

separately for men and women, to the heights y of groups of individuals of weight x, where t and T are heights at birth and maturity, and a would vary by locality (Quetelet, 1835, vol. 2, pp. 23–28; 1842, pp. 61–62).[15]

This use of modern notation captures one aspect of Quetelet's analysis quite well, but it goes far beyond it in another. It captures his intention in his later work of representing individual behavior as a deviation from a population mean and of subjecting only these means to scrutiny for possible lawlike behavior. But it goes far beyond him by suggesting that Quetelet might have encompassed several categorizations within one formal analysis. A modern statistician might represent the "types" μ_{ij} as $\mu_{ij} = \mu + \alpha_i + \beta_j$ and attempt, using all available data, a simultaneous determination of the "laws" described by the μ, α_i's, and β_j's. There is not even a hint of such an intention in Quetelet. He made separate determinations of his equivalent of μ (he called it a "constant cause") and perhaps also of the equivalents of the α_i's *or* the β_j's (he called them "variable causes" and entertained the possibility that they were regular, as with the growth curves, or periodic), but he showed no inclination toward a simultaneous unwinding of multiply categorized data. He was, like the astronomers before Laplace, unable to take the conceptual step of combining measurements taken under a variety of conditions into one analysis. He could write of marginal relationships, and within categories he could seek relations across other variables (such as male height across age); but he could not go further. His elimination of individual variation was at bottom reduced to an assumption of *ceteris paribus*—that all other things are equal.

Quetelet was able to conceive of the application of the theory of probability within homogeneous cells. For example, he wrote, "Everything being equal, the calculus of probabilities shows, that in the direct ratio to the number of individuals observed, we approach the nearer to the truth" (Quetelet, 1835, vol. 1, pp. 13–14; 1842, p. 7). But his appreciation of the multitude of potential causes that would prevent everything from being equal, and his lack both of an analysis of multiply classified data and of a

15. His method of fitting these curves did not involve an algorithm such as least squares but was, like his choice of the family of curves, ad hoc.

sufficient empirical base to determine which categorizing labels need be allowed for, prevented any practical quantification of uncertainty.

The Analysis of Conviction Rates

In dealing with some physical measurements, such as height and weight, Quetelet had large amounts of data, and he did, for example, determine at least roughly how age affected growth. But in other, more socially interesting areas such as crime, he was less well situated. He did make one interesting attempt to measure both the importance and uncertainty of social effects, in work that can be used to reveal his level of understanding and the problems he and others faced. The passage in question was published as the final chapter of his 1835 book (pp. 294–327, vol. 2), although most of it had appeared in another journal in 1832.[16] It was presented more as a cautious example of a method than as a substantive investigation; the method was not to be repeated in Quetelet's work. He stated his purpose as follows:

> In researches of this nature, it will always be necessary to reproduce the original documents carefully, give their source, and any information bearing on their value . . . It will be equally desirable, when numbers are used and results deduced from them, to calculate their probable errors carefully. It is not enough to possess materials; it is also necessary to know their value. One of the greatest defects of existing statistics is, that they present all the numbers collected indiscriminately on the same line, and make them concur to produce a single result, without taking account of either their importance or probable value . . . There is another question that deserves no less attention. It is not sufficient to recognize that an effect depends on several causes; it is extremely important that we be able to assign the proper degree of influence to each of these causes. In concluding this work, I shall demonstrate the possibility of finding a suitable measure for such a purpose. (Quetelet, 1835, vol. 2, pp. 295–296; 1842, p. 103)

Quetelet's example involved the conviction rate for the courts of assize in France. Table 5.1 gives the aggregated data, which covered the six years 1825–1830. Quetelet based his analysis upon what he called the repression of crime, and what we shall call the conviction rate, that is, the fraction of those accused who are convicted. On the basis of the data in Table 5.1, he claimed that there was a slight, but nonetheless apparent, tendency for

16. It was first published as "Sur la possibilité de mesurer l'influence des causes qui modifient les élémens sociaux. Lettre à M. le Dr. Villermé" in vol. 7, 1832, of *Correspondance mathématique et physique*, pp. 321–346. It is also found in Quetelet (1842, pp. 103–108), although the translated passages that I have quoted deviate from that translation in some respects, in order to be more faithful to the original. It was republished again in Quetelet (1869, vol. 2, pp. 407–428).

Table 5.1. Quetelet's data on the conviction rate in the French courts of assize.

Years	Accused	Convicted	Conviction rate
1825	7,234	4,594	0.635
1826	6,988	4,348	0.622
1827	6,929	4,236	0.610
1828	7,396	4,551	0.615
1829	7,373	4,475	0.607
1830	6,962	4,130	0.593
Average	7,147	4,389	0.6137

Sources: Quetelet (1835, vol. 2, p. 298; 1842, p. 103).

Note: Quetelet labeled the last two columns *condamnés* and *répression.* The figures for 1825 differ from those of Poisson (Table 5.4), who used corrected data, and the rate for 1827 should read 0.611.

this rate to decrease annually. Quetelet argued that these conviction rates, themselves averages based on many individuals (and hence not so susceptible to "accidental causes"), were influenced by both "constant causes" and "variable causes." Except for the influence of the variable causes the annual rate should remain unchanged at the average value, 0.6137.[17] Before Quetelet turned to the analysis of the variable (from year to year) causes, he sought to separate and measure the respective influences of the constant causes.

Quetelet interpreted his rates as probabilities. Thus an individual of whom nothing more is known than that he is accused of a crime has a 0.614 chance of conviction. If something more is known, however, something related to a "constant cause," this chance changes. "But if we learn the additional fact, that the accusation is for a crime against persons, the probability of conviction is altered; indeed, experience proves that the conviction rate for crimes against persons is less than that for crimes against property" (1835, vol. 2, p. 299; 1842, p. 104). Quetelet's data allowed him to get a grip on these and several other potential "constant causes:" sex of accused, age of accused, education of accused, and whether or not the accused appeared to stand trial. Not all classifications were available for all six years. The data are given in Table 5.2 and summarized in the second column of Table 5.3.

17. This was the average of the rates; he apparently did not compute the rate based on the totals, 4389/7147 = 0.6141. Throughout his analysis Quetelet employed the marginal *rates* as his data, without weighting them when averaging, say, by the number accused in that year or category.

Table 5.2. Quetelet's conviction rates, broken down by year and state of accused.

State of the accused	Conviction rates, by year						
	1825	1826	1827	1828	1829	1830	Average
Accused of crime against person	0.46	0.51	0.50	0.47	0.46	0.46	0.477
Accused of crime against property	0.66	0.67	0.65	0.66	0.65	0.64	0.655
Male	—	0.63	0.62	0.63	0.62	0.61	0.622
Female	—	0.60	0.60	0.57	0.57	0.54	0.576
Under 30 years of age	—	0.64	0.64	0.64	0.62	0.61	0.63
Above 30 years of age	—	0.60	0.58	0.58	0.59	0.58	0.586
Appeared to stand trial	—	0.49	0.45	0.46	0.50	0.48	0.476
Failed to appear to stand trial	—	0.93	0.97	0.97	0.97	0.96	0.96
Unable to read or write	—	—	—	0.63	0.63	0.62	0.627
Able to read and write imperfectly	—	—	—	0.62	0.60	0.58	0.60
Able to read and write well	—	—	—	0.56	0.55	0.52	0.543
Has a superior education	—	—	—	0.35	0.48	0.37	0.40

Sources: Quetelet (1832; 1835, vol. 2, p. 305; 1842, p. 105).

Note: The numbers in each class were not given. Quetelet's *répression* is translated "conviction rate" here. Misprints from the 1835 and subsequent publications are not given here. The table giving these data was omitted from Quetelet (1869).

Quetelet was not content simply to note in general terms which "cause" appeared most influential and which combination appeared most beneficial, although he did do both of these.[18] Rather he went on to ask two important questions: First, how can we measure the importance of the deviations from the average of 0.614? Second, can we be sure these deviations are not purely fortuitous? Quetelet viewed these as different questions, as we do today. We might phrase them as, How can we measure the scientific and the statistical significance of the observed deviations?

Quetelet's answer to the first question was a simple one: The "importance" of a deviation from the average is the deviation divided by the

18. Because his data did not appear to include multiway classifications, his comments about the best position for an accused to be in (a well-educated female over thirty, appearing voluntarily to answer a charge of a crime against persons) may be open to dispute. However, Quetelet was far from the last to tacitly assume independence in a situation that would not warrant it.

Table 5.3. Quetelet's analysis of the relative degree of influence of the state of the accused upon the conviction rate (probability of conviction).

State of the accused	Probability of conviction	Relative degree of influence	Difference from average	
			Less	Greater
Has a superior education	0.400	0.348 [0.349]	0.125	0.200
Appeared to stand trial	0.476	0.224 [0.225]	0.056 [0.055]	0.050
Accused of crime against person	0.477	0.223	0.035 [0.036]	0.069
Able to read and write well	0.543	0.115 [0.116]	0.042	0.031
Female	0.576	0.062	0.062	0.042
Above 30 years of age	0.586	0.045 [0.046]	0.027 [0.010]	0.024
Able to read and write imperfectly	0.600	0.023	0.033	0.033
Without any designation	0.614	0.000	0.034	0.034
Male	0.622	0.013	0.019	0.013
Unable to read or write	0.627	0.022 [0.021]	0.011	0.005
Under 30 years of age	0.630	0.026	0.032	0.016
Accused of crime against property	0.655	0.067	0.018 [0.023]	0.039 [0.023]
Failed to appear to stand trial	0.960	0.563 [0.564]	0.031	0.010

Sources: Quetelet (1832; 1835, vol. 2, pp. 300, 303, 307; 1842, pp. 104, 105; 1869, vol. 2, pp. 410, 412, 414).

Note: Quetelet headed the columns *État de l'accusé, Probabilité d'être condamné, Degré relatif d'influence de l'état de l'accusé sur la répression,* and *Degrés relatifs de l'importance de l'écart maximum de la moyenne; en moins; en plus.* The figures in brackets are corrections of the original publication. In addition, the figures in the last two columns were incorrectly reversed in the original.

average. In modern symbolism (again, Quetelet used no notation) we could write this as

$$\mathrm{P}(C) = \text{conviction rate, in general}$$

$$\mathrm{P}(C|S) = \text{conviction rate, accused in state } S$$

$$\frac{|\mathrm{P}(C|S) - \mathrm{P}(C)|}{\mathrm{P}(C)} = \text{importance of, or relative degree of influence of, state } S.$$

The results of Quetelet's calculations are given in the third column of Table 5.3. Quetelet's argument for the appropriateness of this measure was short and vague: "It is in essentially this same manner that the first mathematicians who studied the application of the theory of probability to facts bearing upon man (Buffon, in particular), have estimated the importance to an individual of a sum of money, by comparing it with what that individual possessed" (1835, vol. 2, p. 301; 1842, p. 104). The measure he proposed is vulnerable to criticism by modern statisticians, and G. Udny Yule was one of the earliest to discuss its drawbacks (see Yule, 1900; Goodman and Kruskal, 1959). The fact that Quetelet lacked a compelling interpretation for his measure may have even been known by him, as he seems to have never used it again. Still, the fact that he even attempted to find a measure that reflected the substantive importance of the deviations, independent of the level of the average, was itself an interesting advance.

Insofar as he applied his measure to column 2 of Table 5.3, Quetelet was merely rescaling the deviations. He did not stop at that, however; and I think the next step he took was far more interesting. A successful analysis of the type Quetelet was attempting would require a means of sorting the influence of different causes, of determining which were stable over time and could be counted as constant, and which were sufficiently variable that they could be dismissed as useless for purposes, such as prediction, to which the analysis might be put. Later techniques such as analysis of variance might answer some of these concerns if the analyst felt that he had identified and measured all relevant factors, but Quetelet could neither enjoy that comfortable feeling nor use a method that was to be developed more than a half-century in the future. He was groping toward an identification of causes, and he reached for the only tool at his disposal, his measure of influence.

About the data of Table 5.2 Quetelet asked:

> We now come to a question of another kind: to what degree can we regard the causes previously enumerated as constant? Before we can say they are absolutely constant, we would have to show that the results they produce continue to be the same from year to year. Now, this is what does not take place: the deviations from the average which we have taken as constant quantities an-

> nually undergo slight modifications that we have attributed to *variable* causes. These modifications are in general very small, when we only take a small number of years into account, but it is still necessary to take account of them. (Quetelet, 1835, vol. 2, pp. 303–304; 1842, pp. 104–105)

Quetelet looked separately at each of the causes for which he had data, and calculated his measure of influence separately for deviations above and below the average. Thus for female accused where the five conviction rates ranged from 0.54 to 0.60, with the average being 0.576, he calculated

$$\frac{0.576 - 0.54}{0.576} = 0.062$$

and

$$\frac{0.60 - 0.576}{0.576} = 0.042$$

The results are given in columns 4 and 5 of Table 5.3.

Quetelet's analysis of these numbers was not a detailed one. He argued, rather vaguely, that the variation in these annual conviction rates was sufficiently slight to be negligible: "the greatest variations undergone by any of the constant causes affecting the conviction rate have scarcely exceeded the values of the intensity of these causes; in other words, even in circumstances most unfavorable to observation,[19] the effects of constant causes have been but little effaced by the effects of variable and accidental causes" (1835, vol. 2, p. 306; 1842, p. 105). Thus Quetelet appears to have argued that because the influence of annual variation on the separately computed rates, as given by columns 4 and 5 of Table 5.3, are mostly smaller than the influences of these causes upon the overall rate, as given by column 3, the effect of annual variation can be ignored. That is, he performed a very crude "analysis of variance," comparing variation within categories with separate measures of the variations between the category levels and the overall average rate.

Quetelet's incomplete and informal analysis-of-variance-like comparison of sources of variation, of variation between and within categories of accused persons, had begun to address the second question he treated. If the variation in influences between causes had been largely a fortuitous consequence of annually variable causes, then presumably these influences would not be so large in comparison with the range over the years of influence within causes. This conclusion, however, supposed that all annual effects were approximately equally well determined. Quetelet recog-

19. Quetelet did not elaborate on why the circumstances could be considered unfavorable; because he was using French data, he may have been referring to the unsettled political climate, leading up to and including the 1830 French revolution.

nized this implicit assumption in his last nod toward a quantification of the uncertainty in his rates.

Not all of the conviction rates were based on a large number of individuals; the number of accused with superior education seems to have been especially small. Quetelet wrote:

> I have always reasoned under the hypothesis that our results were based on so large a number of observations that nothing accidental (*contingent*) could affect the value of the averages: but this is not the case here. Some results are deduced from small numbers of observations, and we know that, all things being equal, *the precision of the results increases as the square root of the number of observations.*[20] This is particularly applicable to the rate of conviction of the accused with superior education. The values obtained there are deduced from a small number of observations, and the deviations from the average of them have consequently been greater. Now, by using the method of least squares, I have found that the precision of the numbers 0.400 and 0.6137, previously obtained for conviction rates in general and for the accused with superior education, is in the ratio of 0.0870 to 0.0075, or as 11 to 1. (Quetelet, 1835, vol. 2, pp. 307–308; 1842, p. 105)

The source of these last numbers is not clear. In his major work on statistical methods, the 1846 *Lettres sur probabilités,* Quetelet indicated that he would judge the accuracy of a mean by its "probable error," a multiple of $(2\Sigma\epsilon^2/N^2)^{1/2}$, where the ϵ's are the deviations from the mean and N is the number of terms averaged (1846, pp. 395–398). If this formula is evaluated for the $N = 6$ *rates* of Table 5.1 (and his quotation of 0.6137 would indeed suggest that it was these rates he considered as his data), it does give 0.0075, a result suggesting that this is how Quetelet did proceed. But the same formula gives 0.0467 for the $N = 3$ rates for the accused with superior education, or a ratio of about 6 to 1. Quetelet's work abounds in numerical errors, an indication that he frequently calculated in haste and lacked the patience to recheck his work. I am inclined to believe that that is what happened here.[21]

Quetelet did not follow through on this close encounter with the use of the techniques of error theory to measure the accuracy of averages of social data. He had taken a tentative step and then withdrawn, having reflected only on the relative accuracy of two means, not on the absolute accuracy of either. He calculated probability-based measures of accuracy,

20. The anonymous reviewer of Quetelet's book for the *Athenaeum* was evidently not a mathematician because he misquotes Quetelet: "All other circumstances being equal, M. Quetelet states that the probability of truth is as the square of the number of observations" (15 August 1835, p. 611). Alternatively, this misquotation could be based on the passage (1835, vol. 1, pp. 13–14; 1842, p. 7) quoted earlier in this chapter.

21. For examples of his errors, see Table 5.3 and our later discussion of the chest measurements of Scottish soldiers. I have been unable to reproduce Quetelet's 0.0870 despite several attempts. If he had divided by N instead of N^2 he should have gotten 0.0808.

but he made no probabilistic use of them. His crude comparison of two sources of variation by use of his measure of influence had gone part way toward separating variable and accidental causes; and despite its drawbacks this method was susceptible to further development. But there too he stopped, content to remark that "in separating what is purely fortuitous (*purement contingent*) in the deviations from the averages, so that we may only consider causes which have a more or less regular influence on the conviction rate, I believe that we may pretty nearly represent their influence by 0.034" (1835, vol. 2, p. 308; 1842, p. 105). That is, only causes whose influence exceeded that of the annual variable causes upon the general conviction rate were to be taken seriously.

I believe there were two elements that prevented Quetelet from continuing: lack of a well-developed and firmly grasped methodology for the disentangling of the causes he considered important and lack of a device or method to determine when he had formulated a meaningful system of classification. He had begun his investigation with only a vague notion of the ways different causes could produce effects. There were constant causes (that did not change from year to year), variable causes (that did), and accidental causes (individual differences). We might now recognize the start of a mixed two-way analysis of variance model in this vague classification. But Quetelet's conceptual understanding was not well developed, and his crude approach to measurements of various sources of variation reveals this.

This lack of a statistical methodology was not an insignificant difficulty, and Quetelet was not the only statistician of the period to encounter it. Indeed, the English statistician William Farr described the problem in 1848 in rather more direct terms than Quetelet ever did. In the *Eighth Annual Report of the Registrar-General of Births, Deaths, and Marriages, in England,* Farr wrote: "The causes that increase, and the causes that diminish marriage differ in energy; they admit of various combinations; they sometimes neutralize each other; and the marriages express the result of all those forces on the public conduct of the people.*" Farr's footnote explained what he meant in this way:

> *Take only three elements, for example—peace, abundance, high wages; and their three opposites, war, dearth, low wages—then there are eight combinations possible:—
>
> peace, abundance, high wages;
> peace, dearth, high wages;
> peace, abundance, low wages;
> peace, dearth, low wages;
> war, abundance, high wages;
> war, dearth, high wages;
> war, abundance, low wages;
> war, dearth, low wages.

> Of the eight possible combinations, there is one in which the causes all operate in increasing, one in which they all operate in decreasing, the marriages. If p, a, w were the factors expressing the effects of the three first elements, and $\not{p}$, $\not{a}$, $\not{w}$ the factors of their three opposites, the eight following formulae would, if there were no other causes in operation, give the results for any year:—
>
> $$\begin{array}{llll} paw + pa\not{w} & + p\not{a}\not{w} & + \not{p}\not{a}\not{w} \\ & p\not{a}w & \not{p}a\not{w} \\ & \not{p}aw & \not{p}\not{a}w \end{array}$$
>
> Should there be six elements and six opposites, the possible combinations would be $2^6 = 64$. These factors themselves would vary, and it is evident that the subject does not admit at present of strict mathematical treatment. (Farr, 1848, pp. xxiii–xxiv)[22]

The discovery that the techniques of least squares could be bent to deal mathematically with such questions was a long way in the future; and even if it had been available, it is doubtful it could have been used. It would have required the further element Quetelet had lacked in 1835, a way of determining the point at which a classification is fine enough to enable individuals in a class to be regarded, for purposes of statistical analysis, as homogeneous. Quetelet made his most lasting contribution to statistical analysis on just this question, with his fitting of probability distributions to empirical measurements.

Poisson and the Law of Large Numbers

Siméon Denis Poisson (21 June 1781 to 25 April 1840) was the principal successor to Laplace, both in interests and position, and his handling of the same data on conviction rates provides an illuminating contrast to Quetelet's work. Poisson was a man of lively energy (described by Mary Somerville, 1873, p. 109, as having "all the vivacity of a Frenchman"), and his main areas of research were in mathematical physics, where he wrote major texts on mechanics and on the theory of heat. His work in these areas was solid but not of a class with his predecessors, Lagrange, Laplace, and Fourier. In all of this work his role was that of a competent and insightful extender rather than that of a bold originator, and the same is true of his work in probability and its applications.

Poisson's major work on probability was a book, *Recherches sur la probabilité des jugements en matière criminelle et en matière civile (Researches on the probability of criminal and civil verdicts),* published in 1837. The book was in large part a treatise on probability theory after the manner of Laplace, with an emphasis on the behavior of means of large numbers of measure-

22. The passage quoted comes in the report by the Registrar-General, George Graham, dated 25 March 1847. For the role of the General Register Office in nineteenth-century statistics and Farr's role as author and architect of the reports, see Cullen (1975, chap. 2).

ments. The latter portion (pp. 318–415) dealt with the subject matter of the title. Some of this material was taken from memoirs Poisson published in the two preceding years. Only a charitable modern reading could identify a new concept in the work; yet the book contains the germ of the two things now most commonly associated with Poisson's name.

The first of these is the probability distribution now commonly called the Poisson distribution:

$$P(X=k)=\frac{\omega^k}{k!}e^{-\omega}; \qquad k=0,1,\ldots.$$

In a section of the book concerned with the form of the binomial distribution for large numbers of trials, Poisson does in fact derive this distribution in its cumulative form, as a limit to the binomial distribution when the chance of a success is very small (Stigler, 1982a). The distribution appears on only one page in all of Poisson's work (Poisson, 1837, p. 206). Although it is given no special emphasis this brief notice did catch the eye of Cournot, who republished it in 1843 with calculations demonstrating the effectiveness of the approximation (Cournot, 1843, pp. 331–332).[23]

The second most common appearance of Poisson's name in modern literature is in connection with a generalization of the Bernoulli law of large numbers. Let m be the number of times an event E occurs in μ trials, and suppose the chance that E occurs on the ith trial is p_i. Then the difference between m/μ and

$$p'=(p_1+p_2+\ldots+p_\mu)/\mu$$

converges to zero as μ increases; that is, the chance that $m/\mu - p'$ differs from zero by more than any given ϵ decreases toward zero as μ increases. The case where all the p's are equal is essentially Bernoulli's law of large numbers, as it had come to be interpreted in Poisson's time, and the statement given is a generalization of this result to inhomogeneous probabilities p_i. As such, it is relevant both to our immediate discussion of possibly variable conviction rates and to subsequent investigations of the stability of statistical series. Now, essentially the statement I have given can be found in Poisson (Poisson, 1837, p. 138, for example). When we make due allowance for the standards of rigor of the time, it is proved later (§94–§96, pp. 246–254) to be a consequence of Laplace's central limit theorem for nonidentically distributed summands. But Poisson's discussion of the "law of large numbers" is cloudy because he gives greater

23. Poisson's 1837 book did not reprint the "Cauchy" distribution $f(x)=[(1+x^2)\pi]^{-1}$; its significance as a case where the central limit theorem breaks down had been grasped by Poisson in 1824 (Poisson, 1824; Stigler, 1974b). Cauchy's association with this distribution dates from 1853.

emphasis to another version he appears to have seen as a further and more useful generalization of this law. This further generalization has caused confusion in many commentaries on Poisson's work, particularly as it can be viewed as only a special case of the first.

Poisson's further generalization was based upon a setting that renders the result a disguised restatement of Bernoulli's law. Poisson supposed that there were a number of possibly mutually exclusive "causes" C_1, $C_2, \ldots, C_\nu$, corresponding to different possible values for the p's, and that when cause C_j occurred, the chance of E would be c_j. Thus the p's would be a sequence of μ of the values c_j. Now, the stochastic behavior of the relative frequency of occurrences of E (that is, m/μ) will be different according to how we treat the sequence of values $p_1, \ldots, p_\mu$. In particular, in Poisson's first generalization of Bernoulli, he took it as a *fixed* sequence. We can describe it in the following terms: Suppose that $\mu\gamma_j$ equals the number of c_j's in the sequence, $j = 1, \ldots, \nu$, and let $\gamma = \gamma_1 c_1 + \ldots + \gamma_\nu c_\nu$. Then $p' = (p_1 + \ldots + p_\mu)/\mu = \gamma$ and m/μ has expectation

$$E(m/\mu) = \gamma$$

and variance

$$\begin{aligned} \mathrm{Var}_F(m/\mu) &= \sum_{i=1}^{\mu} p_i(1 - p_i)/\mu^2 \\ &= \sum_{j=1}^{\nu} \gamma_j c_j(1 - c_j)/\mu. \end{aligned}$$

In his further generalization, the one he was to stress through application and the one that would lead to confused and contradictory commentaries, he took the p's as a *random* sequence, where on each trial i (independently of all other trials) p_i takes the value c_j with probability $\gamma_j, j = 1, \ldots, \nu$. In this case, $Ep_i = \gamma$ for all i, and m/μ still has expectation

$$E(m/\mu) = \gamma.$$

But now the variance is

$$\mathrm{Var}_R(m/\mu) = \gamma(1 - \gamma)/\mu.$$

The two settings are similar in that the expected number c_j's in the sequence of random p's is γ_j; but they differ in the variability of the ultimate sequence of events and, hence, of m. The variance for the random setting is always greater than that for the fixed setting:

$$\mathrm{Var}_R(m/\mu) - \mathrm{Var}_F(m/\mu) = \sum_{j=1}^{\nu} \gamma_j(c_j - \gamma)^2 \geq 0.$$

Although the law of large numbers can be proved in either of the two settings (and Poisson did both), the statement in the random setting is no more than an elaborately camouflaged statement of Bernoulli's law. The sequence of occurrences of E for the random setting is stochastically indistinguishable from that where the chance of E is a fixed value γ on each trial.

Poisson's proof of this second generalization of the law amounts to a tortured derivation from the first generalization rather than a much simpler proof *ab initio.* The proof occupies pages 277 – 299 and consists of treating m/μ as subject to two independent sources of variation: one due to the p's, one due to variation given the p's. He derived his end result (p. 297) from the two approximating normal distributions, arriving at a statement of the normal approximation to the distribution of m/μ for the random setting. We can admire his technical expertise, although the result could come directly (and more easily) from an appeal to De Moivre's limit theorem.

Poisson's discussion is particularly unclear because it can be read as suggesting that the two generalizations are in fact one, as they are announced in terms that could suit either:

> All manner of things are subject to a universal law that we may call the *law of large numbers.* It consists of this: if we observe a very large number of events of the same nature, dependent upon the constant causes and upon causes that vary irregularly (sometimes in one way, sometimes in another, but not in a deterministic sense), we will find the ratios between the numbers of these events are approximately constant. (Poisson, 1837, p. 7)

A close reading of Poisson's proofs indicates that at a mathematical level he distinguished between the two statements. For example, on page 254 he referred to them as "two general propositions," and they were presented separately in his summary of results (pp. 306 – 316). But at a conceptual level the distinction was blurred. To have the stability that would make the result statistically useful, Poisson needed the framework of the second generalization, and it would appear that he allowed the complicated formulation of this second generalization to convince him that it retained the inhomogeneity of the first. He wanted, and seems to have believed he attained, the impossible — simultaneous inhomogeneity and stability. Without the randomization between causes, however, the stability could not be guaranteed; and with it, inhomogeneity at any but a metaphysical level was lost. The synthesis he thought he had achieved was a false one. A perspicacious contemporary of Poisson, I. J. Bienaymé, took the second of these generalizations (the one most stressed by Poisson) as the law of large numbers and argued correctly that it did not go beyond Bernoulli (Bienaymé, 1855). Poisson's work was further criticized by J. Bertrand in his 1889 *Calcul des probabilités* (pp. xxxii, 94). A more recent discussion of early models for statistical series can be found in Heyde and Seneta (1977, chap. 3).

Poisson and Juries

Poisson's work is relevant to Quetelet's through an application to models for the formation of juries' verdicts and to related analyses of the data on 1825–1830 French conviction rates. Poisson did not originate the application of probability to the formation of juries' opinions. The temptation to consider individual jurors' decisions as analogous to tosses of a weighted coin dated back more than a century, and the topic had already been well developed in work of Condorcet and Laplace. Poisson's model for criminal trials was a slight extension of Laplace's, and it fit in quite neatly as an example of the "random setting" described above—Poisson's second generalization. Poisson supposed that there was a chance k that the defendent was guilty (so guilt and innocence are the two "causes," C_1, C_2, with $\gamma_1 = k$, $\gamma_2 = 1 - k$; and k may, in a large population, be thought of as the proportion of all defendents who are guilty). Now, during the period 1825–1830 in France a guilty verdict could be delivered by a majority of seven of the twelve jurors, and I shall restrict my discussion of Poisson's analysis to this case. In actual fact, a major aim of Poisson and others was to determine the effects of jury size and definition of majority upon correctness of verdict.

Poisson supposed that jurors decided a case independently of one another and that each had a chance u of reaching a correct verdict. Then if

$$B(i, n, u) = \sum_{l=0}^{i} \binom{n}{l} u^l(1-u)^{n-l}$$

is used as modern notation for the chance that no more than i of n jurors reach a correct verdict, the chance of conviction for a guilty defendent is $c_1 = B(5, 12, 1-u)$, the chance of conviction for an innocent defendent is $c_2 = B(5, 12, u)$, and the overall chance of conviction for a randomly selected defendent is

$$\gamma = kB(5, 12, 1-u) + (1-k)B(5, 12, u).$$

We can now summarize the relevant portion of Poisson's analysis of this model: Given the results of a large number μ of criminal trials in which m of the accused were convicted,[24] estimate γ by m/μ. Further, if

$$P = 1 - \frac{2}{\sqrt{\pi}} \int_{\alpha}^{\infty} e^{-x^2}\, dx,$$

24. Poisson used the notation a_5 for our m, to signify that a_5 is the number of trials in which no more than five of the jurors voted for acquittal; he had R_5 where we have γ.

where α is positive but very small relative to $\sqrt{\mu}$, P gives "the probability corresponding to certain limits of the unknown" γ (Poisson, 1837, pp. 370–371):

$$(1) \qquad \frac{m}{\mu} \pm \alpha \sqrt{\frac{2m(\mu - m)}{\mu^3}}.$$

(That is, $P = P\{|Z| \leq \sqrt{2}\, \alpha\}$, where Z has a standard normal distribution.) Similarly, he gave "probability limits" for the difference $m/\mu - m'/\mu'$ of the conviction rates for two different, independent series of trials, corresponding to the same probability of conviction γ, as (Poisson, 1837, p. 371)

$$(2) \qquad \pm \alpha \sqrt{\frac{2m(\mu - m)}{\mu^3} + \frac{2m'(\mu' - m')}{\mu'^3}}.$$

Here as elsewhere in Poisson's work the conceptual distinctions most modern readers (and even some contemporary readers) look for were missing. The analysis was appropriate for the random setting, but some of Poisson's prose suggests that he intended it to apply more generally. And the exact interpretation to be given the probability limits is unclear. Poisson supposed that his model reflected the possibility that the defendents came from different groups, some more likely to be guilty than others, and that the jurors were drawn from a pool of citizens, some more likely than others to deliver a correct verdict. The variables k and u then represented average probabilities of guilt and correct verdict. Thus he hoped here, as he had hoped in his earlier theoretical development, to allow for the diversity of human propensities and the inhomogeneity of the French nation while still maintaining the stability that permitted statistical analysis. On one level he was successful, but the success was limited. It amounted to no more than a realization that the model of Bernoulli trials was sufficiently broad to encompass this diversity, as long as the diversity was conceived of as occurring randomly in each court case and as long as all jurors were drawn at random from the same pool. The frame of reference within which his analysis was valid remained one of homogeneous Bernoulli trials, albeit of rather complicated construction.

The intervals Poisson gave were of the type Laplace had introduced to express the uncertainty of an estimate. They were based on normal approximations for large numbers of trials and were not overtly Bayesian, although many of the statements describing them had a Bayesian flavor. The first we might now interpret as an asymptotic P% confidence interval for γ, the second as the acceptance region for a $1 - P$ level two-sided test of the hypothesis that the two series had the same value of γ. In both cases the interpretation would be anachronistic, although in both cases it would catch the spirit of Poisson's intention. We shall be in a better position to judge this intention if we look at the way in which the intervals were applied.

Poisson's models went only slightly beyond those of Laplace, but his application went much further. The data he considered were, with one minor difference, the same as those considered by Quetelet. The source was the French government series, *Compte général de l'administration de la justice criminelle en France,* and in Poisson's case the data covered the years 1825–1833. Poisson analyzed the data for 1825–1830 separately because of a change in the law after 1830 that redefined a majority decision. Through 1830 a jury could convict the accused with a majority of seven (of a total of twelve jurors) and the concurrence of the judge; from 1831 a majority of eight (of twelve) was required. Quetelet's earlier analysis included only the years 1825–1830 and is thus directly comparable to Poisson's, though Quetelet's figures for 1825 differed from those of Poisson. Quetelet had given the numbers of accused and convicted as 7,234 and 4,594; Poisson gave them as 6,652 and 4,037, respectively. Thus for Quetelet the 1825 conviction rate was 0.635, for Poisson it was 0.607. The explanation for this discrepancy is that in preparing the report for 1827 the minister of justice (Count Portalis) had discovered that in the report for 1825 the figures given had been augmented by those for accused condemned in absentia, and he had provided the needed correction. Apparently Quetelet missed this change (announced in a footnote on p. v of the *Compte général* for 1827), but Poisson did not.

Poisson did not cite Quetelet in his book, but the similarity of some of the questions they asked, the fact that they used the same data, and the wide publicity Quetelet's work (which had been published in Paris) had received, make it implausible that he had not read Quetelet.[25] Their modes of analyses were quite different, however.

Table 5.4 gives some of Poisson's data (a fuller republication of these data can be found in Gelfand and Solomon, 1973). Poisson's analyses of these data differed from Quetelet's in three respects. First, Poisson's main concern was the decision-making of juries in the context of his model, including a determination of the unknown u and k. Only secondarily did he want to see how they might vary under different circumstances. Quetelet had no probabilistic model and was solely interested in the strength of influence of different characteristics upon the conviction rate. Second, Poisson dealt directly with the counts, aggregated over years, whereas Quetelet took the yearly rate as his unit of analysis and averaged these rates. Third, Poisson employed the apparatus of Laplace for the measurement of the uncertainty of the overall rates he computed, whereas Quetelet's only attempts in this direction were of a tentative, qualitative nature.

Poisson began his analysis with the overall conviction rate for France.

25. That Quetelet and Poisson were in contact at this time is clear, because Poisson wrote to Quetelet in March 1836 on this topic (Quetelet reprinted the letter in 1869, vol. 2, pp. 258–259).

Table 5.4. Poisson's data on French conviction rates for all of France, for the department of the Seine, for crimes against persons, and for crimes against property.

	Year	Number accused	Number convicted	Conviction rate
All of France	1825	6,652	4,037	0.6068
	1826	6,988	4,348	0.6222
	1827	6,929	4,236	0.6113
	1828	7,396	4,551	0.6153
	1829	7,373	4,475	0.6069
	1830	6,962	4,130	0.5932
	Total	42,300	25,777	—
Department of the Seine	1825	802	567	0.7070
	1826	824	527	0.6396
	1827	675	436	0.6459
	1828	868	559	0.6440
	1829	908	604	0.6652
	1830	804	484	0.6020
	Total	4,881	3,177	—
For crimes against persons	1825	1,897	882	0.4649
	1826	1,907	967	0.5071
	1827	1,911	948	0.4961
	1828	1,844	871	0.4723
	1829	1,791	834	0.4657
	1830	1,666	766	0.4598
	Total	11,016	5,268	—
For crimes against property	1825	4,755	3,155	0.6635
	1826	5,081	3,381	0.6654
	1827	5,018	3,288	0.6552
	1828	5,552	3,680	0.6628
	1829	5,582	3,641	0.6523
	1830	5,296	3,364	0.6352
	Total	31,284	20,509	—

Source: Poisson (1837, pp. 371–378).

Note: The data disagree with those of Quetelet (Table 5.1) in the figures for all of France for 1825 (Poisson's figures reflect a correction announced in a footnote to the 1827 report); also, 0.6068 for that year should read 0.6069.

(Where we use m, Poisson used a_5; here and later we shall stick to our earlier, simpler notation.)

I take for μ the sum of the number accused during the six years, and for m the corresponding number of convictions. We will have

$$\mu = 42300, \qquad m = 25777;$$

from which the limits (1) become

$$0.6094 \pm \alpha(0.00335).$$

If we take $\alpha = 2$, for example, we will have

$$P = 0.9953$$

for the probability, quite close to certainty, that the unknown [γ] and the fraction 0.6094 do not differ by 0.0067, one from the other. (Poisson, 1837, p. 372)

Poisson's statement is as clear an application of Laplacian methods to the uncertainty of social data as we could hope to find. Even the arithmetic is correct. There is no reluctance, no equivocation, no qualification. The interpretation intended appears to be modeled after the informal fiducial arguments from the theory of errors in astronomical observations rather than being an unambiguously Bayesian statement about an unknown quantity: It is the uncertainty of the difference between estimate and unknown that is being measured.

Poisson's underlying model depended upon the assumption that the behavior of juries remained stable over time. He proceeded to check that assumption by comparing the rate of 1825–1827 with that for 1828–1830. He found the rate for 1825–1827 to be

$$\frac{m}{\mu} = \frac{12{,}621}{20{,}569} = 0.6136$$

and that for 1828–1830 to be

$$\frac{m'}{\mu'} = \frac{13{,}156}{21{,}731} = 0.6054;$$

the difference between them is 0.0082. He calculated the "limit of this difference" from (2), again correctly, to be $\pm\alpha(0.00671)$, and noted that with $\alpha = 1.2$ the limits were ± 0.00805, nearly the same as the difference. This interval corresponded to a probability of $P = 0.9103$, so $1 - P = 0.0897$, and "it should thus be a bet of nearly 10 to 1 that the difference of the two ratios m/μ and m'/μ' will fall between the limits ± 0.00805. Although the observed difference taken without regard to sign, ± 0.0082, exceeds this slightly, the probability of such a difference will be approximately P, which is not so large as to support a belief that there has been a notable change in the causes" (Poisson, 1837, p. 373). That is, where Quetelet (with a higher 1825 rate of 0.635) had discerned a downward trend in conviction rates, Poisson's test leaves him concluding that such an instability is not supported by the data. The point here is the difference in their approaches to the data, not the difference in the conclusions. In fact, applying Poisson's method with Quetelet's data for 1825 would give P =

0.9998, a result lending strong support to a hypothesis of instability. But Quetelet gave no such test.

Poisson continued to make other comparisons. He compared the conviction rate in Paris (the department of the Seine, see Table 5.4) with that for all of France and found a significant difference: "the fraction for France is smaller by 0.0416, or about a fifteenth of its value; now, the limits (2) and their probability P renders such a deviation altogether unlikely, unless we have a particular cause for the department of the Seine that makes convictions easier than in the rest of France" (1837, p. 375). Curiously he erred in not noting the fact that the count for the Seine was included in that for France and, hence, that the rates were not independent. As we have seen, Laplace made a similar error in a situation where the dependence was not notationally explicit. This problem was clearly noted later, however, by Cournot (1843, p. 177), who cited Bienaymé (1840) as calling attention to it.

Poisson also compared the rate in Paris for 1830 with that for the earlier years, finding a difference that he attributed to the Revolution of 1830.

Poisson's comparisons included some of those that Quetelet had made. He compared the rate for crimes against persons with that for crimes against property (1837, pp. 377–379; see Table 5.4). For the rate of crimes against persons he found the limits $0.4782 \pm \alpha(0.00675)$;[26] for the rate of crimes against property, $0.6556 \pm \alpha(0.00380)$. Taking $\alpha = 2$ ("a probability near certainty") the intervals were quite distant, a finding that he felt supported his contention that "these two large classes have presented annual rates very different from each other." Quetelet had reached the same conclusion, but without a supporting calculation and with no attempt to measure the potential variability in the rates—and hence the chance of error in the conclusion.

The remainder of Poisson's investigation focused on the effect of the change of the definition of majority in 1831 and upon the determination of the probability of guilt k and the chance of a correct decision u from estimates of the probability of conviction γ. Gelfand and Solomon (1973) have presented a summary of Poisson's results and the methods he used in this early attempt to identify the components of a mixture of distributions.

What led Poisson to use the calculus of probabilities to weigh his evidence, to judge the accuracy of the rates he calculated, when Quetelet did not? And, once Poisson had apparently broken through whatever psychological barrier might have existed, why did Quetelet not follow suit and adopt the methodology as his own? These questions have no definitive answers, but some of the reasons seem clear.

Poisson brought an altogether different background in empirical

26. The coefficient of α should be 0.00673.

science to his study, and he sought the answers to questions very different from those of Quetelet. Poisson's experience was that of an experimental physicist in control of his environment, with relatively few influential factors to worry about at any one time. Quetelet's background in census studies had, as we have seen, given him a different outlook. By 1832, when Quetelet first treated these data, he had already acquired a measure of the near paranoia that attacks social scientists of some experience who are entering into untested areas. The number of potential causes of instabilities, or the number of sources of variation, is not even limited by the imagination; indeed, a fear of unimagined causes makes the scientist reluctant to jump far ahead. In these modern times, we might accept Poisson's analysis as valid, but we know now that the actual empirical experience he brought to his investigation was insufficient to justify its tacit assumptions. He could not really defend his choice of so few factors for consideration in his analysis; and Poisson's aggregation of counts over several years masked the year-to-year variations that even Quetelet caught with his use of yearly rates as his basic data. Indeed, Poisson's analysis required the "random setting" we have described, a setting that in turn leads easily to the combination of counts because all variation is subsumed in the binomial variation of individual trials. If the assumptions of stability and independence are valid, the analysis will be also; but Poisson did not demonstrate this validity.

In addition, Poisson's objective was different from that of Quetelet. Quetelet's subject was society; his aim was to study his average man in different circumstances. Poisson's focus was on the jury, and, more particularly, on how the conviction rate for a stable series could determine the proportion actually guilty and the correctness of the verdicts. Poisson's model gave these quantities symbols and an objective value that Quetelet's average man had not yet developed. Poisson's isolation of his ultimate objective in terms of simple symbols such as γ, u, k, quantities stable over time and not dependent upon variable causes, was reflective of his limited experience with social data and his unverified assumptions, but it did much to ease his way to an application of probability. When Quetelet looked into the difference between conviction rates for crimes against persons and against property, he was considering just one of many ways of classifying the data. When Poisson made the same comparison he was in effect comparing γ and γ', two mixed binomial probabilities. From there it was an easy step to a Laplacian analysis borrowed from gaming and astronomy, where the assumptions implicit in the model had received a better test.

Once Poisson's book appeared, it rapidly came under Quetelet's notice, but with no apparent effect. Quetelet reviewed the book at the end of the 1837 volume of his journal *Correspondance mathématique et physique.* The review was polite and laudatory, but Quetelet's opening restatement of his own approach, with an emphasis upon the measurement of a multitude of causes and their intensities, might be read as an implicit criticism of Poisson's more limited approach by one who wished no rift among those bold enough to attempt a quantification of social science:

> All sciences of observation follow the same course. One begins by observing a phenomenon, then studies all associated circumstances, and finally, if the results of observation *can be expressed numerically* [Quetelet's italics], estimates the intensity of the causes that have concurred in its formation. This course has been followed in studying purely material phenomena in physics and astronomy; it will likely also be the course followed in the study of phenomena dealing with moral behavior and the intelligence of man. Statistics begins with the gathering of numbers; these numbers, collected on a large scale with care and prudence, have revealed interesting facts and have led to the conjecture of laws ruling the moral and intellectual world, much like those that govern the material world. It is the whole of these laws that appears to me to constitute *social physics,* a science which, while still in its infancy, becomes incontestably more important each day and will eventually rank among those sciences most beneficial to man.
>
> For the intelligent observer, and those who have not allowed themselves to be captured by narrow prejudices, there are a large number of moral phenomena that are well established by statistical documentation for which these assertions will not seem strange. But in general, such facts have not been judged or evaluated with the necessary precision. We are unaware of the importance we should attach to them, and of the probability we should associate with their repetition. It is for the calculus of probabilities to trace for us a suitable course that permits such delicate judgments. The time seems to have come when this important branch of mathematics will take its proper place. Each day the great works of Pascal, Leibniz, Condorcet, Bernoulli, Laplace, etc. receive new applications producing interesting and useful results. The work that M. Poisson has published could not have appeared at a more opportune time. The author's name is deservedly celebrated, and will remind friends of science of new terrain, promising an ample harvest of discoveries to reward their investigations. Of the work under notice, the major part is comprised of those principles of the calculus of probabilities that are of greatest importance in the observational sciences. In the last chapter, M. Poisson presents general applications to the decisions of juries and the judgments of tribunals. This special example is treated very thoroughly, and it sketches out the course to follow in a multitude of other, analogous studies. The facts that result from statistical documents relating to crime alone will be of the highest interest to philosophers and legislators, but this study will continue to encounter much opposition and prejudice, and we owe our thanks to the most eminent men of science for giving such generous support and hastening the time when it must surely triumph. (Quetelet, 1837, pp. 485–486)

Despite this seeming endorsement of Poisson's methods (at least in some cases), Quetelet did not use them. In the second edition of his 1835 book, a work that appeared in 1869 under the slightly altered title *Physique sociale, ou Essai sur le développement des facultés de l'homme,* he republished his earlier treatment of conviction rates with essentially no changes (vol. 2, pp. 407–428). And in his 1846 *Lettres sur probabilités* he explained the use of methods such as those Poisson presented but did not apply them to social data. Nor was Quetelet alone in his abstention from this use of an available methodology. With a few isolated and transitory exceptions, the application of probability to the measurement of uncertainty in the social sciences was unknown before the 1870s, exceedingly rare before the 1890s, and uncommon until well into the twentieth century.

Comte and Poinsot

To appreciate the deep conceptual barriers that had to be overcome in the course of developing and adopting modern statistical methods, we may ask why available methods were seemingly ignored. Actually, they were not totally ignored. In some cases they were actively and derisively attacked and in others thoughtfully discussed and dismissed as inappropriate. The best-known and most dogmatic of the attacks came on various grounds from the French sociologist and positivist philosopher Auguste Comte and his followers. Poisson's first public discussions in 1835 and 1836 of his application of probability to judicial statistics elicited a hostile response from the mathematician Louis Poinsot, who had attended Comte's lectures a decade earlier. Poinsot considered Poisson's work a "false application of mathematical science . . . This singular idea of a calculus applicable to things where the ignorance and passion of men are intermingled in an imperfect light is dangerously illusory in several senses" (discussion published with Poisson, 1836b, p. 380). "The application of this calculus to matters of morality is repugnant to the soul. It amounts, for example, to representing the *truth* of a verdict by a *number,* to thus treat men as if they were dice, each with many faces, some for error, some for truth" (discussion published with Poisson, 1836a, p. 399).

Comte's own lectures had evolved into printed form by 1842, in his six-volume *Cours de philosophie positive* (1830–1842); and in the fourth volume (first published in 1839) he argued for the autonomy of sociology and roundly criticized "the vain pretension of a large number of geometers that social studies can be made positive by a fanciful subordination to an illusory mathematical theory of chances" (Comte, 1877, vol. 4, p. 366; 1896, vol. 2, pp. 267–268). Comte went on to castigate Condorcet and Laplace for "a gross abuse of the credit which justly belongs to the true mathematical spirit . . . Is it possible to imagine a more radically irrational conception than that which takes for its philosophical base, or for its principal method of extension to the whole of the social sciences, a supposed mathematical theory, where, symbols being taken for ideas (as is usual in purely metaphysical speculation), we strain to subject the necessarily complicated idea of numerical probability to calculation, in a way that amounts to offering our own ignorance as a natural measure of the degree of likelihood of our various opinions?" (Comte, 1877, pp. 368–369; 1896, pp. 268–269). Comte's strictures were aimed primarily at the nonempirical works of Condorcet and Laplace, but he added a footnote taking approving notice of Poinsot's recent "memorable academic discussion" of Poisson's work. Without mentioning Poisson by name, Comte noted that Poinsot, "with the sagacious philosophical lucidity that usually characterizes him, had undertaken to warn common mathematics against a momentary new invasion of this antiquated aberration, then being resurrected in a sort of scientific fracas by a much less rational analyst" (Comte, 1877, p. 368). Comte's criticisms were later echoed by John Stuart Mill, who referred to these works of Condorcet, Laplace, and Poisson as "the real opprobrium of mathematics" (Mill, 1846, vol. 2, p. 76). This phrase was later sharpened in a French translation by J. Bertrand to "le scandale des Mathématiques" (Bertrand, 1889, p. 327).

Comte's caustic view of a lack of a role for probability in the social sciences was never well argued by him, and it now reads as an unexamined and naive initial reaction, primarily taken because it was consistent with his more general philosophical position (he had rejected the founding of the social sciences on any other science as early as 1819) (Baker, 1975, pp. 377–378). But another perceptive philosopher who was not philosophically opposed to the quantification of social science also dismissed the possibility of using probability to measure the uncertainty inherent in social data. This man was the remarkable mathematical economist, A. A. Cournot.

Cournot's Critique

Antoine-Augustin Cournot (28 August 1801 to 30 March 1877) is best known today for his 1838 book, *Recherches sur les principes mathématiques de la theorie des richesses,* a work that was ignored at the time it was published but that came to be recognized as a remarkable anticipation of many later concepts of mathematical economics. That book introduced abstract quantitative reasoning into the study of economic questions, but made no attempt at the analysis of empirical questions and no use of probabilistic methods. In 1843 Cournot published his *Exposition de la théorie des chances et des probabilités* (the preface indicates that the book had existed in outline form as early as January 1836). The *Exposition* was not the pioneering work Cournot's book on mathematical economics had been, but it found a larger audience. In the preface he announced two aims for the work: "to bring the rules of the calculus of probabilities to those who have not cultivated higher mathematics, rules without which we cannot get a clear idea of the precision of measurements made in the sciences of observation, of the worth of numbers furnished by statistics, or of the conditions leading to the success of commercial enterprises," and "to correct the errors, remove the qualifications, and dissipate the obscurities from which even the most able geometers' works on this most delicate subject are not exempt."

Cournot's *Exposition* was immensely wise without being profoundly original. He was proudest of the contribution the book made to the philosophical understanding of probability, and indeed he did achieve a much higher level of clarity than did any of his predecessors in his discussions of distinctions between subjective and objective probability. His mathematical development shared the same elegant clarity, as he not only spelled out the difference between the random and fixed settings Poisson had discussed for Bernoulli trials, but also proved that the intervals for the random setting were always larger (and hence more conservative) than those for the fixed setting. Letting $p_1, p_2, \ldots$ be the different values of the probabilities and taking k_i as the fraction of time p_i occurs in a sequence of values in the fixed setting ($k_i = \gamma_i$ in Poisson's earlier notation), Cournot (1843, pp. 137–138) had

$$p = k_1p_1 + k_2p_2 + \text{etc.}$$

for the average value of the probabilities in the fixed setting and, equally well, as the constant value of the probability in the corresponding random setting. He then multiplied out $p(1-p)$, replacing k_1^2 by $k_1(1 - k_2 - k_3 - \text{etc.})$ and similarly for $k_2^2, k_3^2, \ldots$); he found

$$\begin{aligned} p(1-p) &= (k_1p_1 + k_2p_2 + \text{etc.})(1 - k_1p_1 - k_2p_2 - \text{etc.}) \\ &= k_1p_1(1-p_1) + k_2p_2(1-p_2) + \text{etc.} \\ &\quad + k_1k_2(p_1-p_2)^2 + k_1k_3(p_1-p_3)^2 + \text{etc.} \\ &> k_1p_1(1-p_1) + k_2p_2(1-p_2) + \text{etc.} \end{aligned}$$

Because the right-hand side is proportional to the lengths of the probability intervals appropriate to the fixed setting and the left-hand side proportional to the length of the probability intervals in the random setting, Cournot's proof was complete. We can modernize Cournot's proof as follows: Let X and Z, Z' be independent random variables where

$$P(Z = i) = k_i = P(Z' = i)$$

$$P(X = 1|Z = i) = 1 - P(X = 0|Z = i) = p_i.$$

Then $p = P(X = 1) = E\{P(X = 1|Z)\}$, and Cournot's aim is to prove

$$\mathrm{Var}(X) \geq E\{\mathrm{Var}(X|Z)\}.$$

Algebraically, he finds

$$\mathrm{Var}(X) = \mathrm{E}\{\mathrm{Var}(X|Z)\} + \frac{1}{2}\,\mathrm{E}[\{\mathrm{E}(X|Z) - \mathrm{E}(X|Z')\}^2|Z < Z'],$$

and the inequality follows since the last term on the right is never negative. In fact, since

$$\frac{1}{2}\,\mathrm{E}[\{\mathrm{E}(X|Z) - \mathrm{E}(X|Z')\}^2|Z < Z'] = \mathrm{E}\{\mathrm{E}(X|Z) - \mathrm{E}(X|Z')\}^2$$
$$= \mathrm{Var}\{\mathrm{E}(X|Z)\},$$

the breakdown he has arrived at is equivalent to

$$\mathrm{Var}(X) = \mathrm{E}\{\mathrm{Var}(X|Z)\} + \mathrm{Var}\{\mathrm{E}(X|Z)\}.$$

Of course, Cournot is only working in the limited context of Bernoulli trials.

Cournot went on to extend the result generally to means formed by taking weighted averages of means from inhomogeneous series of observations. Cournot was working in the context of large samples and limiting normal distributions, but at least at a formal level he had clearly established the now-famous inequality, that a mean taken with simple random sampling always has at least as high a variance as one taken by proportionally stratified sampling (Cournot, 1843, chap. 7, pp. 135–152). Cournot gave a great deal of credit for other mathematical developments he discussed (although apparently not this inequality) to I. J. Bienaymé. Heyde and Seneta (1977) discussed Bienaymé's work extensively.

At an abstract level, Cournot must be counted as highly sympathetic to the quantification of the social sciences. Whatever prejudices against such an idea Comte and Poinsot may have had, Cournot did not share them; his work in economic theory is ample testimony to this. Hence it may appear ironic that nearly the opposite was true at an empirical level. Having announced in the preface to his 1843 *Exposition* that the calculus of probabilities was essential if we are to determine "the worth of numbers furnished by statistics" and having devoted several chapters to mathematical techniques for the measurement of uncertainty, Cournot then went on to discuss at length how and why the techniques were of no practical relevance with the "statistics" of social data. His argument was clear and logical—quite in contrast to the dogmatic statements of Comte—and all the more forceful in that it came from a partisan of mathematical reasoning in the social sciences.

Cournot's argument was, basically, that an unlimited number of ways of classifying social data existed and any probability analysis that did not allow for the selection of categories after the collection of data was, in a practical sense, meaningless. Because in Cournot's view this was not possible in general, probability was irrelevant to statistics (1843, §111 – §114, pp. 190 – 199). Cournot went into detail in the context of the example of the proportion of male births in a population: Cournot allowed the probabilistic comparison (via a significance test) of these proportions for some classification schemes (such as legitimate – illegitimate, or rural – urban) that could be regarded as natural and of interest a priori. But there would always be an indefinite number of other ways of proceeding, also of interest to a social scientist. Even a scientist of only average curiosity could classify births by birth order, by parents' age, profession, wealth, or religion, by season of the year, by whether it was a first marriage for both parents, and so forth. For any *single* such classification, say, a dichotomization of births into class *a* and class *b*, Cournot could accept the calculation of either an a priori or an a posteriori probability. If n_1 represents the number of the m_1 class *a* births that are male, and n_2 the number of the m_2 class *b* births that are male, he would take

$$t = \frac{\delta m_1 m_2 \sqrt{m_1 m_2}}{\sqrt{2[m_1^3 n_2(m_2 - n_2) + m_2^3 n_1(m_1 - n_1)]}}$$

$$\left(t = \frac{\delta}{\sqrt{2\left[\dfrac{\hat{p}_1\hat{q}_1}{m_1} + \dfrac{\hat{p}_2\hat{q}_2}{m_2}\right]}}, \quad \text{where } \hat{p}_i = 1 - \hat{q}_i = n_i/m_i \right)$$

$$\mathrm{P} = \frac{2}{\sqrt{\pi}} \int_0^t e^{-x^2}\, dx, \quad \text{and } \Pi = \frac{1 + \mathrm{P}}{2}.$$

Let $x_1 = \mathrm{P}$(male birth for class *a*); $x_2 = \mathrm{P}$(male birth for class *b*). He would then interpret P as the a priori (prior to the observation of the data) probability that the difference of the ratios, $n_1/m_1 - n_2/m_2$, would exceed (in magnitude) a given positive δ, there being no intrinsic difference between the *a*'s and *b*'s (that is, $x_1 = x_2$). On the other hand, once $n_1/m_1 - n_2/m_2 = \delta > 0$ was observed, Cournot would take Π as the a posteriori probability that $x_1 > x_2$ (Cournot, 1843, §100, §110; pp. 175 – 176, 190). His formulas agree with what one would find from the large sample normal approximation, supposing in the case of Π that $x_1 - x_2$ is a priori uniformly distributed; and it is clear from Cournot's text that this is what he intended. That is, treating n_1 and n_2 as independent random variables, respectively distributed binomial (m_1, x_1) and binomial (m_2, x_2), then in terms of normal approximations to the distributions in question,

$$\mathrm{P} = \mathrm{P}\left\{ \left| \frac{n_1}{m_1} - \frac{n_2}{m_2} \right| \leq \delta \,\middle|\, x_1, x_2; x_1 = x_2 \right\},$$

and

$$\Pi = \mathrm{P}\left\{ x_1 > x_2 \,\middle|\, \frac{n_1}{m_1} - \frac{n_2}{m_2} = \delta \right\}.$$

Even though Cournot would accept this logic for a *single* classification, he nevertheless recognized that as the number of possible ways of dividing the population increased the relevance of P and Π for evaluating the differences between classes disappeared. As Cournot put it, "It is evident that as the number of divisions increases without limit, it is *a priori* more and more probable that, by chance alone, at least one of the divisions will produce ratios of male births to total births for the two classes that are sensibly different." Because social statistics data admitted to many possible, even plausible, categorizations and because usually the sorting out of potentially informative ways of classifying the data leaves no trace, there is no fixed rule that can be used to evaluate whether or not observed differences might be due to chance alone. "In a word, the probability that we have called Π corresponding to a deviation δ will lose its objective meaning for someone unfamiliar with the examination that revealed the deviation; depending upon the view such a person has of the *intrinsic value* of the characteristic that served as the basis for the corresponding categorization, the same size deviation may lead to many different judgments" (Cournot, 1843, §111, pp. 192–193). An investigator, Cournot argued, might accept probabilistic evidence that the ratio of male births in the summer months differed from that of the winter months, but only because he carried an external belief into the investigation, that such a division was a priori likely to be meaningful. Exactly the same difference between births on odd-numbered and on even-numbered days would be dismissed out of hand (although even such a hypothesis could be validated by future, independent data).

For Cournot, the evaluation of empirical evidence, the judgment of whether or not an observed difference could be attributed to chance, depended on two elements. One was susceptible to precise mathematical evaluation and an objective interpretation—the a priori probability of such a difference, given by P. The other element, however, the a priori judgment of how "natural" it is to focus on one particular division from among the infinite number possible, was more elusive. It was a conjectural judgment that could not be objectively resolved by an enumeration of chances. Because this latter element could only be quantified by means of judgments external to the data and because it was so crucial to the analysis of the statistics of society, mathematical probability lost relevance there. Cournot put it this way:

> There is always a variable element entering into our judgment in inspecting statistical tables. It escapes precise measurement, and we will be well protected by concluding that the mathematical theory of chances is immaterial (*indifférente*) to the statistician. It is evident that the importance of the deviation δ, as given by observation, depends both on the size of the deviation and on size of the numbers used [that is, sample size]; but how does it depend on them? The theory of chances is able only to teach us to calculate the ratio P

corresponding to the deviation δ. As for the probability denoted above by Π, it lacks an objective meaning in statistical applications. It does not at all measure the chance of truth or error of a given judgment. (Cournot, 1843, §113, pp. 196–197)

Cournot's rejection of the a posteriori meaningfulness of probability was not based on reservations about a priori assumptions on the unknown chances x_1 and x_2, such as might occur to a modern critic, but rather upon an acute awareness of the even more devastating effect that selecting a hypothesis after the data is at hand could have upon the conclusion. For emphasis, he supposed that one of the eighty-six departments of France exhibited a much larger ratio of male births than did all of France treated as a whole. For the corresponding a posteriori probability Π to have an objective meaning, Cournot argued (pp. 197–198), it would be necessary that we know that the department in question was selected at random, as if its name were drawn from among eighty-six tickets in an urn and not merely singled out as the department with the largest male birth ratio among the eighty-six. But if the departments were selected at random or specified by the statistician prior to the analysis of the data, Cournot would have no difficulty accepting the a posteriori probability Π.

Cournot's strictures were thus the converse of those Keverberg had presented to Quetelet sixteen years earlier. Keverberg had warned of the need to subdivide and cross-classify social populations, to avoid lumping nonhomogeneous groups together. Cournot spoke to the deceptions that too fine a subdivision could produce, by making even homogeneous groups appear significantly different, when measured by the yardstick of probability. The two arguments may be taken as a rational articulation of the reservations of many at that time and as the prime contemporary statements explaining the lack of early application of probability to the measurement of uncertainty in social science. Neither man could realistically solve the problem he posed, much less resolve the dilemma they presented together. Keverberg's suggestion of a complete enumeration was matched in some ways by Cournot's that data be gathered in such enormous quantities that questions of the type he raised would become unimportant (Cournot, 1843, §120, p. 208). The few such applications of probability Cournot made himself (Cournot, 1843, §170, pp. 308–310) were heavily qualified and lacked force.

The central problems were as follows: How could a social scientist determine that he had sufficiently subdivided his data to permit a useful analysis? And, how could he guard against a selection effect that could come with too fine a categorization? The solutions did not come early. Indeed, they cannot be said to have been completely solved today. Nevertheless, in the 1840s, shortly after Cournot's book appeared, Quetelet took an influential step toward a solution to the first. His step involved the fitting of probability distributions to data; and although it cannot be counted successful, it left its mark on all subsequent work on the problem.

The Hypothesis of Elementary Errors

Quetelet's fame in the history of statistics rests primarily upon two hypnotic ideas: the concept of the average man, and the notion that all naturally occurring distributions of properly collected and sorted data follow a normal curve. Both ideas were flawed and have been more often attacked than embraced. Nonetheless, both were boldly provocative and have been vastly influential on the development of statistics. The average man, whom we discussed earlier, still entrances the popular mind; and despite frequent debunking of the naive notion of an actual "average man," the concept still persistently recurs in social research. Quetelet's enthusiasm for the normal curve was no less influential, although in a different way.

Quetelet may have first encountered the "normal" curve[27] in his 1823 visit to Paris. At that time it was a centerpiece of Laplacian probability theory. In April 1810 Laplace had read to the Academy of Sciences his famous memoir presenting the central limit theorem for arbitrary error distributions (Chapter 4). The main memoir had extended Laplace's own techniques and those of Lagrange to prove a generalization of De Moivre's approximation theorem; in the supplement that Laplace added before publication he had summarized this result as follows (see Figure 4.2):

> If we suppose that in each observation, positive and negative errors are equally likely (*faciles*), the probability that the mean error of n observations will be within the limits $\pm rh/\sqrt{n}$ is equal to
>
> $$\frac{2}{\sqrt{\pi}}\sqrt{\frac{k}{2k'}}\int \exp\left[-\frac{k}{2k'}r^2\right]dr$$
>
> where h is the length of the interval within which the error of a single observation can fall.[28] If we denote the probability of the error $\pm x$ by $\phi(x/h)$, k is the integral $\int\phi(x/h)\,dx$ taken from $x = -\frac{1}{2}h$ to $\frac{1}{2}h$; k' is the integral $\int(x^2/h^2)\cdot\phi(x/h)\,dx$ taken over the same interval; π is half the circumference of the unit circle; and e is the number whose hyperbolic logarithm is one. (Laplace, 1810, p. 559)

27. Quetelet did not use the term *normal.* Rather he variously referred to it as "la courbe de possibilité" (Quetelet, 1846, p. 386), "la loi de possibilité" (Quetelet, 1848, p. 85), and, in his last published book, "la courbe binomiale" (Quetelet, 1870, p. 269).

28. The main memoir makes it clear that the limits are intended to be $-rh/\sqrt{n}$ and $rh/\sqrt{n}$, as is correct. The statement given here corrects an obvious misprint (rh/n for $rh/\sqrt{n}$) in both the 1810 and 1898 printings. (The 1810 version is shown in Figure 4.2.) The occurrence of so conspicuous an error in the original suggests that the supplement, which appears at the very end of the volume, was added too late for its author to correct in proof. Indeed, the title page of the volume is dated August 1810, only four months after the main memoir was read.

We can express this in modern notation: Letting ϵ_i, $i = 1, \ldots, n$, stand for the n independent errors, Laplace supposed that the ϵ_i symmetrically distributed over $[-h/2, h/2]$ with density $\phi(x/h)$. If we take h fixed and $\phi(x/h)$ as a normalized density, then Laplace's k will equal 1, $k' = E(\epsilon_i^2/h^2) = \sigma^2/h^2$, and $k/k' = h^2/\sigma^2$. His conclusion is that

$$P\left\{|\bar{\epsilon}| \leq \frac{rh}{\sqrt{n}}\right\} = \sqrt{\frac{2}{\pi}} \cdot \frac{h}{\sigma} \int_0^r \exp\left[-\frac{h^2u^2}{2\sigma^2}\right] du,$$

where $\bar{\epsilon} = (1/n)\Sigma_{i=1}^n \epsilon_i$ is the mean error. Laplace's application of this result in his supplement amounted to an explanation of how, if each measurement were itself a mean of a large number of more basic observations, the measurements themselves would have the distribution supposed by Gauss. If individual measurements were themselves averages of large numbers of independent components, their distribution would be (at least approximately) of the form $\phi(x) = (h/\sqrt{\pi})e^{-h^2x^2}$. This argument was the basis for what later came to be called the hypothesis of elementary errors; it justified the assumption of Gauss's form by the hypothesis that the errors were derived from more elementary random components.

The combined prestige of the derivations of Gauss and Laplace was sufficient to enshrine $\phi(x)$ in a special niche in the early theory of errors, but there was at least one attempt to make an empirical check on its adequacy as an error distribution. In 1818 Frederick Wilhelm Bessel published *Fundamenta Astronomiae,* a catalog of the positions of 3,222 stars, computed on the basis of about 60,000 individual observations made by the English Astronomer Royal James Bradley between 1750 and 1762. Bessel's introduction included an investigation of the distribution of the residual errors of observations of the declinations (essentially, celestial latitude) and right ascensions (essentially, celestial longitude) of three groups of observations of a few selected stars. In a group of γ observed errors, Bessel wrote, there should be (by Gauss's theory of errors)

$$\frac{2\gamma}{\epsilon\pi} \int \exp\left[-\frac{1}{\pi}\frac{\Delta\Delta}{\epsilon\epsilon}\right] d\Delta$$

residual errors between Δ and Δ', where the integral is evaluated between these limits and ϵ is the mean error observed. (It is not clear whether Bessel calculated the mean error ϵ directly from his data, or by first evaluating the mean square error or the probable error. He gave all three and the theoretical relationship among them.) Bessel noted that this integral could be found from Kramp's tables (Kramp, 1799), but, where $\Delta' - \Delta$ was small, he thought the following approximation sufficient:

$$\frac{\gamma(\Delta' - \Delta)}{\epsilon\pi}\left\{\exp\left[-\frac{1}{\pi}\frac{\Delta\Delta}{\epsilon\epsilon}\right] + \exp\left[-\frac{1}{\pi}\frac{\Delta'\Delta'}{\epsilon\epsilon}\right]\right\} \tag{3}$$

Bessel's comparisons are given in Table 5.5. He found a general agreement with what would be expected from the hypothesized error distribution, although he did note a slight tendency toward more large errors than the theory would predict. These, Bessel suggested, could be due to unanticipated aberration of starlight or to chance movement of the instrument —in other words, sources of variation not anticipated by the theory.

Bessel's investigation was a lonely and unrepeated first look at an empirical distribution. If it caught public notice at all, it was as added support for the assumption of Gauss's special form of error distribution. It seems to have occurred to no one, not even Bessel, that such an exercise would be a useful part of many statistical studies. No one at that early date seems to have realized that comparisons of this type could have implications far beyond those they held for the theory of errors, where the primary role was the validation of the method of least squares, a method that even in 1818 had no real competitors for complex problems. There was, however, one aspect of Bessel's comparison that went beyond this narrow purpose and foreshadowed later use of a similar device by Quetelet. Bessel had shown a tentative willingness to learn about the mechanism generating the data from the characteristics of the departures from the assumed form, to learn of the existence of sources of variation other than simple measurement error.

The Fitting of Distributions: Quetelismus

> The theory [of errors] is to be distinguished from the doctrine, the false doctrine, that generally, wherever there is a curve with single apex representing a group of statistics — one axis denoting size, the other axis frequency — that the curve must be of the "normal" species. The doctrine has been nicknamed "Quetelismus," on the ground that Quetelet exaggerated the prevalence of the normal law. (Edgeworth, 1922)

The feeling that Laplace's central limit theorem generated in Quetelet was much the same as that it generates in neophytes today: a sense of wonder that out of chaos comes order. A single component of a measurement, call it error if you will, could be totally unpredictable; but the average of many such components, no one or few of them dominant (that is, all subject to the same general dominating forces), will be subject to a law of remarkably specific form. In the mid-1840s it seems to have occurred to Quetelet that this remarkable property might be used to give a solution to the problem, posed by Keverberg, of deciding when a group of observations could be treated together as homogeneous. If a collection of variable measurements were in fact homogeneous (that is, susceptible to the same dominant causes, differing only in the more minor and random aspects that Quetelet would term accidental causes), then Laplace's theorem

Table 5.5. Bessel's comparisons of the distributions of the absolute values of three groups of residual errors.

Range (seconds)	Frequency of errors: Observed	Calculated	Recalculated
	A. 300 Observations of declinations		
0.0–0.4	66	65	64.63
0.4–0.8	58	60	60.01
0.8–1.2	55	53	51.72
1.2–1.6	28	41	41.39
1.6–2.0	27	30	30.75
2.0–2.4	23	21	21.21
2.4–2.8	10	13	13.58
2.8–3.2	15	8	8.08
3.2–3.6	8	5	4.46
3.6–4.0	4	2	2.28
Above 4.0	6	2	—
	B. 300 Observations of right ascensions		
0.0–0.1	114	107	107.22
0.1–0.2	84	87	86.80
0.2–0.3	53	57	56.87
0.3–0.4	24	30	30.14
0.4–0.5	14	13	12.91
0.5–0.6	6	5	4.47
0.6–0.7	3	1	1.25
0.7–0.8	1	0	0.28
0.8–0.9	1	0	0.05
	C. 470 Observations of right ascensions		
0.0–0.1	94	95	94.39
0.1–0.2	88	88	88.50
0.2–0.3	78	78	77.81
0.3–0.4	58	64	64.15
0.4–0.5	51	49	49.58
0.5–0.6	36	35	35.94
0.6–0.7	26	24	24.42
0.7–0.8	14	16	15.56
0.8–0.9	10	9	9.30
0.9–1.0	7	5	5.21
Above 1.0	8	5	—

Source: Bessel (1818, pp. 19–20).

Note: The calculated values were given by Bessel as found from the approximation (3); these values were recalculated from this formula as a check, using the values of ϵ given by Bessel; namely, A, $\epsilon = 1.16$; B, $\epsilon = 0.1687$; C, $\epsilon = 0.3119$. In only one case (0.8–1.2 in A) do the calculated and recalculated values differ by more than can be reasonably attributed to rounding.

would tell us to expect the observations to follow the normal law $\phi(x)$, supposing the accidental causes sufficiently numerous. What was true of astronomical observations would also be true of heights of men, of birth ratios, and of crime rates. Now, if homogeneity implied that observations would follow the normal law, then why not use this as a device for discerning homogeneity? Simply examine the distribution of a group of measurements. If they fail to exhibit this form, then this is surely evidence of lack of homogeneity—or at least evidence that the primary inhomogeneities are not in the nature of a large number of accidental (independent, random, of comparable force and size) causes. If they do exhibit this normal form, then this is prima facie evidence that the group is homogeneous and susceptible to statistical analysis as a group, without distinguishing the members of a group by identifying labels. In particular, it will be meaningful to take and compare the averages of such groups.

There are, of course, many pitfalls in the course just outlined. First of all, it is a hypothetical step in that the conclusion of homogeneity does not follow necessarily from the fact that observations are normally distributed. We know now that many situations Quetelet might well have regarded as inhomogeneous are quite capable of producing that same distinctive distribution. This fact was not widely known in Quetelet's time, however, if indeed it was known at all. A second difficulty involves the fitting of a normal curve to data. Bessel had apparently had no problem there, but his method (even if it had been widely noticed by social scientists) would not carry over readily to some of the grouped and truncated data sets found in nonastronomical problems. A third point involved judging the adequacy of the fit: How could a statistician determine when a departure from normality was sufficiently striking to conclude that the data were inhomogeneous? Quetelet's encounters with these problems were all the more difficult because of their tentative nature. He was groping for solutions to problems that had never been well articulated, and his view was further obscured by the attractive glitter of his average man.

Quetelet's approach to the fitting of normal curves was critically tied to Laplace's theorem. To Quetelet, the curve was the end product of a mechanistic model—a binomial urn scheme—and he did not consider it outside the context of this model. Bessel had taken a very different view, treating the curve $\phi(x)$ as a distinct mathematical entity, Gauss's error curve, and therefore had moved easily and naturally to an analytic approximation of its integral. In contrast, Quetelet considered it in the terms of the binomial approximation theorem of De Moivre and Laplace. Indeed, the connection between the curve $\phi(x)$ and the idea of sums of independent accidental causes was crucial to Quetelet's application, and it is natural that he should use this binomial mechanism to find the approximating curve.

Quetelet's method was explained in detail in his 1846 book, *Lettres à S.A.R. Le Duc Régnant de Saxe-Cobourg et Gotha sur la Théorie des Probabilités appliquée aux sciences morales et politiques*, in letters 19–21 and accompanying notes. This book is really an original, if elementary, treatise on probability and social statistics, written in the form of a series of letters to the Belgian king's two nephews, Ernest (the duke to whom the book was dedicated) and Albert (who by 1846 was husband to Queen Victoria of Great Britain). Quetelet had tutored the two in the 1830s, and in writing his book as a series of letters he was adopting a form that had been used with great success by Euler in 1768, with *Letters to a German Princess*, a popular exposition of physical science.

We can best gain an understanding of Quetelet's method by following it through in one of his examples. The table in Figure 5.3 shows Quetelet's first major example, an example he presented after illustrating the method's use with astronomical data. This table is one of the most frequently republished data sets in the nineteenth-century statistical literature. It purports to show the frequency distribution for the chest measurements of 5,738 Scottish soldiers; thus 3 of the soldiers are said to have chests 33 inches in circumference and 1,079 had 40-inch chests. The data were extracted from the *Edinburgh Medical and Surgical Journal* (1817) and, in fact, were given erroneously. Although the errors have no important bearing on the explanation of the method, they do exemplify Quetelet's tendency to calculate somewhat hastily, without checking his work. The original data consisted of separate tables of measurements for the members of eleven different local militia, cross-classified by height and chest circumference. None of the eleven tables gave marginal totals for the chest measurements, hence Quetelet would have had to do quite a bit of summing to get his data. A careful recomputation of the totals shows that there were actually 5,732 soldiers, distributed as in Table 5.6. The data were collected by an army contractor, presumably charged with clothing the various militia. These original data were also discussed by Baxter (1875). It is interesting that Quetelet ignored the entire cross-classification, as it is just the kind of table that in a different intellectual climate might have suggested a *bivariate* normal distribution.

Returning to Quetelet's version of the data, we see that the third column of Figure 5.3 gives the frequency distribution renormalized to sum to one, although the entries are given without decimal points or leading zeros, as parts of 10,000. Thus the entry "5" opposite 33 inches means that the 3 soldiers out of 5,738 with 33-inch chests were a fraction 0.0005 of the total. Columns 4 through 7 give the details of his calculations and the final column gives the relative frequency distribution as would be found from a fitted normal distribution. Quetelet's comparison then would be between columns 3 and 8.

MESURES de la POITRINE.	NOMBRE d'hommes.	NOMBRE PROPORTIONNEL.	PROBABILITÉ d'après L'OBSERVATION.	RANG dans LA TABLE.	RANG d'après le CALCUL.	PROBABILITÉ d'après LA TABLE.	NOMBRE D'OBSERVATIONS calculé.
Pouces.							
33	3	5	0,5000			0,5000	7
34	18	31	0,4995	52	50	0,4993	29
35	81	141	0,4964	42,5	42,5	0,4964	110
36	185	322	0,4823	33,5	34,5	0,4854	323
37	420	732	0,4501	26,0	26,5	0,4531	732
38	749	1305	0,3769	18,0	18,5	0,3799	1333
39	1073	1867	0,2464	10,5	10,5	0,2466	1838
			0.0597	2,5	2,5	0,0628	
40	1079	1882	0,1285	5,5	5,5	0,1359	1987
41	934	1628	0,2913	13	13,5	0,3034	1675
42	658	1148	0,4061	21	21,5	0,4130	1096
43	370	645	0,4706	30	29,5	0,4690	560
44	92	160	0,4866	35	37,5	0,4911	221
45	50	87	0,4953	41	45,5	0,4980	69
46	21	38	0,4991	49,5	53,5	0,4996	16
47	4	7	0,4998	56	61,8	0,4999	3
48	1	2	0,5000			0,5000	1
	5738	1,0000					1,0000

Figure 5.3. Quetelet's analysis fitting a normal distribution to data on the chest circumferences of Scottish soldiers. Column 1 gives the chest circumference in inches; columns 2 and 3 give the frequency and relative frequency distributions for 5738 individuals, columns 4–7 give the details of Quetelet's calculations (see text), and column 8 gives the fitted relative frequency distribution. (From Quetelet, 1846, p. 400.)

It might seem curious at first glance that Quetelet in no way used a table of the normal distribution in deriving his fitted values in column 8, although his book included such a table, republished from Cournot (1843). Quetelet tells us (1846, p. 387) that he had not seen Cournot's book when he derived his method, presumably in 1837, but other tables (such as Kramp's) and analytic approximations to the normal integral (such as Bessel's, or Laplace's continued fraction expansion) were easily available. In fact, what Quetelet did was to construct his own table based on the binomial distribution itself rather than on the approximating normal dis-

Table 5.6. Distribution of heights and chest circumferences of 5,732 Scottish militia men.

Height (inches)	Number of men with chest circumference (inches) of—																Total no. of men in height class
	33	34	35	36	37	38	39	40	41	42	43	44	45	46	47	48	
64–65	1	7	31	69	108	154	142	118	66	17	6	3	0	0	0	0	722
66–67	1	9	30	78	170	343	442	337	231	124	34	12	3	1	0	0	1,815
68–69	1	2	16	34	91	187	341	436	367	292	126	70	13	3	2	0	1,981
70–71	0	1	4	7	31	62	117	153	209	148	102	40	16	7	0	0	897
72–73	0	0	0	1	9	7	20	38	62	65	45	43	18	7	1	1	317
Total no. of men in chest-size class	3	19	81	189	409	753	1,062	1,082	935	646	313	168	50	18	3	1	5,732

Source: Edinburgh Medical and Surgical Journal (1817, pp. 260–264).

tribution. The binomial distribution he chose was the symmetric binomial distribution with 999 trials (and thus 1,000 possible outcomes; Figure 5.4). That Quetelet took such a tack reveals his conceptual view of the comparison he made. His computations may at first seem laborious (although he devised an interesting shortcut), but the result was that by this method he could bring the mechanism of Laplace's theorem into full view whereas an analytic approach could not. The accidental causes (here represented by the individual independent trials) that conspired to produce the marvelous curve $\phi(x)$ were not allowed to recede into the invisible background. Rather they were kept permanently in view by forcing all calculations to be in terms of the binomial distribution itself, even though Quetelet recognized (and even demonstrated, pp. 387–388) that there was no important difference between the symmetric binomial distribution with 999 trials and the approximating curve $\phi(x)$.

A portion of Quetelet's binomial table is given in Figure 5.5. The table is given in terms of the possible outcomes of the experiment: Draw 999 balls, with replacement, from an urn containing equal numbers of white balls and black balls. Because the distribution of possibilities is symmetric,

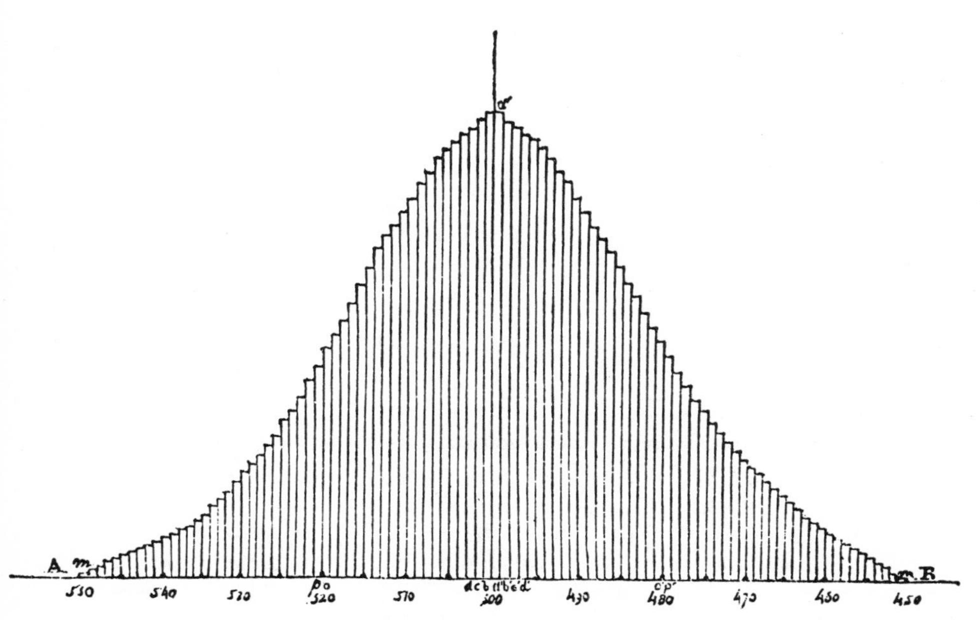

Figure 5.4. Quetelet's 1846 rendition of a symmetric binomial distribution with 999 trials, based upon the table in Figure 5.5. (From Quetelet, 1846, p. 396.)

GROUPES DE				RANG des GROUPES.	ÉCHELLE de possibilité. — PROBABILITÉ du tirage de chaque GROUPE. *Table* A.	ÉCHELLE de précision. — SOMMES des probabilités à partir du groupe le plus probable. *Table* B.	ÉCHELLE de possibilité. — PROBABILITÉ relative du tir. de chaque GROUPE. *Table* C.
499 boules blanches	et	500	noires. .	1	0.025225	0.025225	1.000000
498	id.	501	id. . .	2	0.025124	0.050349	0.996008
497	id.	502	id. . .	3	0.024924	0.075273	0.988072
496	id.	503	id. . .	4	0.024627	0.099900	0.976285
495	id.	504	id. . .	5	0.024236	0.124136	0.960789
494	id.	505	id. . .	6	0.023756	0.147892	0.941764
493	id.	506	id. . .	7	0.023193	0.171085	0.919429
492	id.	507	id. . .	8	9.022552	0.193637	0.894040
491	id.	508	id. . .	9	0.021842	0.215479	0.865882
490	id.	509	id. . .	10	0.021069	0.236548	0.835261
489	id.	510	id. . .	11	0.020243.	0.256791	0.802506
488	id.	511	id. . .	12	0.019372	0.276163	0.767956
487	id.	512	id. . .	13	0.018464	0.294627	0.731958
486	id.	513	id. . .	14	0.017528	0.312155	0.694860
485	id.	514	id. . .	15	0.016573	0.338728	0.657008
484	id.	515	id. . .	16	0.015608	0.344335	0.618736
483	id.	516	id. . .	17	0.014640	0.358975	0.580364
482	id.	517	id. . .	18	0.013677	0.372652	0.542197
481	id.	518	id. . .	19	0.012726	0.385378	0.504516
480	id.	519	id. . .	20	0.011794	0.397172	0.467576
479	id.	520	id. . .	21	0.010887	0.408060	0.431609
478	id.	521	id. . .	22	0.010008	0.418070	0.396815
477	id.	522	id. . .	23	0.009166	0.427236	0.363366
476	id.	523	id. . .	24	0.008360	0.435595	0.331407

Figure 5.5. A portion of Quetelet's table of a symmetric binomial distribution for 999 trials. The table lists outcomes ranked by distance from the center, for one side only. The first column gives the outcomes (499 white balls and 500 blacks, etc.), the second column gives the rank, the last column (Table C) gives the relative probabilities (as a ratio to the central term), the third column (Table A) gives the probabilities themselves, and the next-to-last column (Table B) gives the one-sided cumulative probabilities $P\{500 \leq \mathrm{X} < 500 + r\}$*, with* X *= number of black balls,* r *= rank. (From Quetelet, 1846, p. 375.)*

Quetelet only gave the half corresponding to the possible outcomes: 499 white and 500 black, 498 white and 501 black, and so forth. The second column of his table ranks these outcomes, assigning the rank 1 to the most probable of those listed, 499 white and 500 black. The table actually continued to the outcome with rank 80, namely, 420 white and 579 black.

Quetelet recognized that tabling such a distribution by actually evaluating the binomial coefficients was impractical because of the large numbers involved. He was aware of Stirling's formula, but he preferred not to use it as he had another method better suited to calculation. Instead the table was based on the fact that the ratios of the successive terms of a binomial distribution are of a particularly simple form. If y_n represents the probability that n black balls and $999 - n$ white balls are drawn, then

$$y_{n+1} = y_n \cdot \frac{999 - n}{(n + 1)}.$$

What Quetelet did, then, was to begin by tabling the relative values of these probabilities. Taking the term corresponding to rank 1 (namely, y_{500}) as unity, he multiplied successively by the factors 499/501, 498/502, 497/503, and so forth to obtain the entries in column C of Figure 5.5. Thus, for example, he found the rank 3 term to be $y_{502}/y_{500} = (499/501)(498/502) = 0.988072$. He recognized that he needed to consider only the central portion of the distribution because the more distant terms were vanishingly small. He stopped with the rank 80 term y_{579}, giving $y_{579}/y_{500} = 0.000003$, although it is evident that he had carried the calculation further in the manuscript than in the printed form.

Quetelet then converted the relative probabilities of column C to probabilities by simply adding all the entries in column C, multiplying by 2 (since he was considering only half the distribution), and dividing each term by the column total. The result was given in column A. Column B gives the cumulative probabilities found by summing entries in column A. Thus, for example, the probability of the rank 5 outcome of 495 white balls and 504 black balls is given as 0.024236, and the cumulative probability of ranks 1 through 5 (that is, the probability of getting between 500 and 504 black balls, inclusive) is given as 0.124136.

Quetelet's comparison was based on a clever use of the cumulative probabilities of column B (which he called the scale of precision) and the column of ranks. It may be easier to follow the application of his procedure to the Scottish chest circumferences if we first present it in a modern notation. Let $S(r)$ represent Quetelet's scale of precision, expressed as a function of rank r. That is, if X is a binomial (999, 0.5) random variable, $S(r) = P\{500 \leq X < 500 + r\}$, for $r = 1, 2, \ldots$, a "one-sided" cumulative distribution. Then, in outline, Quetelet would have manipulated a set of grouped data (such as the Scottish data of Figure 5.3) as follows:

(1) Split the grouped relative frequency data in half, at the median.
(2) Compute upper and lower cumulative relative frequencies for the data, cumulating outward from the median.
(3) Transform these cumulated relative frequencies to a rank scale by the inverse of the function $S(r)$.
(4) Match (by trial and error and inspired guesswork) a list of fitted ranks to these transformed frequencies, so that the fitted ranks varied linearly with the original measurement scale.
(5) Transform the fitted ranks by $S(r)$ to obtained fitted frequencies.

The rationale for the procedure is essentially the same as that underlying the modern use of the probability integral transformation, or normal probability plots. If the data are exactly as would be expected from a binomial distribution, the application of the inverse of $S(r)$ in step 3 should produce a list of ranks that vary exactly linearly with the measurement scale (inches of circumference, in the example). In practice they do not vary linearly, but it is easier to construct a list of ranks that do vary linearly (step 4), and hence do correspond to an exact binomial distribution, than it is to construct a fitting (nonlinear) binomial distribution directly.

Let us follow this procedure through for the Scottish data of Figure 5.3. Quetelet began by converting the raw frequencies of column 2 to the relative frequencies of column 3, as we noted earlier. Now, the sum of the relative frequencies for classes 33 through 40 is 0.6285; that is, it exceeds 0.5 by 0.1285. To cut the relative frequency distribution into two equal portions (recall that Quetelet's table was designed to correspond to one-half a symmetrical distribution), he divided the frequency for 40-inch chests into two parts, $0.1882 = 0.0597 + 0.1285$, and he associated 0.0597 with the lower part of the distribution, 0.1285 with the upper part. Starting with these values he calculated the upper and lower cumulative distributions, working outward from the middle. This completed step 2 and gave him column 4. (The reader attempting to verify these calculations will find that Quetelet made small errors in computing the relative frequencies for classes 39 and 40 and rounded incorrectly for classes 42 and 46.)

Column 5 of Figure 5.3 was computed from column 4 by using the tabled values of the scale of precision and crude linear interpolation. For example, the upper cumulated relative frequency of 0.2913, corresponding to the class 41, agrees fairly closely with the value at rank 13 in Figure 5.5 namely, 0.294627, so Quetelet assigned a rank of 13 to this value. Similarly, the lower cumulated relative frequency at class 39 of 0.2464 falls roughly midway between the entries 0.236548 at rank 10 and

0.256791 at rank 11; it was assigned rank 10.5. This completed step 3, albeit in a fairly rough manner.

The next step (4), the construction of a corresponding set of theoretical ranks, proceeded in a similarly rough manner. Quetelet knew (indeed this was the whole point behind his transformation) that if the observed frequencies were *exactly* distributed as if from a suitably grouped binomial distribution, then the ranks should increase linearly with the measurement scale, in this case, inches. Quetelet was not totally explicit about the procedure for finding such a sequence of ranks to approximate those in column 5. What he did was apparently ad hoc. He recognized that the first differences of the fitted rank sequence should be constant, so he found the differences of the ranks in column 5 and chose an integer or simple decimal fraction that approximated as many as possible, with particular attention given to the middle of the distribution. In the present case these differences were, for the lower half of the distribution, 9.5, 9, 7.5, 8, 7.5, 8; for the upper half, 7.5, 8, 9, 5, 6, 8.5, 6.5. The constant difference 8 must have seemed like a convenient choice; there is no indication that he followed any more formal procedure, such as averaging first differences, although he may have tried more than one choice. He then repeatedly added this difference to the ranks for the central class, completing step 4 and arriving at column 6. (The entries 50 and 61.8 in column 6 appear to be misprints for 50.5 and 61.5.)

Quetelet's calculations and step 5 were then completed by finding the entries in column 7 from column B of Figure 5.5 (by using linear interpolation) and converting the cumulated relative frequencies to the relative frequencies of column 8, expressed as parts in 10,000.

In some respects the procedure is cumbersome, particularly in comparison with that of Bessel. A large number of calculations are involved, and several entries into a table are needed, with frequent interpolation. In other respects, though, it is very easy; and it requires no advanced mathematical training. All the computations are quite simple (addition, subtraction, and division), no complicated exponential functions are involved, and there is no need for a table of logarithms. Also, the procedure adapts easily to grouped or truncated data. Interestingly, Quetelet's calculations have a modern (and quite elementary) graphical counterpart. The procedure is essentially equivalent to plotting the cumulated data on a normal probability scale, fitting a straight line to the points by eye (making sure the line agrees with the data at the median), and finding the fitted values from that straight line. With practice, the method becomes easy, almost natural, and the first appearance of cumbersome and laborious calculation is dispelled. But it failed the ultimate test—it did not become widely adopted, nor was it frequently imitated. I shall examine the reasons for this later.

After completing the calculations, it remained for Quetelet to evaluate the fit and examine its implications. Here he fell back on intuitive impressions and an exaggerated faith in the accidental cause – binomial mechanistic model. The result was a series of analogies that were to have a powerful (and potentially misleading) effect upon many readers and, indeed, upon Quetelet himself. This is exemplified in the case of the Scottish soldiers. There Quetelet accepted the fit as self-evident from the table and proceeded to an analogy: The fitted distribution would have a probable error (median deviation) of about 1.312 inches or 33.34 millimeters (mm) in the original scale, and the distribution was what one might expect from measurement error alone. "I now ask if it would be an exaggeration to claim it an even bet that someone engaged in taking measurements of the human body would err by about 33 mm in measuring a chest of more than a meter in circumference? Well, accepting this probable error, 5738 measurements taken on the same person would certainly not group themselves with more regularity (as regards to order of magnitude) than the 5738 measurements taken of the Scottish soldiers. And if we were given the two series of measurements without being told which was which, we would be at a loss to tell which series was taken on 5738 different soldiers, and which had been obtained on one and the same person, with less ability and rougher judgment." Quetelet went on to develop this idea further: The distribution found was "as if the chests measured had been modeled on the same type, on the same individual, an ideal if you wish, but one whose proportions we can learn from sufficiently prolonged study" (Quetelet, 1846, p. 137). This ideal was the average man, an entity Quetelet now thought took on "the character of a mathematical truth."

Quetelet's analysis was this, then: Because the measurements were distributed as they would be if nature were aiming for an ideal type, this must be essentially the truth. The distribution he had found gave proof to the existence of his average man (or at least average Scottish soldier); the measurements in hand were those of the average man, plus the effects of accidental causes. His earlier concept took on a new, mathematically demonstrated reality. He had demonstrated not merely homogeneity but (apparently) immutability, for Quetelet believed that this curve could *only* arise from a concurrence of accidental causes. "This symmetry in the data only exists and can only exist in so far as the elements that concur to give the mean can be traced back to one single type" (Quetelet, 1846, p. 216). The argument was powerful even though incorrect, and Quetelet was not able to resist its temptation. Like Pygmalion, he was, at least to a degree, taken in by his own creation.

There were two important consequences of this argument as far as Quetelet's work was concerned. The first was that this renewed emphasis on a real (if ideal) average man distracted both Quetelet and others and shifted their attention from those aspects of the work that were of greater potential importance. Having created the average man, Quetelet tended to sit back and admire it rather than continue to study the relations of different average individuals to explanatory social variables, and Quetelet's critics could focus on it as a fanciful fiction. The second consequence was that the tool he had created was too successful to be of use for its original purpose. The fitted distributions gave such deceptively powerful evidence of a stable homogeneity that he could not look beyond them to discover that further subclassification could produce other distributions of the same kind, that some normal distributions are susceptible to dissection into normal components. The method was lacking in discriminatory power; too many data sets yielded evidence of normality. Few new classifications were uncovered; the primary use of the method was as a device for validating already determined classification schemes.

There were exceptional cases where the fitting of a normal (or binomial) distribution produced new information, or at least new theories, but these were rare. One was presented by Quetelet in the 1846 *Lettres;* it involved a distribution of what were given as the heights of 100,000 French conscripts. These data, and Quetelet's treatment of them, are given in Table 5.7. Here Quetelet noticed the excess of conscripts in the shortest class and the deficit in the next class. He might have explained this in any number of ways, such as inhomogeneity in the French population, but instead he put forth an explanation that showed both great cleverness and a blindness to the possibility that human measurements in a group he a priori expected to be homogeneous might not be normally distributed. The shortest class, Quetelet told us, was exempt from service, and the surplus in this class was evidence of fraud! By one means (official indulgence) or another (self-imposed shrinkage) about 2,200 men had avoided conscription (Quetelet, 1846, pp. 144–145, 401–403).

Another example of what was viewed at the time as a successful dissection and subclassification based on a fitted distribution was presented in 1863 by Adolphe Bertillon. Bertillon, in treating different measurements of the heights of French conscripts after the manner of Quetelet, found an anomalous pattern in the distribution of heights of 9,002 young men measured between 1851 and 1860 in the department of Doubs, in eastern France (Figure 5.6). Bertillon found that the distribution did not exhibit the usual symmetrical shape, with a single most frequent group; rather it had two modal values (see Table 5.8). Moreover, this effect was constant in time: If he broke the data into two sets, 1851–1855 and 1856–1860, both groups had the same pattern; if he subdivided by year, nine out of the ten years had the pattern. As Bertillon put it, "This is therefore not accidental, and, being constant, it *necessarily* depends on a constant cause." Bertillon hypothesized that the population of Doubs consisted of two human types,

Table 5.7. Quetelet's analysis of 1817 data on the heights of French conscripts.

(1) Height (in meters)	(2) Number of men	(3) Relative freq.	(4) Cumulative freq.	(5) Rank	(6) Fitted rank	(7) Prob.	(8) Fitted no. of men	(9) Difference (2)–(8)
Less than 1.570	28,620	0.2862	0.5000	—	—	—	26,345	+2,275
1.570–1.597	11,580	0.1158	0.2138	9	10.0	0.23655	13,182	−1,602
1.597–1.624	13,990	0.0980	0.0980	4	4.2	0.10473	14,502	−512
		0.0419	0.0419	1.6	1.6	0.04029		
1.624–1.651	14,410	0.1441	0.1860	7.5	7.4	0.18011	13,982	+428
1.651–1.678	11,410	0.1141	0.3001	13.5	13.2	0.29814	11,803	−393
1.678–1.705	8,780	0.0878	0.3879	19	19.0	0.38539	8,725	+55
1.705–1.732	5,530	0.0553	0.4432	25	24.8	0.44166	5,627	−97
1.732–1.759	3,190	0.0319	0.4751	31	30.6	0.47355	3,189	+1
Above 1.759	2,490	0.0249	0.5000	—	—	—	2,645	−155
	100,000						100,000	

Sources: Quetelet (1846, pp. 401–402; 1849, pp. 277–278).

Note: The entries in columns 5 and 7 are derived from Quetelet's table (Figure 5.5). The seventh and eighth entries in columns 8 and 9 are here corrected from the original, where they read 5,527 and 3,187 in column 8 and +3 and +3 in column 9. Quetelet apparently obtained the data from Villermé (1829), who cited as his source an 1817 "brochure" by Hargenvilliers titled "Recherches et considerations sur la formation et le recrutement de l'armée en France."

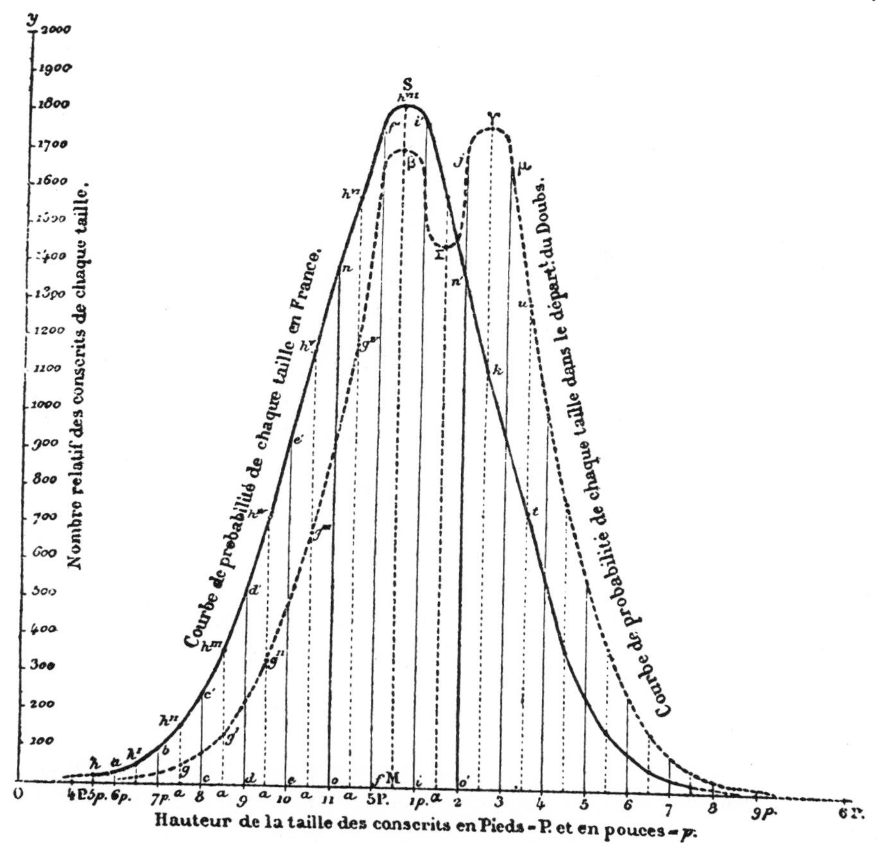

Figure 5.6. Adolphe Bertillon's depiction of the curve of heights of conscripts for all of France (solid line) *and for the department of Doubs* (dotted line), *purporting to show a bimodal distribution. (From Bertillon, 1876, p. 306.)*

one short (average height, 5′) and one tall (average height, 5′3″). Subsequently, his colleague Lagneau (1870) found that the inhabitants of Doubs were primarily of two different races, the Celts and the Burgundians[29] (Bertillon, 1876, p. 307). On their face, Bertillon's investigations bears an uncanny resemblance to later work by Weldon, which inspired one of Karl Pearson's first statistical works (Pearson, 1894). But there is a later ironic

29. The 1837 *Penny Cyclopaedia* (article, "Doubs") described them differently: "The inhabitants of the mountains are tall, robust, and healthy; sober, economical, gentle, willing to oblige, hospitable, and true to their word, but untaught and credulous: those of [illegible] plain are neither so robust, nor temperate, nor obliging." (vol. 9, p. 106).

Table 5.8. Relative frequency distribution of the heights of 9,002 French conscripts from the department of Doubs, 1851–1860.

Height	Relative frequency
Below 4′10″	577
4′10″ to 4′11″	637
4′11″ to 5′0″	1,116
5′0″ to 5′1″	1,766
5′1″ to 5′2″	1,457
5′2″ to 5′3″	1,777
5′3″ to 5′4″	1,313
5′4″ to 5′5″	820
5′5″ to 5′6″	291
5′6″ to 5′7″	153
5′7″ to 5′8″	64
5′8″ to 5′9″	17
5′9″ to 5′10″	9
5′10″ to 5′11″	3
	10,000

Source: Bertillon (1863, p. 238).
Note: Livi (1896) found that the apparent drop in the class 5′1″ to 5′2″ was due to Bertillon's incorrect conversion of measurements grouped by centimeters to measurements grouped by inches.

twist to the story. Livi (1896), puzzled by the fact that Bertillon's phenomenon did not persist in later years, took a careful look at Bertillon's investigation and discovered that the double hump was a simple artifact of the analysis. The heights had been originally recorded to the nearest centimeter, and Bertillon had converted the class boundaries to inches, then grouped the data by inch. This had the interesting effect of putting three of the original centimeter-wide classes in each of the classes 4′11″ to 5′0″, 5′0″ to 5′1″, 5′2″ to 5′3″, and 5′3″ to 5′4″, but only two of the original classes corresponded to 5′1″ to 5′2″. This anomalous grouping alone produced the effect, as Livi was able to demonstrate convincingly.

Thus, even Bertillon's apparently constructive use of Quetelet's method was illusory. The fact that even a mistaken finding could elicit such convincing explanations from anthropologists may be partially attributable to the lack of power of Quetelet's technique: A large number of real inhomogeneities lurked below the surface, and with even the most specious incentive they would burst forth. But so limited was the power of the technique that even false phenomena were rare. In almost all cases in this period where normal (or binomial) distributions were fit to frequency data, the fit was found to be good (or at least acceptable). This was true of Quetelet's own work, much of which was summarized in his last book *Anthropométrie*, published in 1870; it was also true of those few others who tried this type of comparison. Some of these workers were Americans, starting with the actuary Ezekiel B. Elliott (1863) and the astronomer Benjamin Apthorp Gould (1869). This and other work was reviewed and summarized by Baxter (1875).

One of the most unusual early applications of Quetelet's method was an 1861 paper, "On typical mountain ranges: an application of the calculus of probabilities to physical geography," by William Spottiswoode, a physicist. Spottiswoode attempted to use Quetelet's own procedure to fit a normal curve to the distribution of direction of orientation of eleven mountain ranges to see whether they corresponded to a common "type." His work was more notable for enthusiasm and wishful thinking than for understanding, however, as he mistakenly ignored the fact that his eleven residuals were not equally spaced. He ended up with a really dismal fit, commenting nonetheless that "the agreement . . . although not perfect, is yet sufficiently approximate to justify the conclusion that, the various mountain ranges which we have been considering do point to a common type, that their directions are not accidental, and that the geologist and the physical philosopher will at least have good grounds for seeking some common agency which has caused their upheaval" (Spottiswoode, 1861). The work had one major effect, though. According to Francis Galton's autobiography, it was this work of Spottiswoode that caught his eye and first interested Galton in the normal law of error (Galton, 1908, p. 304). This is a tribute more to the keenness of Galton's eye than to the quality of Spottiswoode's work, however.

Quetelet's method failed to solve the problem that was the main stumbling block to the advancement of statistics in the social sciences. It did not provide the key to evaluating and finding useful ways of classifying data for analysis, except in very specialized situations. It was, as I have argued, too successful in revealing patterns—in almost all cases where it was tried it revealed the same pattern. It was not sensitive to the more subtle types of inhomogeneities, such as age or diet, for which a method of analysis was needed. It could not even detect the presence of widely differing racial types. This is not to say Quetelet's work on this was without influence. To the contrary, it helped create a climate of awareness of distribution that was to lead to a truly major advance in statistical methods over the period 1869 to 1925, in works of Francis Galton, Francis Y. Edgeworth, Karl Pearson, and Ronald A. Fisher.

Quetelet's work also fostered a growing belief in the explanatory potential of stochastic statistical models. The striking fits he found convinced his contemporaries that something like the "central limit effect" was at work. Even though we might now see the argument as non sequitur, in 1850 the dramatic law he found in unusual places was proof that an averaging of random causes was at work. And even though he failed to provide the real payoff he aimed for, he succeeded in showing that nature could be counted on to obey the laws of probability.

Nowadays Quetelet is only known for his fits of symmetric distributions, and it is true he gave these the most emphasis. But they were not the only stochastic models he used; for example, he fit asymmetric binomial distributions to skewed meteorological data (Quetelet, 1846, pp. 174–183; 1852), and he was well aware that many other distributions (such as changes in the price of wheat) were skewed, a circumstance he thought due to curious accidental causes acting unequally in two directions (Quetelet, 1846, pp. 166–169).

6. *Attempts to Revive the Binomial*

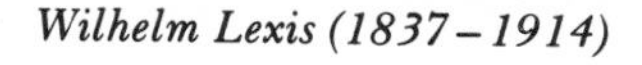

Wilhelm Lexis (1837–1914)

John Arbuthnot (1667–1735)

THE CENTRAL conceptual problem that nineteenth-century statisticians encountered in extending statistical methodology from astronomical to social data was the isolation of social data into homogeneous classes or categories. For investigators to apply the techniques of the theory of errors to social data, they had to perceive the data within classes as amenable to such analyses: Variation within the class had to be seen as analogous to the random fluctuations affecting independent observations of a single astronomical phenomenon. It was not acceptable to analyze as homogeneous data whose variation was thought to be in large measure due to factors that were important to the study at hand. If mortality rates were different for different areas or occupations, then it was not thought proper to aggregate data from different areas or occupations in a study of mortality rates. If crime rates changed from year to year with known changing social conditions, then crime rates should not be based on data aggregated across years in a comparison of different countries, at least not if the calculus of probabilities was to be used to judge the significance of the comparison. On the other hand, if data on the sex ratio at birth was seen as homogeneous in time and location—subject to the same random fluctuations in one year as in the next, in one country as in the next—then the application of probability to the evaluation of hypotheses concerning this ratio would be natural and accepted. It was not necessary that the data *be* homogeneous to the eyes of a later century; it was essential that they be *perceived* as homogeneous by the investigator himself and be thought likely to be so perceived by his intended audience.

The judgment of homogeneity could be made on external grounds or could be based on evidence internal to the data. The nineteenth-century awareness of a vast multitude of potentially influential factors made an external judgment of homogeneity difficult, if not impossible. Keverberg's admonition made this point forcefully to Quetelet, and it must have been felt keenly by all objective investigators of the time. Too much was known about too many potential factors to make an a priori judgment of homogeneity widely acceptable in the study of important matters.

There were three apparent ways in which this impasse could be overcome: (1) Exhaustively study factors external to the data, with the hope of isolating all important influences and thereby permitting a homogeneous categorization; (2) gather vast sets of data, thereby rendering all residual variation negligible and permitting analyses without resort to the quantification of uncertainty; (3) develop a methodology for the internal evaluation of a class's homogeneity. Much of the development of empirical research in the nineteenth century can be viewed as employing the first two of these approaches: the founding of large numbers of statistical societies, the publication and proliferation of statistical journals devoted entirely to empirical research, the growth of national records offices and national statistical bureaus, the introduction of international statistical congresses, and the development of censuses. The third approach—the development of tests of homogeneity internal to the data at hand—is more central to our present purpose, however.

Lexis and Binomial Dispersion

As mentioned in Chapter 5, Quetelet initiated an internal evaluation of homogeneity by fitting normal-like binomial distributions to his data. In effect, he used Laplace's normal approximation theorem in reverse: If the data had a distribution like one that would arise as the sum of a large number of independent errorlike effects, then, by analogy to the treatment of astronomical observations, they could be treated as if they *had* arisen in such a manner, as if they *were* homogeneous. Quetelet had thus sought to validate his analysis by a probability model. If the data were well classified and the only residual variation was due to independent and individually unimportant effects omitted from the analysis, then the distribution within the classes should, following Laplace, be approximately normal. Just so, the reasoning then went, data that exhibited this necessary consequence of being well classified must themselves be well classified. As we have already noted, Quetelet's approach lacked power. Too many empirical distributions passed his test, and the normal distribution came to seem an almost ubiquitous law of nature rather than an effective test of data classification. Quetelet did not fully realize what we know now (although he may have sensed it): Too many alternative hypotheses, including many requiring further or different classification, lead to the same normal distribution.

Quetelet's fitting of distributions was to find many followers, but perhaps his most direct intellectual descendant was the German statistician–economist Wilhelm Lexis (1837–1914). Lexis's early studies were in science and mathematics. He graduated from the University of Bonn in 1859 with a thesis on analytic mechanics and a degree in mathematics.[1] In 1861 he went on to Paris to study social science, and he subsequently held positions at many European universities: Strassburg in 1872, Dorpat (later Yuriev, in western Russia) in 1874, Freiburg in 1876, Breslau in 1884, and finally, Göttingen in 1887. By 1875 Lexis had written an introductory book on the theory of population; and his most important statistical work consisted of several articles published between 1876 and 1880 while he was at Freiburg, all on population and vital statistics.

Lexis brought to social statistics a mathematical background much like that of Quetelet (this fact may help to explain the great attraction he felt to Quetelet's conceptual approach). It seems likely he first directly encountered Quetelet's work during his years in Paris. Citations in his early work mentioned Quetelet (using the Parisian spelling Quételet) but no other mathematical statisticians, and much of his data (for example, in Lexis, 1879) came from Belgian compilations edited by Quetelet. Some of Lexis's statistical work was built rather directly on Quetelet's use of the normal curve. He fitted a normal curve to empirical distributions of age at death, noting an observed surplus of early deaths, which he classed as "premature deaths." Those deaths that fit the curve he classed as "normal deaths" (*Normalgruppe der Sterbefälle,* Lexis, 1877, p. 45) (Figure 6.1). But Lexis is best known today for a method of analysis he invented, even though he described it as a generalization of Quetelet's conceptual approach to a different probability model (Lexis, 1880, p. 482).

Lexis was primarily concerned with data presented as series (over time)

1. For an account of Lexis's life, see Heiss (1978).

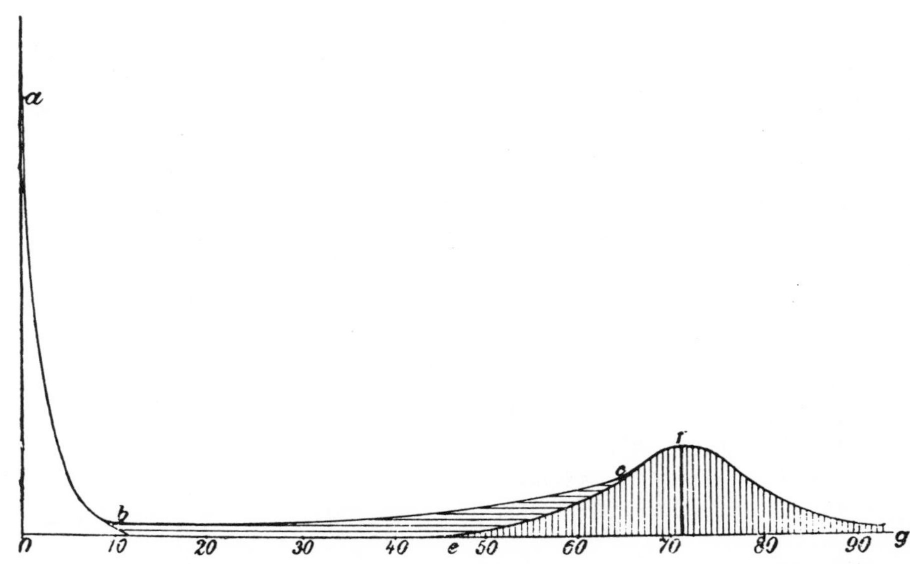

Figure 6.1. Lexis's drawing of a mortality curve, showing infant mortality and "premature deaths," that is, those corresponding to the horizontally shaded area bc *in excess of a normal curve chosen to fit later deaths, or "normal deaths." (From Lexis, 1903, p. 88.)*

of rates. Suppose, to take an example from Lexis (1879), we have data gathered over the twenty-five years from 1835–36 through 1859–60 on the number of infant deaths in Bavaria, classified by sex, by age (that is, stillbirth, aged 0–1 year, or aged 1–2 years), and by whether the birth was legitimate or illegitimate. Then for each category (for example, legitimate stillbirths) we can calculate for each year a male death rate and a female death rate: If a equals the number of male deaths in the category and b equals the number of female deaths in the category, then $a/(a+b)$ and $b/(a+b)$ give the fractions of male and female deaths for the category and these numbers times 1,000 give the respective death rates (male and female deaths per 1,000 deaths). In this example Lexis found it more convenient to look at the (for his purposes, equivalent) ratio 1,000 a/b, the number of male deaths per 1,000 female deaths for the category. He then wished to compare these ratios across categories, perhaps aggregating over years. For example, was the ratio of male to female stillbirths for legitimate births different from that for illegitimate births? Did either ratio change over time?

At a naive level these questions were easy to answer and required a statistical technology that went little beyond that available to Bernoulli and

De Moivre (and certainly not beyond that available to Laplace and Poisson). If the stillbirths are considered as independent trials, like draws with replacement from an urn of tickets labeled M and F, then the question becomes, Does the "legitimate" urn have the tickets in the same ratio as the "illegitimate" urn, and do the ratios change in time? The application of the technology of Bernoulli–De Moivre–Laplace–Poisson to the problem hinged on whether or not the investigator could bring himself to adopt this naive view. At an earlier time there was no question that the sex at birth was like a draw from an urn, and the probability calculus was applied without compunction. By Lexis's time the same process was perceived quite differently, and a different, more skeptical analysis was called for.

Arbuthnot and the Sex Ratio at Birth

The comparison of early eighteenth-century analyses of the sex ratio with those of Lexis in the late nineteenth century illuminates the conceptual difficulties involved and demonstrates the role statisticians' perceptions of their data played in determining which analyses were to be viewed as appropriate. Perhaps the earliest brush between mathematical probability and social statistics was in a short note published by John Arbuthnot in the *Philosophical Transactions* of the Royal Society in 1710. Arbuthnot (1667–1735) was physician to Queen Anne from 1709 until her death in 1714, but he is better known to posterity as a witty satirical writer—a collaborator of Jonathan Swift and creator of the prototypical Englishman John Bull. Arbuthnot had earlier, in 1692, translated into English and slightly extended Christian Huygens's short tract on probability, *Ratiociniis in Aleae Ludo* (1657).[2]

Arbuthnot's 1710 note was called "An argument for Divine Providence, taken from the constant regularity observ'd in the births of both sexes." Of prime interest to us are the terms in which he framed his problem: He claimed to demonstrate that divine providence, not chance, governed the sex ratio at birth. To prove this point he represented a birth governed by chance as being like the throw of a two-sided die, and he presented data on christenings in London for the eighty-two-year period 1629–1710. Under Arbuthnot's hypothesis of chance, for any one year male births will exceed female births with a probability slightly less than one-half. (It would be less than one-half by just half the very small probabil-

2. Huygens's tract (Huygens, 1657; 1659) was the first printed work on probability; Arbuthnot's translation (which appeared anonymously) and extension to a few games of chance not considered by Huygens was apparently the first work on probability published in English. See Todhunter (1865) for a discussion of subsequent editions of Arbuthnot's work. Todhunter overlooked one other early English translation of Huygens, in the article on "Play" in vol. 2 of John Harris's *Lexicon Technicum* (1710).

ity that the two numbers are exactly equal.) But even when taking it as one-half Arbuthnot found that a unit bet that male births would exceed female births for eighty-two years running to be worth only $(1/2)^{82}$ units in expectation, or

$$\frac{1}{4\ 8360\ 0000\ 0000\ 0000\ 0000\ 0000},$$

a vanishingly small number, "From whence it follows, that it is Art, not Chance, that governs."

Arbuthnot had thus rejected his simple model for chance in favor of divine providence, but he did not question the basic appropriateness of a model of this type for the problem at hand. He rejected chance but not the representation of chance by independent trials. Indeed, independent identical trials is the only model for chance he felt constrained to consider. Anything else, for Arbuthnot, was divine providence (and beyond the realm of mathematical modeling).

Arbuthnot's paper excited a lively debate, engaging men such as Nicholas Bernoulli (in Montmort, 1713, pp. 388–393), who argued that chance could as well be represented by an unbalanced die (or an urn of unbalanced constitution), but who also did not question the relevance of these simple probability models to birth statistics. Over succeeding years the regularity of many types of vital statistics, including those for births, was frequently remarked upon. Derham (1713, 1754, pp. 171–179) and Süssmilch (1741) took the regularity as apparent without probabilistic calculation and as evidence of divine order. 'sGravesande (1774, vol. 2, pp. 221–248) (the only scientist in history whose name begins with an apostrophe) and Laplace (1781; 1812, pp. 377–384) did apply probability calculations to the sex ratio, the latter as a formal example of the application of his methodology rather than as part of a serious investigation in vital statistics.

Buckle and Campbell

In the mid-1870s two men (Lexis and Émile Dormoy) independently proposed almost the same method for analyzing series of rates. By this time the intellectual climate had changed dramatically from that of a century before, when Laplace had almost playfully subjected birth data from London and Paris to the intricacies of his developing probability calculus. Even in the 1830s and 1840s the world as viewed by statisticians had been a stable (although increasingly complicated) place. Quetelet and others had pointed out what were claimed to be vast regularities in increasingly extensive statistical compilations: births, deaths, crimes, suicides, marriages

—all seemed sufficiently constant over time in sufficiently many societies to warrant remark. Their apparent regularity or stability was only over time and they were seen to be sufficiently variable over other characteristics that the Keverberg admonition was heeded and no post-Laplace probability analyses of the uncertainty of differences between categories were undertaken. But the apparent temporal stability was exciting, and it was seized upon by some as a major discovery.

Henry Thomas Buckle, a forerunner of modern scientific historians, presented Quetelet's findings in the first volume of his *History of Civilization in England* and claimed that they held the promise of making history a science, by proving "the regularity with which mental phenomena succeed each other" (Buckle, 1857, p. 19). He saw the statistics on crime as telling a powerful story for all moral conduct: "if we can in any period detect a uniformity and a method in the vices of a people, there must be a corresponding regularity in their virtues"; and "if it can be demonstrated that the bad actions of men vary in obedience to the changes in the surrounding society, we shall be obliged to infer that their good actions, which are, as it were, the residue of their bad ones, vary in the same manner; and we shall be forced to the further conclusion, that such variations are the result of large and general causes, which working upon the aggregate of society, must produce certain consequences, without regard to the volition of those particular men of whom the society is composed" (Buckle, 1857, pp. 20–21). To Buckle, as to many of his time, the regularity of statistical phenomena could reveal the laws of society, just as the regularity of physical phenomena had revealed the laws of nature to an earlier generation. The number of observations needed to discern a social law was larger, but the principle was the same. The vast statistical compilations of the nineteenth century could make social physics a reality.

It is hard to judge why these hopes began to come unraveled in the 1870s. A primary reason seems to have been that the importance of the enterprise had the effect of increasing the closeness with which the data were scrutinized. This examination, together with the increased discriminatory power furnished by increasingly large data sets, revealed instabilities, trends, and changes that had not appeared at first glance. As the glow of the first appearance of regularity faded and as statisticians' eyes became adjusted to a darker reality, previously unsuspected patterns emerged, not all illusory. With the increasing awareness of the lack of constancy in nature and society came the need to reevaluate old assumptions and an accompanying need for a methodology for this reevaluation.

A first whisper of discord was heard as early as 1859 when a Cambridge-trained, Scottish lawyer, Robert Campbell, published a short note in the *Philosophical Magazine,* "On a test for ascertaining whether an observed degree of uniformity, or the reverse, in tables of statistics is to be looked upon as remarkable."[3] The test, Campbell said, was directly inspired by his reading of Buckle; he felt that before one embarked upon Buckle's road, eliciting consequences from an assumed regularity, the apparent regularity should be put to the test. "For supposing the observed uniformity to be not more than that which might be expected from events, the occurrence of which to individuals was conceived of as perfectly fortuitous, the whole argument would resolve itself into a pure metaphysical question, from which it would be hopeless to expect any practical issue" (Campbell, 1859, p. 360). Campbell's test was a conditional one—he took as given that a favorable event (for example, a male birth) occurs ab times in a years (or an average of b times a year) and that the number of times it could occur (that is, the total number of births in a years) is na (or an average of n per year). The question then was, Is the observed series closer (or farther) from the perfect regularity of a series "$b, b, \ldots, b$" than we should expect? His reference for this judgment was to be the hypothesis that the na events were distributed at random over the a years, the ab favorable cases falling where they may. Campbell wanted, in effect, to look at each single year in isolation and see whether its departure from the most probable number of favorable cases, b, was sufficiently small or large to warrant surprise. The test was based upon the combinatorial probability (the hypergeometric distribution) appropriate to his situation. He did not really carry through his program, although he made some tentative calculations involving data on suicides to illustrate feasibility—without drawing any conclusions. He showed no awareness of a need to allow for the multiplicity of hypotheses (one for each year) being considered.[4]

Campbell's test did not attract much attention, perhaps because it was cumbersome to apply and because it did not emphasize application to the question that came to be the major concern of statisticians. Campbell would have considered a homogeneous random distribution among the years to be an *unremarkable* regularity and not worthy of attention. In fact, he was still operating in a world that accepted stability as the norm, and he had read Buckle (and by inference, Quetelet) as claiming even more—a super regularity, one even greater than could be expected in a sequence of homogeneous independent trials. Campbell did not emphasize the appli-

3. Campbell had read his note at a 20 September 1859 meeting of the British Association at Aberdeen, in whose *Report* an abstract appears. He also published a slightly enlarged version in the 1860 volume of the *Assurance Magazine, and Journal of the Institute of Actuaries.*

4. With a minor difference Campbell's test was anticipated by Bienaymé (1840), who would, however, have divided the series into two parts (rather than consider individual years) and considered the distribution of favorable cases among the two parts. Bienaymé worked no example. Both Bienaymé and Campbell are discussed by Heyde and Seneta (1977, §3.5).

cation of his mathematical calculations to the possibility that was more to the point—that deviations from the average may be larger than allowed for by his random allocation of favorable cases.

The Dispersion of Series

When Lexis turned to the investigation of statistical series in 1876–1879, it was in a more skeptical intellectual climate. Hence, a broader view of potential hypotheses came more naturally. It also came nearly simultaneously in two quarters, to Lexis and to a Frenchman, Émile Dormoy. Although Dormoy's work came first (in 1874), it lacked the foundation of a serious empirical study—a foundation that Lexis's later work had—and it attracted no wide notice until after Lexis's work became known.

Lexis's first publication on statistical series came in 1876 and concerned birth and death statistics. His fullest discussion of his theory, however, was in an 1879 article, "On the theory of the stability of statistical series," published in the *Jahrbücher für Nationalökonomie und Statistik,* the premier German journal of economics and statistics of the time. Lexis wished to compare statistical series (we now call them time series) for different places, for different classes of people, and so on. If the series were "stable" (a word we shall return to), he compared their levels to see whether they differed significantly (using a probabilistic criterion). If they were not stable, he either compared their instabilities in some way or reclassified the data to form stable series. Lexis succeeded in producing a sometimes useful measure of stability and a method of analysis that could be used in those rare cases that passed his test of stability. But at a more fundamental level—as an attempt to provide a full analysis of statistical series—his work was a failure. Nevertheless, the way in which he adapted Quetelet's conceptual approach to a different probability model and used it to explore the application of probability to social data is revealing. Although he achieved only a limited success in addressing one narrow aspect of the problem, the reasons why it was limited had much to do with the directions taken by subsequent work.

Lexis's limited success came in testing what he called the stability of the series. He worried that series of ratios could be "periodic," "undulatory," "evolutionary" (*evolutorische*—a word that would not have been encountered twenty years before), or that in a more stable environment they might be "oscillatory" ("fluctuate entirely incoherently within a certain range"). Within this latter group he proposed as a baseline for stability the "typical" series: "Their individual values are an approximation of a constant underlying value, differing from it only by random deviations" (Lexis, 1879, p. 61). These are vague words, and the short description of the type of application he had in mind was no more clear: "Does the percentage of marriages that are first marriages fluctuate less in Alsace-Lorraine than in France or England? Does it fluctuate more or less than do other population ratios, such as the general marriage rate?" (Lexis, 1879, p. 61). But the technical discussion that followed dispelled much of this vagueness.

In a modernized version of Lexis's notation, we can describe the model he considered as follows: Suppose that in the ith year, for $i = 1, \ldots, n$, there are g cases to be considered—say, g births or g marriages. Of these, X_i are considered favorable—say, X_i male births or X_i first marriages. Then the series under consideration is $Y_1, Y_2, \ldots, Y_n$, where $Y_i = X_i/g$. (Lexis permitted the g's to vary slightly over time but only at the expense of introducing an uncertain element of approximation into the results.) The question was, Is the series of rates $Y_1, Y_2, \ldots, Y_n$ "stable," and/or is one such series "more stable" than another? Lexis defined stability in terms of a binomial sampling model. Suppose the ratio Y_i is viewed as approximating an underlying probability v_i; if the individual events (births or marriages) are independent, we could describe X_i as then having a binomial (g,v_i) distribution:

$$P\,(X_i = k) = \binom{g}{k} v_i^k(1 - v_i)^{g-k}, \qquad k = 0, 1, \ldots, g.$$

Suppose we decide to measure the variability of a series by the "probable error" of the terms in the series, that is, by 0.6745 times the standard deviation.[5] Then the probable error of series of rates could be estimated by

$$R = 0.6745\sqrt{\textstyle\sum(Y_i - \bar{Y})^2/(n-1)},$$

treating the Y's as simple measurements in the manner of the method of least squares.[6] On the other hand, if the series did behave like a stable binomial series with $v_1 = v_2 = \ldots = v_n$, then the calculations appropriate to that model would suggest estimating the probable error (if the common value v were known) by

$$0.6745\sqrt{v(1-v)/g}.$$

5. Recall that the probable error played the role of the standard deviation in much of the nineteenth-century work on the theory of errors. Introduced by Bessel before 1820, it was that multiple of the standard deviation that would correspond to the distance from the mean to a quartile if the distribution was in fact normal: Normally distributed errors are as likely to be within one probable error of the mean as not.

6. Lexis said the estimate of the probable error based on the sum of squared errors was theoretically preferable to that based on the sum of absolute errors, a fact known for the normal distribution since Gauss (1816) (see Stigler, 1973a,b).

Because v must be estimated, we would use

$$r = 0.6745\sqrt{\bar{Y}(1-\bar{Y})/g}.$$

Lexis tested stability by comparing the two different estimates of probable error. He noted that if the series is stable and the variability is in line with the binomial sampling model, then we expect $R \cong r$. On the other hand, if the v_i's vary from year to year (but the binomial sampling model holds within individual years), then we expect

$$R = \sqrt{r^2 + p^2},$$

where p is the probable error of the v_i's:

$$p = 0.6745\sqrt{\sum(v_i - \bar{v})^2/(n-1)}.$$

Lexis gave no source for the formula relating R, r, and p, but it seems based on one that, as we have seen, was known to Poisson and Cournot. Lexis's version would only hold in an asymptotic sense, because it neglected a factor $n/(n-1)$ before r^2 and it treated the Y_i's as identical to the v_i's in describing p. It is clear, however, that he only intended it for large sample use; and his later discussion showed he was aware both of the approximations he was making and of the correct "small sample" formula.

Lexis referred to R as measuring the total fluctuation of the series and r and p as the unessential, or the random (or normal-random, *normal-zufällige*), and physical components, respectively (Lexis, 1879, p. 72).[7] He recalled an analogous situation from his training in mechanics: "The 'unessential' and the 'physical' components are, therefore, related to the observed total fluctuation as a force resulting from two opposing forces meeting at right angles" (Lexis, 1879, p. 66). His comparison was accomplished by calculating the ratio $Q = R/r$. (It is tempting to suppose that Lexis selected the symbol Q in honor of Quetelet.) If Lexis found that (approximately) $Q = 1$ (a situation he referred to as "normal dispersion"), he would conclude that for all practical purposes the physical component was negligible (that is, $p \cong 0$) and the series stable. If he found $Q > 1$ ("supernormal dispersion"), the series was not stable in the sense he used the word—the physical component p made an important contribution to the variability of the series and the rates Y_i could not be regarded as estimating the same, constant underlying probability v. The physical component itself could be estimated by $p = r\sqrt{Q^2 - 1}$, which Lexis noted was approximately $rQ = R$ when Q is large.

What if $Q < 1$ ("subnormal dispersion")? Under the model Lexis used,

7. In later work (for example, Lexis, 1886) he referred to r as the modulus determined by the combinatorial method and R as the modulus determined by the physical method.

with the individual counts X_i binomial and independent, subnormal dispersion was impossible, except to a small degree due to sampling variation. He readily allowed that subnormal dispersion could arise; for example, if a law specified the percentage of the population that must serve in the military at any one time, then the service rate would be "subnormally" disperse, that is, more stable than binomial variation would permit. Lexis claimed that this could only happen when the ratios in the series were subject to normative laws or compensatory reactions, not when the series were of unrelated items (and "series of unrelated items constitute the principal subject of statistical analysis"; Lexis, 1879, p. 71). As Lexis put it, "It would be inconceivable that the maximum stability expressed by the condition $R = r$ would be exceeded in an unrelated series representing a mass phenomenon where we could detect neither a compensatory interrelation nor the effect of an enforced law" (Lexis, 1879, p. 69). Thus to Lexis normal dispersion represented the extreme in stability.

Unknown to Lexis, an analysis and measure of dispersion much like his had been presented earlier by Émile Dormoy, a French actuary. Dormoy presented his *coefficient de divergence* in 1874 as part of a review of probability in the *Journal des Actuaires Francais,* a review he reprinted without change four years later as the first three chapters of his *Théorie mathématique des assurances sur la vie* (1878). Dormoy's coefficient was in most essentials the same as Lexis's. It too was a ratio of two quantities that would, for a stable binomial model, estimate the same quantity. Dormoy, however, used a numerator proportional to $\Sigma|Y_i - \bar{Y}|$ in place of Lexis's R. Dormoy compared his coefficient with 1, as had Lexis, but he lacked Lexis's decomposition of the dispersion and he considered values both larger and smaller as possible. Coefficients smaller than 1 were possible, Dormoy noted, if the series exhibited a positive serial dependence ("if the trials already made influence those which are to be made, in such a way that they make the past result more probable"; Dormoy, 1874, pp. 452–453). On the other hand, he saw values greater than 1 as due to negative serial dependence.

Dormoy, like Lexis, examined birth and marriage series; he also looked at crime, suicide, accidental death, and rainfall data. He found his coefficient to be generally larger than 1, the sole exception being the sex at birth, where he found (as Lexis was to find later) that the stable binomial model was well satisfied in all years, in all countries. He concluded that "such an event depends only on chance" (Dormoy, 1874, p. 455).

Dormoy does not seem to have seen his coefficient as a general tool for statistical investigations; rather (insofar as we can judge from his brief comments) his analyses were intended to explore the limitations to be encountered in applying probability in the real world of insurance. Apparently he never returned to his coefficient. Lexis learned of Dormoy's work by 1886, although only of the 1878 publication, for Lexis still politely claimed priority (Lexis, 1886). No real dispute developed, however, because Dormoy remained silent and Lexis eventually took note of the earlier publication (Lexis, 1903, p. 130; von Bortkiewicz, 1930).

Lexis's Analysis and Interpretation

Lexis had a formal structure that Dormoy lacked, namely, his decomposition of the total fluctuation R into a random component r and a "physical" component p; and he used that structure to explore and analyze statistical series. The method of analysis was somewhat ad hoc, but it showed good sense and a reasonably good intuitive grasp of the behavior of data. We can distill Lexis's sequence of examples and commentary on procedure into a few simple rules: First compute Q. If $Q < \sqrt{2} = 1.41$, the difference from 1 may well be due to error; so either look at other series (gather more data) or take $p = 0$ and treat the series as stable and subject to normal dispersion. If $Q > 1.41$, the physical component can be estimated by $p = r\sqrt{Q^2 - 1}$. The value $\sqrt{2}$ was selected because it is for this value that the two components p and r are equally large. Above this value, p dominates; below it, r does.

In the first case ($p = 0$, or normal dispersion), Lexis would have used the estimate of the probable error for binomial variation, r, to evaluate the significance of different comparisons. For example, he might have used it to test whether the rate of stillbirths was the same for legitimate and illegitimate births. Having once established stability, he would have been able to make similar comparisons of ratios for similar, but less extensive series.

In the second case ($p > 0$, or supernormal dispersion), Lexis would have attempted a further analysis. At the least, he would have compared the values of p for different series; but in some cases he would have tried to reduce the variation in the series to normal dispersion, perhaps through reclassification of the data. One suggestion he made (but does not seem to have followed up on) was to use techniques borrowed from least squares to eliminate externally caused variation. For example, the true probabilities might be linearly related to time, population, or the price of bread:

$$v_t = a + bt,$$

$$v_t = a + b \cdot \text{(population at time } t\text{)}, \qquad \text{or}$$

$$v_t = a + b \cdot \text{(price of bread at time } t\text{)}.$$

Then, Lexis suggested, the series Y_t could be fit to t (or population or price) by least squares. If v_t represented the fitted values, this gives the breakdown[8]

$$\sum(Y_t - \bar{Y})^2 = \sum(v_t - \bar{Y})^2 + \sum(Y_t - v_t)^2.$$

If this equation is multiplied through by $(0.6745)^2/(n-1)$, then "the expression on the left of the equation becomes exactly R^2. The first term of the sum on the right coincides with the square of the probable physical error in our previous cases and the second term on the right equals approximately r^2" (Lexis, 1879, p. 84). He did not suggest using this equation for sifting through possible models in an anticipation of an analysis of variance; rather, he suggested trying to repeat the previous analysis, but for the residuals from the fitted line. For R^2 he used $(0.6745)^2/(n-2)$ times the sum of squared residuals; for r he used the same value as before, supposing that the fitted values did not vary further from 0.500 than 0.350 or 0.650. In this way he hoped to be able to reduce a supernormal series to a linear trend on binomial probabilities. It appears from his comments that it was the computational difficulty that inhibited his application of this approach: "The attempt to find such an equation even in a simple case requires rather laborious work and is of doubtful value."

Lexis realized that the interpretation of his measure Q was made difficult by its sensitivity to sample size. He noted that Q could be so near 1 that the difference could be due to chance alone, and yet p might be large: $Q = 1.1$ might correspond to $p = 0.447r$. He observed that r varied with g; in fact, r decreased toward 0 as g increased, whereas p did not, and so for sufficiently large aggregates (large g) any nonzero p produced a large Q and $R \cong p$. On the other hand, small g sometimes led to an r sufficiently large to mask even moderately large p. Because of these uncertainties, Lexis only expressed confidence in the stability of a series if he found Q near 1 with a large number of observations. Thus, although he lacked a formal procedure for evaluating the significance of his measure Q, he had a vague understanding of the relationship between its power for detecting alternatives to stability and the number of observations available.

Why Lexis Failed

In some respects Lexis's procedure bears an uncanny resemblance to a modern analysis of variance, but this resemblance is quite misleading, in

8. Lexis's notation differed slightly: With V for our $\bar{Y}$ and (v_t) for our Y_t he wrote, suppressing the subscripts, $\Sigma\,[(v) - V]^2 = \Sigma\,(v - V)^2 + \Sigma\,[(v) - v]^2$. In a footnote (which was expanded in the reprinting of Lexis, 1879, in Lexis, 1903, pp. 196–197), he presented this relationship for the full linear model. Formally it was known in the literature of error theory, but Lexis did not cite this literature.

regard to both the way Lexis viewed it himself and the limitations inherent in the procedure that helped keep it from enjoying widespread popularity as a technique for the analysis of social data. To Lexis, the purpose of his measure Q was the same as that of Quetelet's earlier analyses: to establish an objective basis for statistical analysis by demonstrating that the data arose as if from a well-understood probability model. For Quetelet the model had been that of a large number of independent influences, and its validation had been an adequate normal distribution for the observed data. For Lexis the model was a sequence of binomial distributions based upon a common probability, and its validation was a "normal dispersion" such as should arise from such a model.

The entire rationale for Lexis's measure Q was inductive: Since for stable binomial series we should find Q near 1, then if Q was found to be near 1 for a large number of observations, this would justify our treating the data as if they were a stable binomial series. As Lexis wrote, "This information [Q near 1] is sufficient to enable one to state with greater or lesser certainty (according to the value of n) that the ratio v has fundamentally the character of a mathematical probability, and that its empirical values are distributed approximately in the same way as can be derived theoretically from the analogy of a proper game of chance" (Lexis, 1879, p. 74). In 1880 after discussing the dispersion of the sex ratio, Lexis wrote, "One could say that the variations in the empirical ratios are the same as if there was a constant proportion of male and female germs in each ovary over the entire range of observation. I do not claim that things are really thus, but that hypothesis accounts most simply for the observed dispersion in the sex ratios" (Lexis, 1880, p. 487). Lexis's analysis was directed to the primary end of establishing an objective meaning for the average level of the series in such a way that the uncertainty inherent in its determination could be measured against the benchmark of a well-understood urn model, the model of binomial sampling.

Lexis's innovative expansion upon Quetelet's conceptual approach avoided one of Quetelet's difficulties, but at a fatal cost to another. Quetelet had been too successful—he found nearly all distributions to be normal, and he thus lost the war by not being able to isolate homogeneous aggregates in the way he had wished to. As a tool for discrimination, Lexis's Q was far more effective. In fact, it was too effective; and that was its principal limitation and the reason it ultimately proved to be sterile as a general methodology for analyzing social data. For, as Lexis found, very few series were in fact stable.

The insistence upon binomial dispersion as a benchmark for stability had the effect of ruling out most interesting series of any social importance. It is almost literally true that the only series that passed Lexis's test were series that, like the sex ratio, were genetically rather than socially determined. The further Lexis found himself from a purely biological level, the further he was from "normal dispersion." And, what is almost worse, the measure had the perverse property of not being able to discriminate in some of those situations when a judgment of inhomogeneity would be most desired.

The first of these situations involved series where the individual terms themselves were inhomogeneous. Lexis's base measure r assumed that within a single year, say, the variation was binomial, and Q was designed to react to variation from year to year. But if the individual years were not binomial — say, if the marriage rate was based on an aggregate population consisting of two social classes with different propensities to marry — then "subnormal" dispersion could easily arise, notwithstanding Lexis's claim that "interrelation" (dependence) was necessary. Of course, the inhomogeneity within years would have to remain roughly constant from year to year, or between-year inhomogeneity could overcome the "subnormal" effect of within-year inhomogeneity. Thus Lexis was ironically in a situation where inhomogeneity of a type he might have wished to detect actually could mask the effect of the instability to which his measure was most sensitive. Lexis could detect unstable homogeneity, but not stable inhomogeneity. It is doubly ironic that Robert Campbell, who had actually looked for subnormal dispersion as proof of regularity, correctly diagnosed the situation twenty years before, writing, "Suppose that in a table of statistics the numbers were found more uniform than should be expected from the whole number of the community. The hypothesis that such phenomena are confined for the most part to a definite section of the community, might in some cases go far to explain such a uniformity" (Campbell, 1859, p. 365).

A second situation where Lexis's measure was ineffective was found by his own most direct intellectual descendant, Ladislaus von Bortkiewicz, in 1898. Von Bortkiewicz noticed that if the empirical ratios were all small, if the events in question were rare events, then even if the number of observations was quite large a considerable instability in the v_i could be masked by the great variability in the counts. Binomial variation gave way to Poisson variation, and the component r dwarfed the component p as a result of the small size of the v_i. Thus for many interesting series concerned with rare events, the methods of Lexis were useless. Von Bortkiewicz's 1898 pamphlet, "The Law of Small Numbers," is best known today for the example in which he showed that the number of Prussian officers killed by horse kicks between 1875 and 1894, although quite variable from year to year and corps to corps, nonetheless would pass Lexis's test and, incidentally, fit the Poisson distribution quite well (von Bortkiewicz, 1898; Newbold, 1927; Winsor, 1947).

Lexian Dispersion after Lexis

Lexis had made a bold attempt to bring the calculus of probability to bear on social data and end the century-long impasse. In the end he was unsuccessful, and the line of work he began gradually withered as a branch of statistics, surviving only as a source of problems for a growing corps of probabilists seeking areas in which to try their developing methods of analysis. Lexis's only lasting success was as a reaction to those who uncritically treated any empirical ratio as a probability. He demonstrated convincingly that the interesting ratios of empirical research did not have the characteristic variation of binomial sampling. But this success was too complete, and he was left without an analysis to replace the one he helped discard. His estimates of "physical" dispersion, although they permitted a rough descriptive comparison of non-"normal" series, did not lead the way to a really useful analysis. For the vast majority of the series Lexis and his contemporaries wished to treat, binomial variation was simply not relevant. Lexis could reject a simple hypothesis of stability, but he could not follow through on the more important question, How are the series unstable? His discussion showed that he understood the nature of the problem but was unable to provide the needed development.

Von Bortkiewicz fell heir to Lexis's approach, and he carried the torch with a fervent teutonic determination, but to no avail. Von Bortkiewicz understood the limitations to Lexis's scheme; but instead of abandoning it, he sought to save it with increasingly complex models and analyses. He, and later Markov and Chuprov, provided more rigorous derivations to replace those of Lexis and even derived moments and analyzed the asymptotic behavior of Lexis's measure Q. Even though the result of this was sometimes good mathematics, the development of the statistical concepts lagged, to the point where von Bortkiewicz's replacement of Lexis's "physical" component p (the probable error of the underlying v_i's) by the coefficient of variation of the v_i's could be cited as a rare advance. The statistical sterility of the post-Lexis theory is exemplified by the fact that it was not noticed until 1924 that Karl Pearson's chi-squared tables would permit the evaluation of the statistical significance of the departure of Lexis's Q from 1; and this was noticed, not by one of Lexis's intellectual descendants, but by Ronald A. Fisher (Fisher, 1928).

Lexis's models did survive as elegant variations on simple binomial models in probability texts, for example, by Keynes (1921), Coolidge (1925, chap. 4), and Uspensky (1937). A full modern treatment can be found in Heyde and Seneta (1977, chap. 3).

Among his contemporaries, Lexis's approach was most perceptively criticized by Francis Y. Edgeworth. From 1885 on Edgeworth made frequent comments on the general inapplicability of Lexis's methods; but his remarks were polite and sometimes oblique and had little effect upon Lexis or von Bortkiewicz. Those replies he did receive (Lexis, 1886; von Bortkiewicz, 1895) were defensive and largely missed the point of the criticism—that limiting full statistical analyses to those few cases that could be bent to conform to binomial variation was not fruitful. In Edgeworth (1885b), he essentially pointed out that an additive two-way analysis of variance model for the underlying probabilities could destroy the efficiency of Lexis's analysis even when the variation was binomial, but the point was not well-expressed and it was lost on Lexis (1886), and even on Keynes (1921, chap. 32). This work of Edgeworth is discussed later and more fully in Stigler (1978a). Von Bortkiewicz took considerable note of Edgeworth's commentary and analysis (von Bortkiewicz, 1895, 1896) and eventually recognized Edgeworth's own analysis, which he then viewed as an extension of Lexis's theory to nonbinomial data (see von Bortkiewicz, 1909, pp. 468–470).

In the end, the approach of Lexis was swept away by a different conceptual approach, already under development in England by 1885. This new line also had roots in Quetelet's work, but it took a very different point of departure. Even though Lexian dispersion did help Edgeworth focus his ideas, its major impact was its negative proof that at last, a century and three-quarters after Jacob Bernoulli's death, the simple urn models of the early years of probability were insufficiently rich to support the needs of a modern statistical analysis.

7. *Psychophysics as a Counterpoint*

Hermann Ebbinghaus (1850–1909)

Gustav Theodor Fechner (1801–1887)

THE GLACIAL PACE with which statistical methods entered into (and eventually reshaped) the social sciences raises the questions, Could this simply be intellectual inertia? Could the resistance to the quantification of uncertainty in the social sciences be accounted for as part of a resistance to new ideas generally and to measurement in particular? Is the transfer of statistical technology from one area to another *always* a slow and difficult process? The answer to all of these questions is "No," and the field of psychology provides dramatic proof and shows how, when conditions are right, statistics can spread rapidly to a new field—indeed, even help to create the field.

The Personal Equation

Experimental psychology as a discipline dates from the middle of the nineteenth century with the works of Fechner, Helmholtz, and Wundt, but many of the central problems predate the emergence of the formal discipline. One of these problems, the study of reaction times, is of particular interest to us because it shows one way statistical methods can be transferred—as part of a larger transplantation. In fact, the study of reaction times was taken so completely from observational astronomy that even the name used there, the "personal equation," was retained.

The fact that different observers were not equally reliable was long known by practicing astronomers. Some—Flamsteed, Bradley—were celebrated for their accuracy; others were ignored or their observations discounted. But the discovery that different individuals working under exactly the same conditions, in the same weather with the same telescope, could persistently and systematically differ from one another, is usually associated with the British Astronomer Royal Nevil Maskelyne (Sanford, 1888; Boring, 1950, chap. 8). A standard part of every observatory's routine in the late eighteenth century was the measurement of time by the determination of the instant at which stars and planets crossed the meridian. A special telescope would be designated for this purpose, one that turned only in the meridian plane (it turned on an axis pointing east and west) and had an eyepiece equipped with a number (say, five) of equidistant fine parallel wires. The middle wire was set to coincide with the plane of the meridian, and the astronomer on duty focused on a particular star and recorded (as exactly as he was able) the times at which the star crossed each of the five wires. The average of the five times then gave the time when the star crossed the meridian. With practice, variability was reduced, and because of the routine nature of the operation, it came to be shared by different observers, affording an opportunity for interindividual comparisons.

In 1796 Maskelyne noticed that his times and those of his assistant, David Kinnebrook, were systematically at variance. Two years earlier they had been "very well in agreement," but in January 1796 Maskelyne noticed that his assistant was setting times down that were consistently eight-tenths of a second slower than Maskelyne's. Now eight-tenths of a second on the clock translates to 12.8 seconds of arc, an astronomical amount in an exact science: It would correspond to an error in longitude at the equator of only a quarter-mile, but even that would be a most unwelcome addition to the many other factors contributing to problems in the determination of longitude. And despite the fact that both Maskelyne and Kinnebrook carefully reexamined their techniques of observing, searching for any cause or irregularity that could account for the difference, they found no explanation. Maskelyne was faced with a genuine anomaly that did not yield to careful scrutiny, and he dealt with it in much the same way that bacteriologists before Fleming dealt with patches of green mold growing in their laboratory cultures. He fired the hapless Kinnebrook, writing, "I cannot persuade myself that my late assistant continued in the use of this excellent method (Bradley's) of observing, but rather suppose he fell into some irregular and confused method of his own, as I do not see how he could have otherwise committed such gross errors" (Sanford, 1888, p. 8).

Twenty years later, Friedrich Wilhelm Bessel was led to Maskelyne's own account of the episode, at about the same time Bessel was preparing *Fundamenta Astronomiae* (1818) (see Chapter 5). Bessel had in that work shown more interest in the empirical characteristics of errors of observation than any other astronomer, and Maskelyne's discovery piqued his curiosity. Could other pairs of observers have similar discrepancies? From 1819 through 1825 Bessel conducted a series of experiments involving himself, Encke, Walbeck, Argelander, Struve, and Knorre. He found that sizable interastronomer differences were the rule rather than the exception. For example, representing each astronomer by his initial, Bessel (1876) found that

in 1820–21	$B - W = -1.041$ seconds	
	$S - W = -0.242$ second	
and hence	$B - S = -0.799$ second	
while in 1823	$B - A = -1.223$ seconds	
	$S - A = -0.202$ second	
giving	$B - S = -1.021$ seconds.	

Several of these differences were even larger than the one Maskelyne had found, $K - M = -0.8$ second. (Bessel's writing the difference in this manner evidently gave rise to the term *personal equation* for this component of variation.) Others confirmed this phenomenon through their own experiments, and by midcentury the determination of the personal equation of individual observers (and allowance for it in the reduction of observational data) was a standard part of observatory routine.

The experiments went further than the simple determination of a constant difference between astronomers. Bessel himself noted that the personal equation seemed to change over time (an observation consistent with Maskelyne's statements on the change in Kinnebrook's discrepancy between 1794 and 1796), and he and others investigated the effects of quite a variety of changes in observing conditions. For example, Rogers (1869) found that his own times were markedly faster when he was hungry. One consequence of Bessel's discovery was the invention and adoption of the chronograph for recording transit times, a device that reduced the size of but did not eliminate the personal equation.

Throughout this period, the statistical methods of Laplace and Gauss played a key role in the investigation of the personal equation. The design of the telescopic eyepiece (with at least five wires rather than only one) was intended to take advantage of the diminution of error that comes from the combination of observations. And Bessel's own conviction—that the differences he found were real and not merely a manifestation of observational error—depended crucially upon his comparisons of the differences and estimates of the observers' "probable errors," a type of significance test. Incidentally, Bessel himself appears to have coined the term *probable error,* or *der wahrscheinliche Fehler* by 1815 (Walker, 1929, p. 186). By the 1830s and 1840s when physiological psychologists began to take notice of the astronomers' work, the probability calculus was already in place as a standard part of the methodology. Indeed, the conception of the statistical variability inherent in the data was essential to the definition of the effects involved.

Fechner and the Method of Right and Wrong Cases

During the period 1850–1880 the study of the personal equation hovered between astronomy and the emerging field of experimental psychology. By the end of that period, astronomers' interest in the problem was waning, primarily as a result of the development of improved chronoscopes that rendered the difference between individuals' reactions less important. The problem eventually was taken over wholly by psychology, where it came to be known as the method of average error. But the study of the personal equation was not the only problem or even the most important one of the early years of experimental psychology, despite its important role in showing the feasibility of quantitative planned experimentation with human subjects. Rather, the telling methodological breakthroughs that led to the development of experimental psychology as a field in its own right came with the study of lifted weights. The landmark of this development was Gustav Theodor Fechner's 1860 book, *Elemente der Psychophysik.*

Fechner was born in 1801 in Halle, Germany; and in 1817 he entered the University of Leipzig to begin an association that continued until his death in 1887. Fechner seems to have been foreordained to launch a new field—the disciplines he encountered at Leipzig were too narrow to encompass his talents and interests. He studied medicine and earned a baccalaureate in 1822; at about the same time he began writing a series of sometimes mystical philosophical pieces on the identity of mind and matter, a practice that was to last throughout his life. Fechner did not continue to a doctorate and hence did not practice medicine. Instead he turned to physics, lecturing from 1824 on; and in 1831 he published research on the galvanic battery that was of sufficient importance to earn him a professorship of physics in 1834. That 1831 work built on Ohm's pathbreaking 1827 treatise on the subject and bore the hallmark of Fechner's later work. Even though it made no use of probability in its analysis, it was an extensive, painstakingly detailed account of a series of multifactor experiments. Everything that could be varied was varied; everything that could be measured was measured; everything that could be recorded was recorded. And in all this mass of detail (the record of the experimental results alone covers about 200 pages) he did not lose sight of overall objectives.

Biographical accounts of Fechner (Boring, 1950, chap. 14; Adler 1966; Jaynes, 1981) tend to associate his turn in the 1850s to *psychophysics* (a term he apparently coined) with a series of neurotic ailments and personal crises between 1839 and 1851 (including a three-year period of partial blindness). The stories are colorful and dramatic (the headlines might read: 5 October 1850, "Physicist sees souls of flowers;" 22 October 1850, "Idea that stimulation can measure sensation comes to philosopher in bed"). As appealing as such stories are as devices for raising the origin of scientific ideas to the level of heroic myth, they do not seem to be essential to an understanding of Fechner's intellectual development. The urge to experiment, the interest in physics and both mind and body, and an ambition to influence human thought—all the essential ingredients were already in place in the 1820s. The Fechner who by 1855 had begun the extensive experimentation that led to his 1860 *Elemente der Psychophysik* was essentially the same Fechner who had devoted two full years to the study of electrical current in 1829–1831.

If Fechner's psychophysics could be said to have an inverse square law, it was the logarithmic law relating sensation and stimulus (*Reiz*):

$$S = C \log R.$$

This empirical law was developed by Fechner, and he christened it Weber's law after E. H. Weber, who had earlier developed a more limited but related hypothesis. Today this law is frequently referred to as the Weber–Fechner law, although such emphasis on Weber's influence may be misleading because it ignores the fact that Fechner's work was in some respects more closely related to Ohm's early investigations of current. Indeed, Ohm's first paper of 1825 had developed essentially the same empirical formula $V = m\log(1 + x)$ for the relationship between the loss of force of current V and length of wire x (Caneva, 1981).

The Weber–Fechner law may have been the showpiece of the 1860 *Elemente* (and the focus of much controversy), but it was not the major contribution in that work. The most influential intellectual innovation of the book, and the subject of most of it, was Fechner's methodology for measuring sensation. Borrowing from his earlier work on electricity, borrowing from astronomy, borrowing from others' (Weber, Vierordt) techniques of experimentation, and improvising and inventing freely, Fechner developed three methods of measuring sensation that were to constitute the core of experimental psychology for more than half a century.

One of Fechner's methods, a forerunner of modern quantal response that Fechner called the method of right and wrong cases, consisted of presenting an individual with a pair of stimuli, asking for a judgment of which was the greater, and, after several repetitions of stimulus and response, counting the numbers of right and wrong answers. A crude version of the method was used in 1852 by Hegelmaier in an experimental investigation that was primarily concerned with the decay over time of the memory of the lengths of lines (Hegelmaier, 1852). Fechner (1860, p. 62) and Titchener (1915, p. 275) have suggested that the idea for the experiment originated with Hegelmaier's teacher, Vierordt. Regardless, it was Fechner who through two major contributions to the method developed it into a basic tool for experimental psychology. He greatly refined the experimental design and provided the method with a conceptual basis that was rooted in probability and that gave the results a generality Hegelmaier's simple counting of cases had lacked. These two contributions are related, and the relationship is most easily seen in the context of Fechner's main application, experiments involving lifted weights.

Fechner devised an apparatus consisting of two containers, each of weight P. One (and only one) also contained an extra weight, D, making its total weight $P + D$. The subject of the experiment (Fechner himself, in most instances) would successively lift the two containers and judge which seemed heavier. After n such trials, the ratio r/n of right guesses to total trials was calculated. Fechner did not insist that a guess be definite, that is, he allowed the judgment "doubtful." He kept count of the doubtful cases and split them evenly between right and wrong before analyzing the data. This device for handling doubtful cases was to prove controversial twenty years later; we shall return to it.

The discussion of the details of the design and analysis of this experiment is among the most extensive discussions in *Elemente.* Even so, Fechner found he had to make frequent reference to a future (never-to-be-written) book, *Methods of Measurement,* where the full development was to appear. In effect, a major portion of *Elemente* is a handbook on experimental design, the most comprehensive treatment of that topic before R. A. Fisher's 1935 *Design of Experiments.* The discussion ranges from general pronouncements ("If possible, experiments should be carried on from the start according to some prearranged plan, appropriate to a given purpose"; Fechner, 1860, p. 67) to specific warnings ("possibly the time elapsed since sleeping or eating may influence the sensitivities that are to be investigated"; p. 70).

Fechner's basic aim was to investigate how different factors affected the individual's sensitivity, that is, the accuracy of his judgment. There was seemingly no end to the factors that might prove important or worth study. Which hand was used might be influential in determining sensitivity, or, if two hands were used, the order in which they were used could change the outcome. "What would happen if one changed one's grip or the mode of attack or the position of the weights within the containers? Would not the speed of lifting each container, the intertrial interval, the sequence, whether the heavier one is lifted first or second, or the height to which they are lifted bring about differences? Would one still find the same results if trials were run with the order of the standard weights *P* ascending from light to heavy or if one were to run them in reverse order? What is the influence of a fatigued versus a nonfatigued arm? What change occurs in the ratio of right and wrong cases as a function of the magnitude of the comparison weight? And so on" (pp. 67–68). And in addition to external factors such as these, there were internal factors beyond control that varied over time and made long series of trials of dubious value, even if all external factors remained constant (p. 69).

This plethora of potentially influential factors is reminiscent of the litany recited to Quetelet by Keverberg in 1827 (see Chapter 5). But Fechner had control over his experiment in a way Quetelet had not, and he proposed to use this control cleverly and systematically to overcome the seeming obstacles. Fechner was not rendered helpless by the situation; rather he recognized it as the opportunity it was: "In the sensitivity of these methods to experimental conditions affecting constant errors lies a proof of their discriminatory power" (p. 77). He would reduce the influence of some internal variability by confining attention to relatively short series of trials, all made at the same time, and he would attempt to allow for what remained with an explicit factorial design. The principle guiding him in his emphasis on design was that "only those measurements of sensitivity may be considered comparable which can satisfy the assumption of an equal play of chance. This assumption demands strict uniformity of both external and internal conditions at the time of experimentation" (p. 65). To see how he defined sensitivity and how he developed a factorial design and analysis to achieve this equal play of chance, we must examine the probability model he employed.

Fechner defined sensitivity as differential sensitivity, that is, as the increase in sensitivity obtained by adding an increment to the base weight *P*. If the same increment *D* in weights gave the same fraction r/n of right cases for two individuals, they were equally sensitive. If different increments *D* and *D'* were required to produce the same ratio r/n, then the individuals' sensitivities were in inverse ratio to their increments. At this level he had not advanced much beyond Hegelmaier. A pair of experiments could demonstrate (approximately) equal sensitivity or suggest an ordering, but the only prospect for actually measuring the unequal differential sensitivity of two individuals was tedious (and impractical) trial and error. And even if such trial and error was successful, there was no reason to believe that the sensitivity did not depend critically upon the levels of increments; that is, it was by no means clear that if both increments D and D' were increased by 20 percent the ratios r/n for the two individuals would remain equal. What Fechner needed was a way of measuring sensitivity for a single individual that did not depend critically upon the level of increment. "In studying the theory of probability, to which my interest in the development of our methods drove me again and again" (p. 84), Fechner found a way.

Fechner's method was based in the theory of errors, and his measure of sensitivity was inspired directly by the precision constant h of Gauss and Laplace (he even used Gauss's symbol h). Fechner let $\theta = \theta(t)$ represent the error function:

$$\theta = \frac{1}{\sqrt{\pi}} \int_{-t}^{t} e^{-u^2} du = \frac{2}{\sqrt{\pi}} \int_{0}^{t} e^{-u^2} du.$$

Taking $t = hD/2$, where D is the increment in weights and h the individual's sensitivity, Fechner specified the relationship between r/n, h, and D (ignoring chance variation) to be given by

$$\frac{r}{n} = \frac{1 + \theta}{2}$$

or

$$2r/n - 1 = \theta(hD/2).$$

In the terminology of the nineteenth-century theory of errors (where the normal curve with standard deviation $1/\sqrt{2}$ was taken as standard), $\theta(h\Delta)$ was equal to the chance an observer of precision h would make an error of magnitude between $-\Delta$ and Δ; we would write $\theta(h\Delta) = 2\Phi(h\Delta\sqrt{2}) - 1$. So (in modern notation) if Z has a normal (0,1) distribution,

$$\theta\left(\frac{hD}{2}\right) = \mathrm{P}(-hD \leq Z\sqrt{2} \leq hD),$$

$$\frac{1}{2}\left[1 + \theta\left(\frac{hD}{2}\right)\right] = \mathrm{P}(Z\sqrt{2} \leq hD),$$

and Fechner's model was tantamount to assuming what we now call a probit model,

$$\mathrm{E}(r/n) = \mathrm{P}(Z\sqrt{2} \leq hD) = \Phi(hD/\sqrt{2}).$$

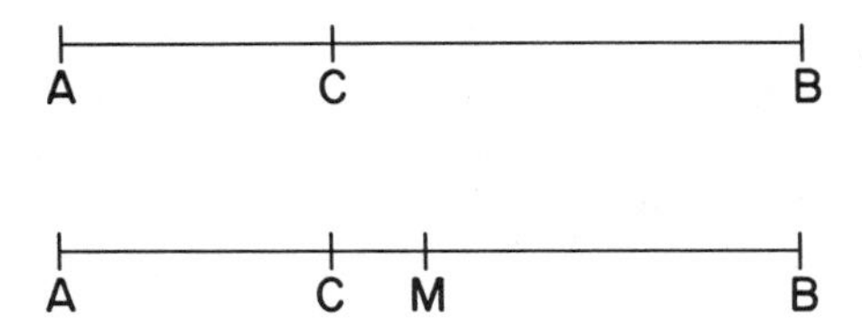

Figure 7.1. An illustration of the 1860 argument Fechner credited to Möbius, justifying the use of the normal error curve in the method of right and wrong cases: Asked to judge whether length AC exceeds CB, the subject forms a mental estimate (making a normally distributed random error) of the distance from C to the midpoint M; if the error exceeds CM, the judgment will be wrong.

This differs from the more general modern model $E(r/n) = P(Z \leq hD + C)$ only in using the information that for $D = 0$ we should have $E(r/n) = 0.5$ and in the use of a differently scaled normal curve as the standard.

Fechner did not simply pull this relationship from thin air; he presented a derivation. The derivation he gave was not his original one (which he termed "somewhat clumsy"), he told us, but "a briefer and more precise" one suggested by his colleague Professor A. F. Möbius (who was clearly a many-sided mathematical talent). Stripped to its essentials, Möbius's argument (which was given in terms of judging the lengths of lines) goes like this: Suppose the subject is presented a line segment AB with an intermediate point C marked and is asked to judge whether AC or CB is longer (Figure 7.1). We may assume that the subject forms an estimate of the distance CM from the point C to the midpoint M of AB and that this estimate errs from the truth by an amount that (from the theory of errors) has a density

$$(h/\sqrt{\pi})e^{-h^2t^2}$$

where h is the precision of the estimate, the individual's sensitivity. If CM, the true distance, is positive, then the judgment will be wrong if it is judged negative, that is, if the error of estimation is below −CM, which happens with probability

$$\frac{h}{\sqrt{\pi}}\int_{-\infty}^{-\mathrm{CM}} e^{-h^2t^2}dt = \frac{1}{\sqrt{\pi}}\int_{-\infty}^{-h\mathrm{CM}} e^{-t^2}dt$$

$$= \frac{1}{2} - \frac{1}{2}\theta(h\mathrm{CM}).$$

Thus the probability of a wrong judgment is $\frac{1}{2} - \frac{1}{2}\theta(h\text{CM})$, and the probability of a right judgment (or the expected fraction of right cases) is $\frac{1}{2} + \frac{1}{2}\theta(h\text{CM})$. In terms of Fechner's lifted weights, $\text{CM} = D/2$, and $\text{E}(r/n) = \frac{1}{2}[1 + \theta(hD/2)]$.

Fechner did not comment upon the implicit assumption that the error distribution was independent of AB and AC (or, for weights, independent of P and D), although he did later provide an extensive empirical investigation of the adequacy of his model.

The practical implementation of Fechner's model was simple enough. He took a table of $\theta(t)$ that Encke had published for astronomers in 1832 (and that was itself derived from Kramp's 1799 table) and recast it as a table relating r/n and hD, giving hD to four places for $r/n = 0.50\ (0.01)\ 1.00$, $r/n = 0.8300\ (0.0025)\ 0.9775$, and $r/n = 0.970\ (0.001)\ 1.000$. The procedure then would be as follows: Given a series of n experiments at the same value of D, record r/n, look up hD in the table (interpolating if necessary), and divide by D to find h. Fechner did not fully develop a method for combining data from experiments with differing D, although he did suggest that an appropriate procedure would calculate separate values of h and then determine "the most probable average of h" from them (p. 159). Müller (1879) later advocated using least squares for this purpose.

Thus far Fechner had a way of determining an individual's sensitivity for a given base weight P and increment D. It remained to test empirically that for a given P the same value of h would be found for different increments D. And it remained to test empirically the Weber–Fechner law: whether (or to what degree) h remained constant as P changed, as long as D/P remained constant. But first, what about the myriad of possibly influential factors that had troubled Fechner earlier? These he sought to deal with through experimental design. Fechner attempted to control for some factors such as practice and fatigue and incorporated others into the analysis.

Fechner explained his factorial design in the context of an example. With himself as subject, he wished to determine his sensitivity when he used different hands to lift the two weights. But he worried about the possible "constant effects" due to whether the heavy weight was in the left-hand container or in the right or to whether he first lifted with his left hand or with his right. He theorized that such a constant effect would be equivalent to a change in the increment: Instead of expecting r/n to be $\frac{1}{2}\ [1 + \theta(hD/2)]$, we should expect the fraction of right cases to be $\frac{1}{2}\{1 + \theta[h(D + M)/2]\}$, "where M is the algebraic sum of all constant influences besides D that also determine the choice of the apparently heavier weight" (p. 93). To permit an analysis he performed the experiment a number of times for each of the four possible combinations of factors and recorded the number of right cases in each instance (see Table 7.1). He

Table 7.1. Fechner's notation for the numbers of right cases for his two-hand lifted weight experiment.

Container with heavier weight	Container lifted first	
	I	II
I	r_1	r_2
II	r_3	r_4

Source: Fechner (1860, pp. 93–94).

then assumed an additive effect model, where p and q represented the constant effects due respectively to order of lifting and location of weight. Specifically, he supposed

$$M = \begin{cases} p + q & \text{I lifted first, weight in I} \\ p - q & \text{I lifted first, weight in II} \\ -p + q & \text{II lifted first, weight in I} \\ -p - q & \text{II lifted first, weight in II.} \end{cases}$$

Fechner would then use his table of $\theta(t)$ to find t_1, t_2, t_3, t_4 from r_1, r_2, r_3, r_4; if n was the number of cases for each combination of factors (so that there were $4n$ cases in all), he thus found t_i so that $r_i/n = \frac{1}{2}[1 + \theta(t_i/2)]$. Then, setting $t = h(D + M)$ in each case, he solved for

$$hD = \frac{t_1 + t_2 + t_3 + t_4}{4},$$

$$hp = \frac{t_1 - t_2 + t_3 - t_4}{4}, \quad \text{and}$$

$$hq = \frac{t_1 + t_2 - t_3 - t_4}{4}$$

whence

$$h = \frac{t_1 + t_2 + t_3 + t_4}{4D}$$

and p and q could be estimated by

$$p = \frac{t_1 - t_2 + t_3 - t_4}{t_1 + t_2 + t_3 + t_4} \cdot D$$

$$q = \frac{t_1 + t_2 - t_3 - t_4}{t_1 + t_2 + t_3 + t_4} \cdot D$$

Fechner did not fully develop his analysis in *Elemente der Psychophysik.* He explicitly deferred many important statistical questions to the unwritten

Table 7.2. Fechner's data, giving the values of r (the number of right cases) for the four main conditions of the two-hand lifted weight experiment ($n = 512$).

	$D = 0.04P$				$D = 0.08P$				
P	r_1	r_2	r_3	r_4	r_1	r_2	r_3	r_4	Sum ($n = 4096$)
300	328	304	328	266	404	358	372	300	2,660
500	352	274	321	288	399	339	364	306	2,643
1,000	334	318	335	309	377	365	410	338	2,796
1,500	346	323	308	344	408	402	399	383	2,913
2,000	296	365	309	373	404	385	439	398	2,969
3,000	244	393	265	433	392	447	390	428	2,992
Sum	1,900	1,977	1,866	2,013	2,384	2,296	2,374	2,153	16,973

Source: Fechner (1860, p. 157).

Note: Fechner counted each actual case as two so that doubtful judgments could be split between right and wrong cases without introducing fractions. Thus the given n's are twice the number of actual trials, and each right case counted for two. These data were based on series of 256 experiments, in each of which two weights were lifted (thus $n = 2 \times 256 = 512$ weights lifted in all). In general, Fechner preferred to analyze data in series of 32 experiments ($n = 64$) to reduce temporal variation in the effects p and q (Fechner, 1860, p. 161).

sequel, *Methods of Measurement.* These included the choice of the magnitude of D ("which one should properly make neither too small nor too large")[1] and "the testing of the significance of the results" (p. 100). But he did present enough detail in his examples so that we can gain a fair idea of how he would have proceeded in general.

Fechner would have focused first on the main effect, h, relying upon his balanced factorial design to cancel out other effects. He would have converted the fraction of right cases to the t scale and estimated hD as described earlier. He would then have looked to see how this quantity varied with such factors as the size of the base weight P. Since D was specified as a constant fraction of P, the Weber–Fechner law implied that hD should remain constant. As the data in Tables 7.2 and 7.3 show, this was not the

1. The choice of D is a nonlinear design problem. In one formulation, rescaling h by $\sqrt{2}$, it amounts to choosing D to minimize the asymptotic variance of the estimate of the sensitivity h,

$$\frac{\Phi(hD)[1 - \Phi(hD)]}{D^2\phi(hD)^2}$$

This is minimized when $hD = 1.575$ or when the expected fraction of right cases is $0.94 = \Phi(1.575)$. This optimum $D = 1.575\ h^{-1}$ of course depends on the unknown h, and the variance rises sharply when too large a D is used. Fechner seems to have arranged his experiments so that the fraction of right cases would be expected to be in the range 0.65–0.85. In later work he suggested 0.84 as a goal (Fechner, 1882, p. 65).

Table 7.3. Values of t ($\times 10^4$) derived from Table 7.2 for the two-handed series ($n = 512$).

	$D = 0.04P$					$D = 0.08P$					
P	t_1	t_2	t_3	t_4	Sum $4hD$	t_1	t_2	t_3	t_4	Sum $4hD$	Total $8hD$
300	2,547	1,677	2,547	346	7,117	5,679	3,692	4,260	1,535	15,166	22,283
500	3,456	624	2,290	1,112	7,482	5,444	2,958	3,932	1,749	14,083	21,565
1,000	2,769	2,181	2,807	1,856	9,613	4,469	3,973	5,971	2,920	17,333	26,946
1,500	3,224	2,363	1,820	3,147	10,554	5,873	5,584	5,444	4,726	21,627	32,181
2,000	1,394	3,973	1,856	4,301	11,524	5,679	4,813	7,558	5,397	23,447	34,971
3,000	−416	5,168	312	7,200	12,264	5,123	8,067	5,034	6,915	25,139	37,403
Sum	12,974	15,986	11,632	17,962	58,554	32,267	29,087	32,199	23,242	116,795	175,349

Source: Fechner (1860, p. 158).

Table 7.4. Fechner's estimated effects of lifting time, as they varied according to order lifted (p = effect due to being lifted first) and location of heavier weight (q = effect due to heavier weight being in container I).

Lifting time (seconds)	Left		Right	
	p	q	p	q
½	6.73	− 3.17	31.49	6.28
1	13.07	−19.46	43.38	3.30
2	12.38	−16.00	38.05	0.36
4	− 7.95	− 3.28	3.43	6.04

Source: Fechner (1860, p. 254).
Note: Fechner lifted both containers with the same hand, and alternated hands.

case—hD increased with P, although the rate of increase diminished as P increased. Fechner finessed this neatly by recalling that the weight of the experimenter's arm should properly be included with that of the base weight P. That is, hD should stay constant only when D increased as a constant fraction of $P + A$. Because A remained constant as P increased, $hD \propto hP = h(P + A) \cdot P/(P + A)$ should increase in proportion to $P/(P + A)$. The effective weight A of the arm was not known, although it could in principle be determined from the data. Fechner did not carry through that analysis,[2] however, being content to note a plausible source of the discrepancy and to observe with satisfaction that for each P the value of hD for $D = 0.08P$ was about twice that for $D = 0.04P$, a finding showing that his model gave results of the right order of magnitude, at least for constant P.

Fechner deferred the full discussion of these data (including the estimation of effects p and q) to *Methods of Measurement.* Indeed, he did present estimates of these effects later, for another experiment testing the relationship between sensitivity and the length of time the weights were held aloft. Working with a base weight of $P = 1000$ grams, experimenting separately for each of his two hands, and apparently averaging results for $D = 0.04P$ and $0.08P$, he found the effects given in Table 7.4. He then interpreted these as additive effects as follows:

> We see from the table that the first container when lifted with the left hand and held for two seconds, appeared 12.38 gm heavier than the one that was lifted second, but when it was held for four seconds it seemed to be 7.95 gm

2. Possibly he discovered that this single correction factor was not enough to do the job. He later discussed more complicated models for incorporating arm weight (Fechner, 1860, p. 164).

> lighter. Lifting with the right hand resulted in changes in the same direction, without going as far as a reversal, however. (Fechner, 1860, p. 254)

Not long after this experiment, Fechner was at least temporarily forced to abandon experiments with lifted weights because he began to experience pain "in the region of the spleen."

Fechner's experimental technique was highly developed for the time, and even a modern critical eye can find little to quarrel with. One major refinement that is missing is randomization. Fechner was extremely systematic in his approach to varying conditions, carefully striving for balance and attempting to diminish temporal effects through the alternation of factor levels and the use of relatively short sessions as units of analysis (thus preserving balance over short as well as long spans of time). The use of careful mathematically rigorous randomization to attain these goals while eliminating other sources of bias was introduced a quarter-century later, by C. S. Peirce in his experiments with lifted weights with J. Jastrow (Peirce and Jastrow, 1885; Stigler, 1978c). Fechner also may have underrated potential difficulties resulting from the fact that the experiments were not performed blindly: He knew which container held the heavier weight. He was not oblivious to this potential difficulty, but he claimed that objectivity and ignorance of the effects of other factors (time order, position) were enough to overcome the problem. In addition, a nonblind experiment did not require an assistant. Peirce and Jastrow also improved upon Fechner in this respect, taking great pains to ensure blindness in all aspects of their experiment.

Fechner's methodology was vastly influential in the following half-century, and for the most part it was adopted with little or no change. Aside from Peirce's introduction of randomization and blindness, only one aspect of Fechner's approach attracted much criticism. From 1879 on, Fechner's practice of dividing doubtful cases between right and wrong excited major debates, both as to what was the correct practice and how sensitivity was to be defined and interpreted; Titchener (1915, pp. 275–318) has outlined the main points of contention. Briefly, the main possibilities were seen to be these:

(1) Follow Fechner's approach and divide the doubtful cases between right and wrong.
(2) Calculate two values of h and average them: one with all doubtful cases treated as right, and one with all treated as wrong. This method was suggested by G. E. Müller in 1879 and later.
(3) Insist that no doubtful answers be allowed, so that all answers became right or wrong. This was called the method of right and wrong

answers and was preferred by Peirce and Jastrow (among others) but scathingly denounced by Titchener (1915, p. 290) as forcing false statements upon the subject.

Curiously, Fechner did not indicate in *Elemente* the frequency of doubtful cases, so we cannot easily judge the severity of the problem. In a later work he gave data in which 10 percent of the cases were doubtful (Fechner, 1882, p. 52, quoted by Titchener, 1915, p. 283). None of the methods was given universal assent (although Müller's system seems to have had a slight edge). Indeed, it is difficult to see how the matter could have been convincingly resolved without a greater consensus on the behavior of sensory processes than was available at the time.

Ebbinghaus and Memory

At a stroke Fechner had created a methodology for a new quantitative psychology. The ingredients were borrowed and the scope of application was limited, but the effect was unmistakable: After Fechner the concepts of the calculus of probability were fundamental to experimental psychology. Probability was not just a tool, for in all of Fechner's methods — and particularly in the methods of right and wrong cases and of average error — it provided the conceptual basis for the very definitions of the quantities measured. And yet something was missing. Even though Fechner could exploit probability explicitly at one level — with his formulation of mathematical models — and could invoke it implicitly at another — with his careful attention to experimental design — he did not take the further step of adapting it to measure the uncertainty in the constants he derived.

It is clear from Fechner's discussions that this next step was not far away. His use of design to permit "an equal play of chance," his factorial analyses, and his insistent control of whatever exterior conditions could be controlled were clearly aimed at reducing the uncertainty of his estimates of sensitivity. He was acutely aware of the need to put the differences that he found to the test, to convince the reader that the constants he found were indeed constants, that the differences he found were indeed differences. Even though he failed to provide a formal apparatus for dealing with those questions, Fechner did indicate plans to provide "practical hints" regarding "the testing of the significance of the results" in the *Methods of Measurement* (Fechner, 1860, p. 100). In the *Elemente*, however, he relied upon informal assessment; and there is no firm indication that he had more in mind.

Although Fechner did not take this next step of measuring the uncertainty in his experimental results, later workers did. A most dramatic and instructive example can be found in the work on memory of Hermann Ebbinghaus—dramatic because Ebbinghaus dealt with higher mental processes (and thus was even further removed from astronomical reaction times than was Fechner), instructive because Ebbinghaus's approach bore a remarkable similarity to Quetelet's.

Hermann Ebbinghaus (1850–1909) had studied philosophy at universities in Halle, Berlin, and Bonn, but it was only after 1875 that experimental psychology caught his attention. From 1875 to 1878 he traveled in France and England, studying and tutoring. While in Paris he chanced upon a copy of Fechner's *Elemente* in a shop selling second hand books (Boring, 1950, p. 387). Upon his return to Germany, he undertook an extensive quantitative study of memory. The experiments started in 1879 and continued through 1885, by which time Ebbinghaus held a post at the University of Berlin. In 1885 his researches culminated in a short and extremely influential book, *Über das Gedächtnis* (translated in 1913 as *Memory*).

Ebbinghaus borrowed two crucial ideas from Fechner: the belief that quantitative study was the only way to give precise expression to the vague notions that had gone before (British associationist psychology, in Ebbinghaus's case) and a reliance upon careful experimental design as fundamental to meaningful experimentation. Ebbinghaus imitated Fechner in performing his experiments upon a single individual (himself) as a means of eliminating large differences between individuals. His analysis was less ambitious than Fechner's—he did not perform factorial experiments and he did not rely upon the fitting of nonlinear response functions to give his measurements validity. What Ebbinghaus did do carefully and self-consciously was to take the single step Fechner had not taken.

Ebbinghaus proposed to adapt the methods of natural science to the study of memory. He was acutely aware of the fundamental difficulties he faced:

> In the first place, how are we to keep even approximately constant the bewildering mass of causal conditions which, in so far as they are of mental nature, almost completely elude our control, and which, moreover, are subject to endless and incessant change? In the second place, by what possible means are we to measure numerically the mental processes which flit by so quickly and which on introspection are so hard to analyze? (Ebbinghaus, 1885, pp. 5–6)

The first was the problem that had long bedeviled social scientists, the problem Keverberg had pointed out forcefully to Quetelet in 1827. And indeed, Ebbinghaus's solution was to have much in common with the one Quetelet had groped for, with the all-important difference that Ebbinghaus had control over *his* experimental material. The second problem—how to measure such an elusive "object" as memory—was new.[3]

Ebbinghaus proposed to measure memory by exhaustively testing himself on his ability to recall series of nonsense syllables taken from a list of 2,300 he had devised for this purpose. For example, he timed how long it took him to learn eight series of thirteen syllables each, syllables such as *zat, bok, sid* (Boring, 1950, p. 383). He repeated that particular experiment ninety-two times in 1879–80, with an average time of 1,112 seconds, or 18.5 minutes (Ebbinghaus, 1885, p. 35). Similarly, he tested long-term memory by determining how much time was needed to relearn a series twenty-four hours after he had first learned it. Ebbinghaus was aware, however, that not all measurements are meaningful, that in many ways the more difficult of the two questions he posed was the first. He wondered,

> Can I bring under my control the inevitably and ever fluctuating circumstances and equalise them to such an extent that the constancy presumably existent in the causal relations in question becomes visible and palpable to me? (Ebbinghaus, 1885, p. 11)

And how could he show whether or not constancy has been attained?

> When, however, we have actually obtained in such manner the greatest possible constancy of conditions attainable by us, how are we to know whether this is sufficient for our purpose? When are the circumstances, which will certainly offer differences enough to keen observation, sufficiently constant? (Ebbinghaus, 1885, p. 12)

To answer these questions, Ebbinghaus in effect retreated to the formulations of Quetelet (although he nowhere cited Quetelet): The test of constancy must be empirical. It is too much to ask that the constancy in experimental averages be absolute or even that deviations be negligible—that "is not necessarily obtained even by the natural sciences" (p. 13). More than stability in large aggregates of experimental data is needed. Ebbinghaus followed Quetelet and others in distinguishing between "constants of natural science" and "statistical constants," although like those others the distinction was blurred and somewhat circular. The constants of the natural sciences were ideally "produced by a combination of causes exactly alike," although "the individual values come out somewhat differently because a certain number of those causes do not always join the combination with exactly the same values" (p. 13). Statistical constants, on the other hand, are averages of separate effects that "arise, rather, from an oftimes inextricable multiplicity of causal combinations of very different

3. Or so Ebbinghaus thought at the time. At least one earlier investigation had been carried out, a fragmentary one in 1876 by the American physicist Francis E. Nipher. See Stigler (1978d).

sorts" (p. 14). The observed constancy in large aggregates was due to the circumstance "that in equal and tolerably large intervals of time or extents of space the separate causal combinations will be realised with approximately equal frequency . . . a peculiar and marvellous arrangement of nature" (p. 14). The intellectual content of Ebbinghaus's distinction was one of degree: With constants of natural sciences the variation of causes was slight, within narrow bands, and compensation was achieved within relatively short series. With statistical constants the variation was large and unpredictable, and only long series could produce the desired compensation. But how was he to test whether or not the first of these situations held for his memory experiments so that the averages of his relatively short series could be meaningful and not merely reflective of unmeasured haphazard variations in underlying causal mechanisms?

Quetelet's answer to this question was also Ebbinghaus's, although Ebbinghaus articulated it more clearly. The answer involved the distribution of the individual measurements. If they followed the "Law of Errors" (the normal distribution), this was prima facie evidence (although not certain proof) that the measurements were of a kind with those made of constants of the natural sciences, and analyses proper to that domain could then be applied. The distribution would validate the average. As Ebbinghaus put it:

> I examine the distribution of the separate numbers represented in an average value. If it corresponds to the distribution found everywhere in natural science, where repeated observation of the same occurrence furnishes different separate values, I suppose—tentatively again—that the repeatedly examined psychical process in question occurred each time under conditions sufficiently similar for our purposes. (Ebbinghaus, 1885, pp. 19–20)

Ebbinghaus recognized that the inference was not a certain one, that the law of errors could in principle be produced by nature "in a more complicated way" than from constant causes. But he thought such a possibility remote and even went so far as to make the unsubstantiated claim that "among all the groups of numbers which in statistics are usually condensed into mean values not one has as yet been found which originated without question from a number of causal systems and also exhibited the arrangement summarized by the 'law of errors'" (p. 18). Ebbinghaus trusted to careful design and repeated experimentation to rule out such unlikely occurrences.

Ebbinghaus's method of testing the closeness of his data to the law of error was a simple one: He computed the average and the "probable error" of the data. He did not explain how he computed the probable error (p.e.), although he expressed a preference for "a theoretically based calculation" over "simple enumeration" (p. 20), a statement suggesting

Table 7.5. Ebbinghaus's table of the theoretical law of error (normal distribution), given cumulatively and based on 1,000 measures.

Within the limits	Number of separate measures
$\pm \frac{1}{10}$ p.e.	54
$\pm \frac{1}{6}$ p.e.	89.5
$\pm \frac{1}{4}$ p.e.	134
$\pm \frac{1}{2}$ p.e.	264
$\pm$ p.e.	500
$\pm 1\frac{1}{2}$ p.e.	688
± 2 p.e.	823
$\pm 2\frac{1}{2}$ p.e.	908
± 3 p.e.	957
± 4 p.e.	993

Source: Ebbinghaus (1885, p. 20).

that he used $0.6745 \cdot [\Sigma(X_i - \overline{X})^2/(n - 1)]^{1/2}$ rather than the median of the $|X_i - \overline{X}|$. His comparison was based upon a short table of the normal distribution (see Table 7.5) giving the chances that a normally distributed quantity would be within $\frac{1}{10}$ p.e., $\frac{1}{6}$ p.e., . . . , $\frac{1}{2}$ p.e., 1 p.e., $1\frac{1}{2}$ p.e., and so on of the mean. From this table he constructed a comparison table (see Table 7.6) giving the actual cumulative distribution of the absolute residuals side by side with a list of the expected cumulative counts calculated from Table 7.5. (For example, to get the calculated count within $\frac{1}{10}$ p.e., $(0.054) \times 84 = 4.5$.) His test was subjective: He simply looked at the two columns and (in all cases that he reported) noted the close agreement, commenting briefly upon the minor discrepancies he found. He checked symmetry by giving the counts for negative and positive deviations separately (Table 7.6).

Ebbinghaus and Quetelet differed slightly in their visual assessment of the fit: Quetelet compared *relative* frequencies, whereas Ebbinghaus compared frequencies. The comparison of frequencies also was used in some earlier anthropometric comparisons (for example, Elliott, 1863; Gould, 1869, p. 251) and has the virtue of permitting an informal subjective allowance for sample size.

Ebbinghaus made many other comparisons but did not report all of them in detail. Of particular interest to us is the fact that his choice of the time to learn *several* series of syllables (rather than a single series) as a unit of analysis was based upon a check of the distributions. That is, the example of Table 7.6 was based upon the times needed to learn groups of *six* series of sixteen syllables each rather than upon *single* series of sixteen

Table 7.6. An example of Ebbinghaus's test of fit to a normal distribution.

		Test of fit		Check of symmetry	
Within the limits	Corresponding deviation	Number of deviations by actual count	Number of deviations calculated from theory	Number of deviations above average	Number of deviations below average
$\frac{1}{10}$ p.e.	±4	4	4.5	—	—
$\frac{1}{6}$ p.e.	±8	7	7.6	5	2
$\frac{1}{4}$ p.e.	±12	12	11.3	7	5
$\frac{1}{2}$ p.e.	±24	23	22.2	13	10
1 p.e.	±48	44	42.0	20	24
$1\frac{1}{2}$ p.e.	±72	57	57.8	28	29
2 p.e.	±96	68	69.0	34	34
$2\frac{1}{2}$ p.e.	±121	75	76.0	37	38
3 p.e.	±145	81	80.0	40	41

Source: Ebbinghaus (1885, p. 37).
Note: These data were based on 84 tests made in 1883–84. Each test recorded a single time needed to memorize six series of 16 nonsense syllables. The average of the 84 times was 1,261 seconds, the probable error (or p.e. $\equiv 0.6745\sigma$) was 48.4 seconds. The deviations were from the average.

each. Because he learned the syllables in groups of sixteen in any case, why aggregate in this way? Why be content with eighty-four units of analysis instead of $84 \times 6 = 504$ units? The reason is that when Ebbinghaus looked at the distribution of the separate single series he found marked departures from the normal curve. The distribution was quite skewed: "The distribution of the arithmetical values above their mean is considerably looser and extends farther than below the mean" (p. 41), the maximum positive deviation being about twice the maximum negative deviation and the mean exceeding the mode and hence producing many more negative deviations than positive ones. Furthermore, the distribution was not smooth but showed several maxima and minima. He also compared the probable error he had computed for the single series with that he had found for the grouped series. For the data of Table 7.6 he found (pp. 44–45) (p.e. based on 504 single series) $= 1.53 \times$ (p.e. based on 84 grouped series), where with independence the 1.53 should be $\sqrt{6} = 2.45$. Thus he effectively found a negative serial correlation between single series (he called it "a kind of periodical oscillation of mental receptivity or attention").

This highlights a curious and revealing limitation to Ebbinghaus's methodology. He validated his use of means by demonstrating that the quantities he had averaged were normally distributed, and he effectively created that normal distribution by using units of analysis that were themselves aggregated sufficiently to provide the needed normal distribution! He justified this procedure by arguing that even with measurements of the constants of the natural sciences some combination of cases is needed to allow compensation for the slight fluctuations of the separate causes (p. 13). But he did not provide any argument that so-called statistical constants could not be similarly bent to fulfil his condition for validation, and he did not observe that the aggregation of groups of series had no essential effect at all upon the averages he had compared, only upon the probable errors he had attached to those averages. He also did not check his eighty-four groups of series themselves for serial correlation.

Ebbinghaus was effectively following the rationale of Quetelet's program, but with this difference: Instead of using the check of normality as proof that a group was homogeneous, and thus using the test of fit to determine when the net had been cast too wide and further subdivision needed, Ebbinghaus used it as a guide to when the aggregation had been insufficient, to when individual, nonnormally distributed causes had been sufficiently smoothed through random compensation to be neglected. From today's perspective neither procedure has compelling merit. Long series of measurements such as Ebbinghaus considered may, if independent, be as validly analyzed separately as in groups. The normality of the aggregates does not contribute to validity. And if they are not independent, aggregation will be useful only when the groups thus formed are less dependent than the single terms, and this was not demonstrated by Ebbinghaus.

Once he had validated his analyses through his distributional check, Ebbinghaus proceeded with few reservations. Nearly every average (and he presented many) was accompanied by an estimate of its probable error, and the subsequent comparisons were performed with this as a yardstick: "It [a difference of averages] attains a value 6 times the probable error. Its existence, therefore, must be considered to be fully proved although naturally we cannot be so sure that its size is exactly what it was found to be in the experiments" (pp. 106–107). "The deviations of the calculated values [of a hypothetical curve describing rate of forgetting][4] surpass the probable limits of error only at the second and fourth values" (p. 78), and so forth. He gave no firm rule, but he seems to have considered a difference

4. The curve was $b = 100K/[(\log t)^c + K]$, where t = time, b = a measure of the amount remembered, and c and K are constants. Ebbinghaus determined c and K crudely, "not involving exact calculation by the method of least squares" (p. 77). Curiously, in 1876 Nipher had verbally described a curve of the decay of memory over time as a "logarithmic one," although he may have had a different relationship in mind (Stigler, 1978d).

of twice the probable error as worthy of note and six times the probable error as proof certain.

It is clear that Ebbinghaus had a keen statistical mind. His inventive comparisons, his critical and careful attention to all manner of potential and actual biases, his awareness of the potential effect of dependence upon his estimates of probable error (as noted earlier and in a footnote on p. 67), his care in design and reporting of details: all of these show a first-rate intelligence at work. But for our purposes the rationale he presented for his methods is the most important aspect of the study. It is revealing both for the form it took and because he saw the need for such a rationale.

Others had not provided such an argument—and others had not taken the step he did. Even though Fechner had founded the field upon probabilistic principles, he had held back from clearly applying them to measure uncertainty. Ebbinghaus had developed a clear argument to justify treating his data sets as sufficiently immune to extraneous uncontrolled causes to merit analysis. Like Quetelet, Ebbinghaus based his rationale upon an inverse to the central limit theorem, upon a belief that a normal distribution necessarily implied that the data formed a homogeneous whole suitable for analysis.

With Ebbinghaus the transfer of at least the core of statistical methodology from astronomy to psychology was accomplished—a scant twenty-five years after Fechner's book. The methods were to receive further development, of course, but this core did not change greatly over the next several decades. The normal distribution continued to play a central role (Boring, 1920). Urban (1910) considered replacing Fechner's "psychometric function" Φ by the cumulative distribution function for the Cauchy distribution [$0.5 - \arctan (ax + b)/\pi$], but an empirical investigation convinced him that Φ gave the better fit. The self-fulfilling validation of analyses by producing normally distributed units of analysis continues to the present day, with the use of normal scaling for test scores.

PART THREE

A Breakthrough in Studies of Heredity

8. The English Breakthrough: Galton

Francis Galton (1822–1911)

THE KEY to the rapid spread of statistical methods to experimental psychology had been the possibility of experimental design, the control of experimental conditions. In the social sciences, where this possibility apparently did not exist, the successful use of probability-based statistical methods did not come quickly; indeed, it may be argued that the development is far from complete today. But beginning in the 1880s there was a notable change in the intellectual climate, as a series of remarkable men constructed an empirical and conceptual methodology that provided a surrogate for experimental control and in effect dissipated the fog that had impeded progress for a century.

Galton, Edgeworth, Pearson

The three principal contributors to this clearing of the air were Francis Galton, Francis Ysidro Edgeworth, and Karl Pearson. Separately, they would rate mention as important figures in the very different fields of anthropology, economics, and the philosophy of science; together they helped create a statistical revolution. Their talents were so different as to be almost incommensurable, yet without the work of any one of them it is doubtful that the effect would have been one-tenth so great.

Of the three, Galton was the idea man, a highly imaginative thinker with considerable energy and driving curiosity. Yet despite a high degree of conceptual insight, earned through years of determined effort, Galton was mathematically backward and unable to extract and develop the full fruit of his own ideas. Edgeworth was the subtle theorist who, perhaps alone among Galton's audience, saw through the vagaries of Galton's prose and was able to translate the ideas to a generalizable mathematical form, eventually permitting a much broader and more fruitful application than Galton had envisaged. Yet Edgeworth too was deficient—his own exposition was incapable of reaching a general audience, and he lacked the will and the interest to build a broad empirical base to prove the methods' worth. It was Karl Pearson, a man with an unquenchable ambition for scholarly recognition and the kind of drive and determination that had taken Hannibal over the Alps and Marco Polo to China, who recognized the power in Edgeworth's formulations of Galton's ideas. Pearson lacked Galton's originality and Edgeworth's depth of understanding, but it was his zeal, with a vital assist from G. Udny Yule, that created the methodology and sold it to the world.

Francis Galton (born 16 February 1822, died 17 January 1911) is a romantic figure in the history of statistics, perhaps the last of the gentleman scientists. Galton studied medicine at Cambridge, where he performed competently and without great distinction. After receiving his inheritance, however, he was able to abandon a medical career, the prospect of which had never really attracted him. Instead, Galton set out to see the world and, in the process, developed a consuming curiosity. The leisure provided by the life of a gentleman did not stifle his energy, rather it nurtured it. During the period 1850–1852 he explored Africa (the major frontier of that era), and in 1853 he received the gold medal from the Royal Geographical Society in recognition of his achievements. In the early 1860s he turned to meteorology and the first signs of his statistical interests and abilities emerged. He circulated a questionnaire to all the weather stations in Europe, asking them to keep a detailed record after his design of all aspects of weather for the month of December 1861. Using the returns Galton constructed intricate weather maps of his own invention and by visual inspection was able to discover the existence of "anticyclones"—counterclockwise movements of air that accompanied sudden changes in pressure. His maps employed ingenious symbols that showed simultaneously the wind direction, temperature, and barometric pressure at each weather station; and Galton's extraction of an unexpected pattern from the chaotic array of data was a remarkable early triumph of the use of graphical methods for the analysis of multivariate data. The results were published in 1863 as *Meteorographica*, and Galton included a cautionary note to those not so carefully self-critical as he: "Exercising the right of occasional suppression and slight modification, it is truly absurd to see how plastic a limited number of observations become, in the hands of men with preconceived ideas" (Galton, 1863, p. 5).

Galton's later interests and publications ranged over psychology, anthropology, sociology, education, and fingerprints, but the dominant theme in his work from 1865 on was the study of heredity. Galton came from a remarkable family. His grandfather was Erasmus Darwin, noted "physician, physiologist, and poet" (Galton, 1869, p. 209), and Charles Darwin was his first cousin. It is impossible to determine whether Galton's own experiences and family or the intellectual content of Charles Darwin's 1859 *Origin of the Species* weighed more heavily in determining his main interest. A recent book by MacKenzie (1981) has emphasized the role of social factors in setting the course of Galton's work (see also Cowan, 1972, 1977). No doubt there is some truth to this position, although it also seems impossible to deny that Darwin's theories opened an intellectual continent more promising and attractive to the explorer Galton than Africa had been.

Galton's Hereditary Genius and the Statistical Scale

The distinguishing feature of Galton's studies in heredity was their statistical nature. Their resemblance to traditional statistical techniques is only superficial, however. He did compile and he did tabulate, but it was his analyses and his conceptual approaches to his compilations and tables that are most interesting to us. As a statistician, Galton was a direct descendant of Quetelet. Nevertheless, even in his first major work on heredity, *Hereditary Genius* (1869), Galton broke away from Quetelet in major respects. In *Hereditary Genius,* Galton sought to show, in effect, that talent runs in families. The book was only moderately successful, perhaps because he failed in that first attempt to formulate that premise in ways that were both surprising and convincing. Nevertheless, the work still holds the key to his later formulation, which was both.

The larger part of *Hereditary Genius* is given over to long lists of famous people and their famous relatives: statesmen, scientists, poets, divines, even oarsmen and wrestlers. Galton looked for famous families and he found them: As a statistical investigation it was naive and flawed, and Galton seems to have realized this. The book is nonetheless remarkable for his adaptation of Quetelet's methodology as a device for measuring talent and the first glimmerings of ideas of regression that he was to develop over the following twenty years.

In his autobiography Galton tells us that he first came upon Quetelet's method of fitting normal curves to data in 1863 (Galton, 1908, p. 304). The excitement of that encounter had not worn off by 1869. Galton not only gave a detailed presentation of Quetelet's method in an appendix, using data taken from Quetelet (1849), he also reprinted two other examples and added one of his own (on examination scores, p. 33). He was fascinated by the appearance of "the very curious theoretical law of 'deviation from an average' " in so many different cases, for heights and chest measurements and even such measures of talent as examination scores. Following Quetelet, he proposed that the conformity of the data to this characteristic curve was to be a sort of test of the appropriateness of classifying the data together in one group; or rather, the nonappearance of this curve was indicative that the data should not be treated together. In *Hereditary Genius* he illustrated the shape of the curve with a dot diagram, each dot representing the height of a man (Figure 8.1). Galton wrote:

> It clearly would not be proper to combine the heights of men belonging to two dissimilar races, in the expectation that the compound results would be governed by the same constants. A union of two dissimilar systems of dots would produce the same kind of confusion as if half the bullets fired at a target had been directed to one mark, and the other half to another mark. Nay, an examination of the dots would show to a person, ignorant of what had occurred, that such had been the case, and it would be possible, by aid of the law, to disentangle two or any moderate number of superimposed series of marks. The law may, therefore, be used as a most trustworthy criterion, whether or no the events of which an average has been taken, are due to the same or to dissimilar classes of conditions. (Galton, 1869, p. 29)

This idea was to recur in Galton's work, although by 1875 he had devised a different way of displaying the data. He ordered the data in increasing order and, effectively, graphed the data values versus the ranks. Figure 8.2 illustrates the ideal curve that the data would, if homogeneous, approximate. Borrowing a term from architecture, Galton called this curve an "ogive"; we now call it the inverse normal cumulative distribution function. In 1883, he wrote, "Whenever we find on trial that the outline of the row is not a flowing curve, the presumption is that the objects are not all of the same species, but that part are affected by some large influence from which the others are free; consequently there is a confusion

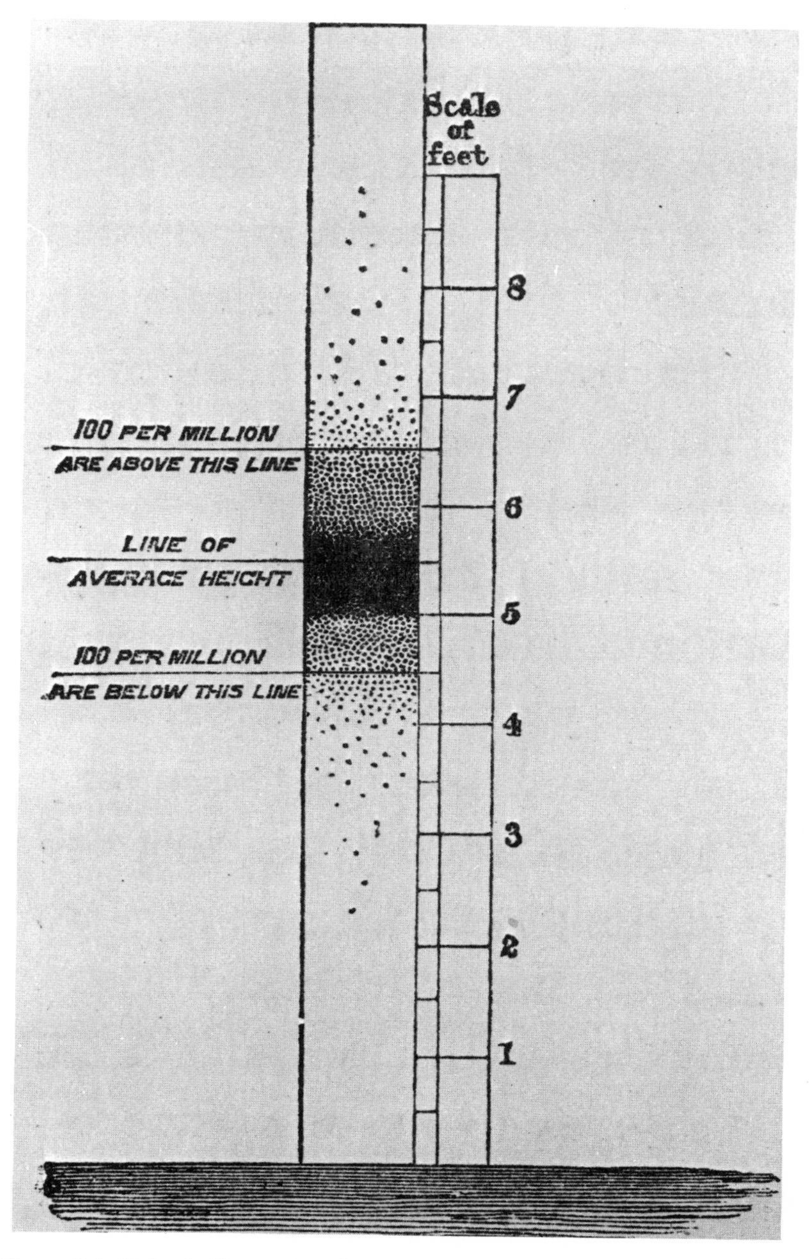

Figure 8.1. Galton's illustration of the "law of deviation from an average," showing the heights of a million hypothetical men. (From Galton, 1869, p. 28.)

A

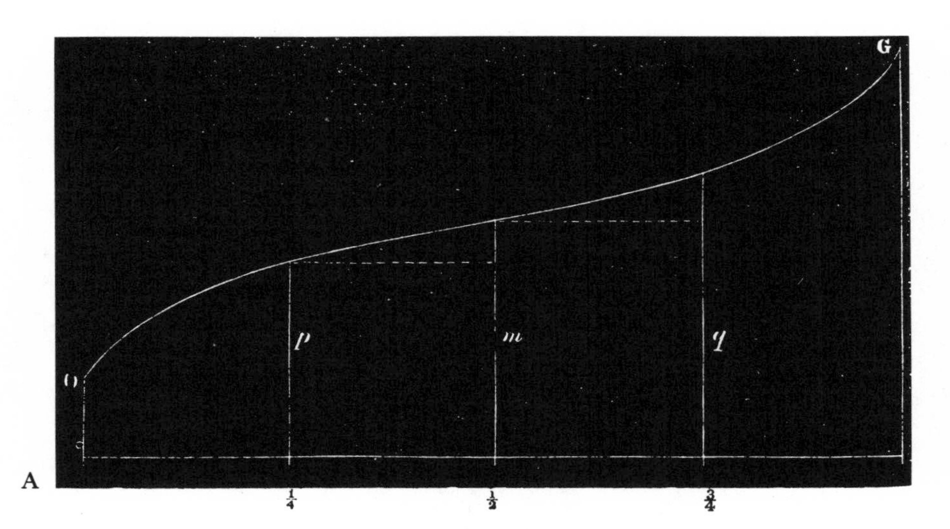

B

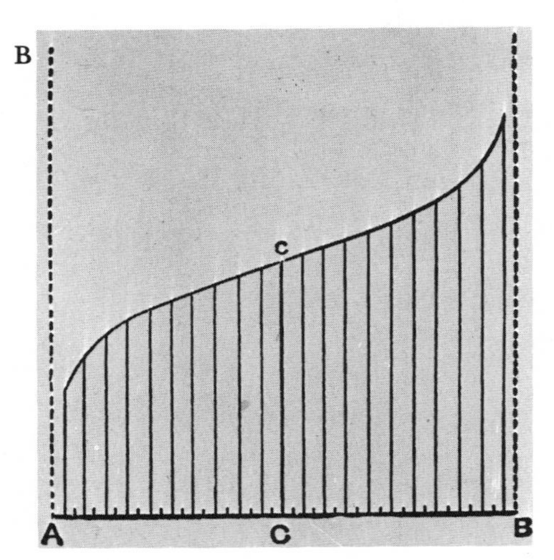

Figure 8.2. Two of Galton's renditions of the ogive. (A) The earlier, from 1875, shows the median and the quartiles (Galton, 1875). (B) The later drawing shows spikes with heights representing 21 equally spaced ideal data values (Galton, 1883, p. 51).

of curves. This presumption is never found to be belied" (Galton, 1883, p. 49).

Even in 1869, though, Galton knew that recognizing a set of data as arising from the same species was the beginning, not the end, of an analysis. At one extreme he argued that "there is sufficient uniformity in the inhabitants of the British Isles to bring them fairly within the grasp of this law" (Galton, 1869, p. 29). Yet it was his basic aim to *distinguish* between Englishmen, to investigate the differences among them and the inheritance of these differences, not to lump them all together as a homogeneous whole. To this end, and in the first of his departures from Quetelet, Galton turned Quetelet's phenomenon to a novel use. If data from the same species arrayed themselves according to this curve and if the unity of the species could be demonstrated by showing that measurable quantities such as stature or examination scores followed the curve, then, once such a species was identified on the basis of measurable quantities, the process could be inverted with respect to qualities that eluded direct measurement! Qualities such as talent or "genius" that were at most susceptible to a simple ordering could, by Galton's method, be assigned a value on a "statistical scale." If a hundred individuals' talents were ordered, each could be assigned the numerical value corresponding to its percentile in the curve of "deviations from an average": The middlemost (or median) talent had value 0 (representing mediocrity), an individual at the upper quartile was assigned the value 1 (representing one probable error above mediocrity), and so on.

Galton later called this method of analysis "statistics by intercomparison" (1874a, 1875), and it was to become the most used (and abused) method of scaling psychological tests (see, for example, Boring, 1920). The argument for the scale was, and remains, weak: It rested solely on analogy. The statistical scale is assumed to be appropriate for talent because it is appropriate for stature; no case is made that the difference in talent between mediocrity and one probable error equals that between one and two probable errors.[1] But the use to which the scale would be put was clear and somewhat ironic. Where Quetelet had used the occurrence of a normal curve to demonstrate homogeneity, Galton based a dissection of the population upon it. Using this inferred scale, he could distinguish between men's abilities on a numerical scale rather than claim that they were indistinguishable.

1. Doubts were evidently raised at the time. In *Natural Inheritance* Galton (1889, p. 56) wrote: "It has been objected to some of my former work, especially in *Hereditary Genius,* that I pushed the applications of the Law of Frequency of Error somewhat too far. I may have done so, rather by incautious phrases than in reality; but I am sure that, with the evidence now before us, the applicability of that law is more than justified within the reasonable limits asked for in the present book."

Galton's use of his scale in *Hereditary Genius* was primarily as a conceptual framework rather than for detailed statistical analyses. It provided a quantitative background for his qualitative tabulations, and it permitted him to make a number of general statements of the comparative abilities of different races, statements that were well in tune with (and in many ways merely reexpressions of) the prejudices of that day. For example, on the basis of the high percentage of illustrious citizens in ancient Athens, he rated the ability of that society "very nearly two grades higher than our own—that is, about as much as our race is above that of the African negro" (Galton, 1869, p. 342). The fact is that in this 1869 book the quantitative trappings were not intrinsic to the arguments because Galton had not yet devised a way to relate them to the main questions he faced, that is, relate them to inheritance.

Conditions for Normality

Galton was faced with a problem. He could recognize the occurrence of the normal curve in his quantitative data, but he could not connect that curve to the transmission of abilities from generation to generation. Indeed, there is a sense in which the classical theory of errors actually impeded his search for a connection. If the normal curve arose in each generation as the aggregate of a large number of factors operating independently, no one of them of overriding or even significant importance, what opportunity was there for a single factor, such as a parent, to have a measurable impact? And why did population variability not increase from year to year? At first glance the beautiful curve Galton had found in Quetelet's work stood as a denial of the possibility of inheritance. It is a tribute to Galton's own genius and to his persistent work over twenty years that he was able to discover a way around this impasse, with his formulation of regression and its link to the bivariate normal distribution.

Some aspects of the notion of regression received crude, qualitative expression in *Hereditary Genius.* For example, Galton wrote of dogs: "If a man breeds from strong, well-shaped dogs, but of mixed pedigree, the puppies will be sometimes, but rarely, the equals of their parents. They will commonly be of a mongrel, nondescript type, because ancestral peculiarities are apt to crop out in the offspring" (p. 64). In one passage he argued against the value of inherited titles as predictors of ability, at least at more than two generations remove (p. 87). In another he found that by selecting eminent men for study he focused on a peak in their families: "The statistics show that there is a regular average increase of ability in the generations that precede its culmination, and as regular a decrease in those that succeed it" (p. 84). He elaborated upon this in the form of a table (Figure 8.3) from which he abstracted an early form of what would become known

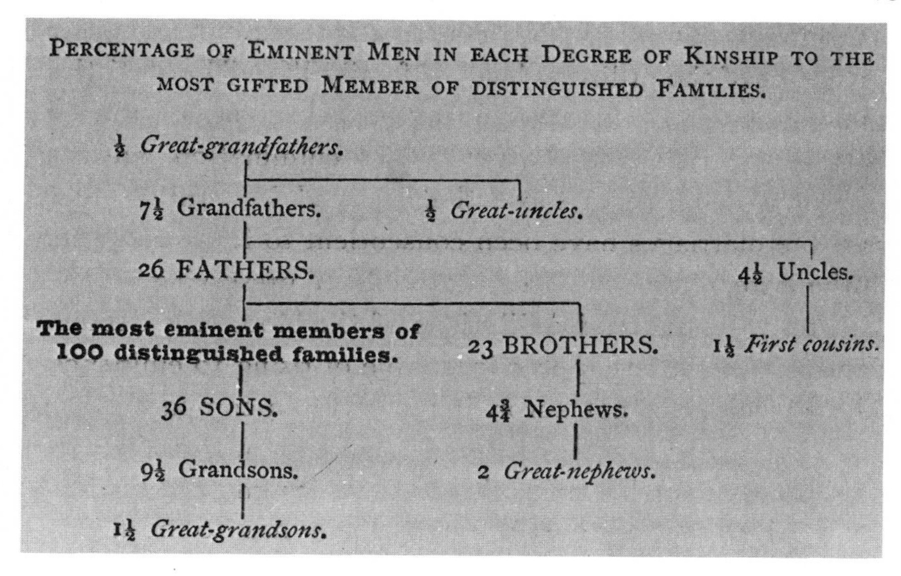

Figure 8.3. Galton's illustration of how the percentage of eminent men declines as the distance from the most eminent man in the family increases. Thus in 36 percent of the cases the son of the most eminent man was also eminent, whereas only 9.5 percent of the grandsons achieved eminence. (From Galton, 1869, p. 83.)

as Galton's law of ancestral heredity: "Speaking roughly, the percentages are quartered at each successive remove, whether by descent or collaterally. Thus in the first degree of kinship the percentage is about 28; in the second, about 7; and in the third, 1½" (p. 83).

Galton's own development of the concept of regression and its probabilistic basis reached full flower in his 1889 book, *Natural Inheritance.* That work, however, was built upon a series of memoirs and experiments dating from 1874 on; some passages are taken verbatim from earlier papers, others were totally new in 1889. From these sources and from Galton's correspondence of the period, it is possible to reconstruct at least the outline of the steps he followed in his crucial reconciliation of inheritance and error theory. There were four basic stages: his initial investigation in 1874–75 of the conditions that would produce the law of error; his experiments with sweet peas, which led to his 1877 formulation of an empirical law of reversion (as regression was first called); his discovery in 1885 of a mathematical framework encompassing regression (after gathering extensive data on human populations); and his elaboration on all of this in *Natural Inheritance* in 1889.

In 1874–75 Galton was thinking deeply about the nature of the law of error. The classical conditions of Laplace implied that data would follow the normal curve, but they were clearly too restrictive for Galton's use. Galton had come to realize that if he were to solve his problem, he would have to demonstrate that Laplace's sufficient conditions were far from necessary for the appearance of the curve and also would have to present a plausible mechanism that could produce the observed distribution. In an 1875 article, "Statistics by intercomparison, with remarks on the law of frequency of error," Galton explained his method of scaling, using the normal curve, and then turned to the conditions that would imply that the normal curve was indeed appropriate. He wrote:

> Considering the importance of the results which admit of being derived whenever the law of frequency of error can be shown to apply, I will give some reasons why its applicability is more general than might have been expected from the highly artificial hypotheses upon which the law is based. It will be remembered that these are to the effect that individual errors of observation, or individual differences in objects belonging to the same generic group, are entirely due to the aggregate action of variable influences in different combinations, and that these influences must be (1) all independent in their effects, (2) all equal, (3) all admitting of being treated as simple alternatives 'above average' or 'below average;' and (4) the usual Tables are calculated on the further supposition that the variable influences are infinitely numerous. (Galton, 1875, p. 38)

Galton knew that the solution to his puzzle involved circumventing these constraints. Indeed, he may have been at least dimly aware of this in 1869, for when he had described the conditions in *Hereditary Genius* (pp. 28–29, 382), the assumption of independence was conspicuous in its absence. In 1875 he addressed the question directly; but although he knew in a vague way what he wanted, he did not yet have a clear conception of how to get it. He wrote that the first three of the conditions "assuredly do not occur in vital and social phenomena; nevertheless it has been found in numerous instances, where measurement was possible, that the latter conform very fairly, within the limits of ordinary statistical inquiry, to calculations based on the (exponential) law of frequency of error. It is a curious fact, which I shall endeavour to explain, that in this case a false hypothesis, which is undoubtedly a very convenient one to work upon, yields true results" (Galton, 1875, pp. 39–40). His explanation was scattered, however; and therefore incomplete. It amounted to two claims that, although true, did not get to the heart of the matter. One was a rearguing of the hypothesis of elementary errors—large influences would frequently, upon closer inspection, be seen to be composed of a large number of smaller influences, and hence the Laplacian conditions could be safely pushed back a stage, out of sight. The other, upon which Galton put more emphasis, was that the number of variable influences did not really need to be "infinitely numerous"; in fact, even for $n = 17$, a binomial distribution was normal for practical purposes.

The Quincunx and a Breakthrough

The 1875 article contained, almost incidentally, one passage that presented the germ of what Galton two years later saw was the key to the whole problem. Consider an influence of "extraordinary magnitude," such as "aspect" as an influence upon the size of fruit, aspect being the location of the fruit as it affects the exposure to sunlight (for example, the southern slope of a hill). This is clearly an important and constant influence upon the size of fruit. Suppose aspect is divided into three classes: large (tending to produce large fruit), moderate, and small. Then the collection of all fruit produced will be a mixture of produce from these three different sources—Why should the total produce follow the normal curve? As Galton put it, "The question is, why a mixture of series radically different, should in numerous cases give results apparently identical with those of a simple series" (Galton, 1875, p. 45). Galton's answer was ingenious: "Now if it so happens that the 'moderate' phase occurs approximately *twice as often* as either of the extreme phases (which is an exceedingly reasonable supposition, taking into account the combined effects of azimuth, altitude, and the minor influences relating to shade from leaves etc.), then the effect of aspect will work in with the rest, just like a binomial of two elements. Generally the coefficients of $(a + b)^n$ are the same as those of $(a + b)^{n-r} \times (a + b)^r$."

We might now describe this as breaking down the factors influencing size of fruit into two categories: aspect and other. Given (or conditional upon) aspect, the other factors obeyed the conditions of classical error theory: They might in particular be supposed to have the net effect of the sum of a large number r of disturbances to the average size for that given aspect. Then if aspect itself was distributed as the sum of $n - r$ disturbances, the net effect, even if $n - r$ is small (it was 2 in the example), is that of $(n - r) + r = n$ disturbances. The overall effect is that the net disturbance is approximately normal, even though the aspect component may not be. The final mixture of sizes is approximately normal because it is an approximately binomial mixture of approximately normal conditional distributions.

In one sense there was a logical difficulty in Galton's argument. It seemingly required that the small number $(n - r)$ of disturbances producing aspect were each of size comparable with the large number (r) accounting for the other factors. Yet if that were true the importance of aspect would be dwarfed by the other factors. Galton dealt with this difficulty by supposing that the factors affecting aspect could themselves be subdivided, justifying what amounted to an assumption that aspect was approximately normally distributed by the observation that such distributions occurred "very commonly." Galton may have felt a bit uneasy about this argument: It came near the end of the article and it received no special emphasis. But if so, by 1877 his view had changed dramatically for it became the centerpiece of his theoretical apparatus. And with its added prominence came a clearer explanation and an empirical illustration.

By 1877 Galton had two crutches to lean on: the results of an extensive series of experiments with sweet peas and a device that later became known as the quincunx. Both were essential to his later success. It is not clear just when Galton devised the quincunx, but it was before 27 February 1874 because he tells us that he showed it at a lecture at the Royal Institution on that date. Indeed, that particular quincunx still survives at the Galton Laboratories (Figure 8.4). As a device for illustrating lectures it must have been marvelously effective, but it was much more than a mere prop to Galton. To Galton it was a means of discovery and proof, a surrogate for his rudimentary mathematics. Figures 8.5 and 8.6 show several of Galton's drawings of the apparatus from 1877.

The quincunx[2] had a glass face and a funnel at the top. Shot was poured through the funnel and cascaded through an array of pins, each shot striking one pin at each level and in principle falling left or right with equal probabilities every time a pin was hit. The shot then collected in compartments at the bottom. The device was an analogue for a binomial experiment; each row of pins subjecting a shot to an independent disturbance, equally likely to be left or right. The total displacement from the point of origin was the sum of as many independent disturbances as there were rows of pins (nineteen in Figure 8.4, twenty-three in Figure 8.6). The resulting outline after many shot were dropped should resemble a normal curve.

One feature not present in the 1874 model that Galton incorporated in his 1877 quincunx was a mechanism for illustrating his 1875 insight. This is perhaps most clearly illustrated in Figures 8.5 and 8.6. Suppose that the shot are intercepted at an intermediate level (AB) and segregated into compartments. The result will be a binomial distribution, which is approximately normal if AB is not near the funnel at the top. Then consider releasing a single compartment, as if it were a funnel. It will produce a small normal-shaped curve at the bottom. The compartments near the center are fuller and will produce higher curves than will the more extreme ones; but except for this difference in scaling they will all be equally disperse (Figure 8.6). The result of releasing *all* of the AB compartments

2. This word is used in *Natural Inheritance,* p. 64, to describe the arrangement of the pins.

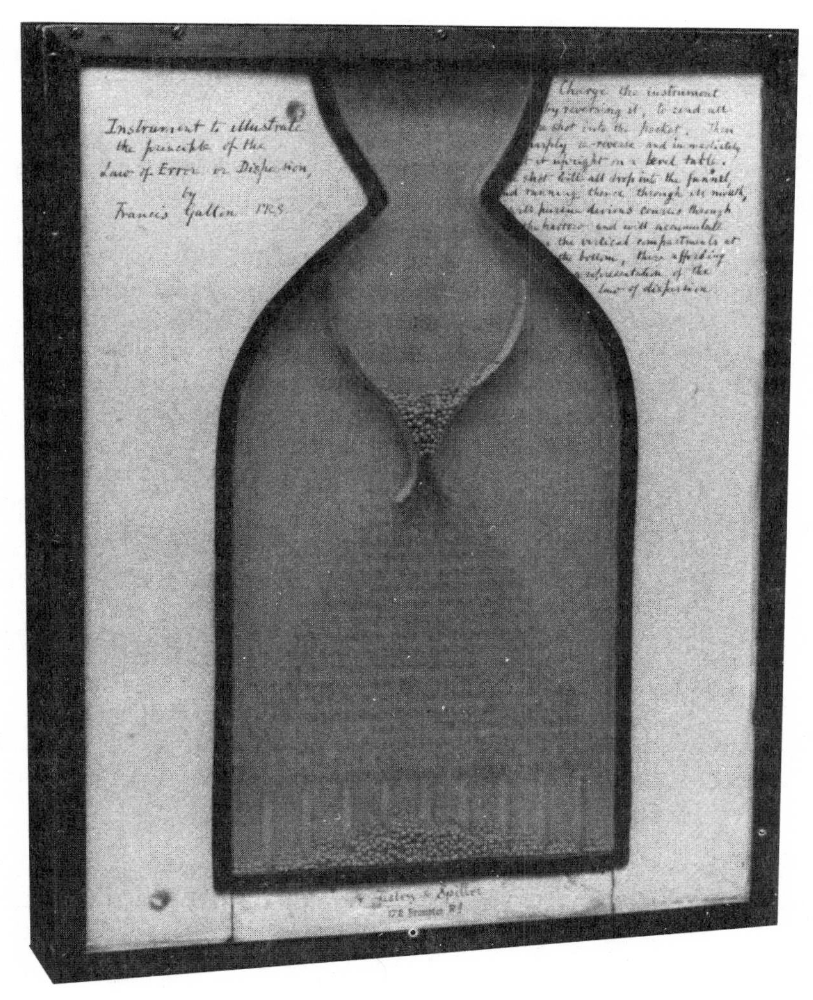

Figure 8.4. The original quincunx, apparently made for Galton in 1873 by Tisley & Spiller. Although it once had an opening at the top through which the shot could be poured, the top is now sealed with the shot inside. The glass has become cloudy with lead dust over the years. The caption, in Galton's handwriting, reads:

Instrument to illustrate
the principle of the
Law of Error or Dispersion
by
Francis Galton F.R.S.

Charge the instrument by reversing it, to send all the shot into the pocket. Then sharply re-reverse and immediately set it upright on a level table. The shot will all drop into the funnel, and running thence through its mouth, will pursue devious courses through the harrow and will accumulate in the vertical compartments at the bottom, there affording a representation of the law of dispersion.

Tisley & Spiller
172 Frampton Rd.

(a4)

(a1)

42 Rutland Gate Jan 12/77

My dear George

How can I thank you sufficiently. I am aghast at the trouble my unlucky memoir gave, and at the great pains you have taken to put clearness into it. I will certainly adopt your suggestions generally, & rewrite the thing.

Let me mention an illustration of one of the processes, (Family Variation) which I think may interest you — You recollect that apparatus of mine with the shot; — well, suppose I want to shew by a modification of it, how

(a2)

it comes to pass that

when the ordinates of an exponic mountain ~~[illegible]~~ subside each of them, into an exponic hillock, as in the sketch, the sum of the hillocks is an exponic curve of larger modulus

In I, I pour shot, & it makes an exponic heap at the bottom.

In II

I have cut the apparatus across at AB, & have interposed a row of vertical compartments A B A' B' with trap door bottoms that I can pull out & in,

(a3)

Shot is poured in

A B A B A' B'

I II

to form a temporary landing for the shot, when I so desire. If these are open, the shot falls through & of course makes an exponic mountain at the bottom of II exactly as it did in I. But if they are closed, they intercept the shot & an exponic mountain (of less modulus) is formed on A'B'

(a4)

Now I open the trap doors, successively the shot in each vertical compartment rushes down and forms its own exponic hillock, & we have already seen what the sum of them will be.

The ratio of the modulus (of the three heaps) is self evident (they vary as √ of the distances which vary directly as the lengths of passage of the shot)

For my R. Instn lecture, I shall simply go into generalities to show what Reversion means & how a law is possible — & shall hang up the formulae but not speak a word about them.

[illegible] F. Galton

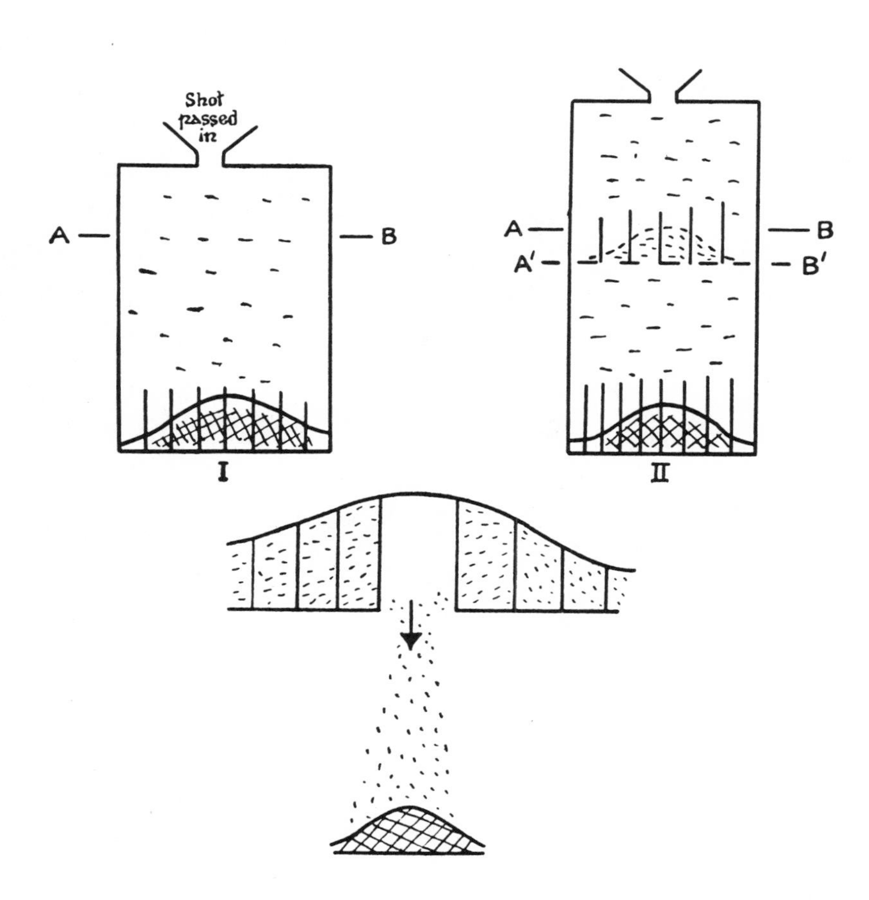

Figure 8.5. (a) Galton's explanation of the working of the two-stage version of the quincunx in a 12 January 1877 letter to his cousin, George Darwin. (From Galton Archives.) (b) Drawings by Karl Pearson, based on Galton's original letter. (From Pearson, 1914–1930, vol. 3B, pp. 465–466.)

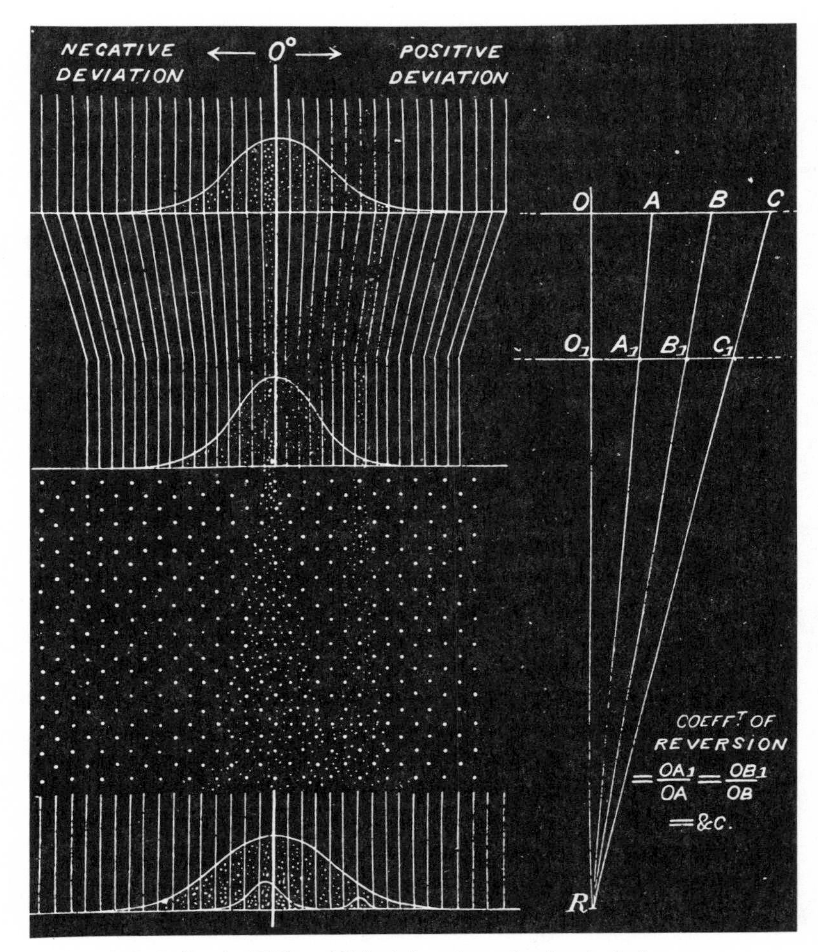

Figure 8.6. Galton's 1877 published drawing, showing both the two-stage quincunx and the law of reversion. (From Galton, 1877.)

will be a mixture of these curves of varying sizes. Because the result of releasing all compartments is the same as if the shot had not been interrupted at AB at all, the resulting mixture must itself be a normal curve!

The quincunx can be related to the 1875 discussion by defining the AB compartments as differing aspects of fruit; the bottom compartments then represent different sizes of fruit. I suspect that it was through reflection upon this device that Galton achieved his 1875 insight. In any case, its use to provide an analogue proof that a normal mixture of normal distributions was itself normal was a stroke of genius. Of course, if we rephrase the result we can make it appear that it was not new. It could be viewed as saying that if one set of normal disturbances is superimposed (added to) another independent set, the resulting sum is normal. This was known to Laplace, and in its binomial form, to De Moivre. But this would be grossly misleading. The crucial point is that Galton's conceptual use of the result was new and ingenious and represents the most important step in perhaps the single major breakthrough in statistics in the last half of the nineteenth century. This concept freed Galton of the restrictions of naive error theory without requiring that he go beyond that theory. He could conceive of his data as a mixture of very different populations, notwithstanding the unity apparent in its normal outline. Of course, he still had a long way to go before this insight was developed into a conceptual and statistical tool. First, it remained for him to relate this insight to heredity. For this he was to use his second crutch, the sweet pea experiments.

Reversion

In 1875 Galton had arranged for several friends to assist him by growing sweet peas from packets of seeds that Galton had carefully separated by weight into seven groups (Cowan, 1972). By 1877, when he addressed the Royal Institution on February 9, he had completed an analysis of these data, the produce of 7 sizes × 10 seeds per size × 7 friends = 490 seeds, with surprising findings. After reviewing the properties of the normal curve for his audience (and reintroducing the quincunx), Galton moved on to the main topic:

> First let me point out a fact which Quetelet and all writers who have followed in his paths have unaccountably overlooked, and which has an intimate bearing on our work to-night. It is that, although characteristics of plants and animals conform to the law, the reason of their doing so is as yet totally unexplained. The essence of the law is that differences should be wholly due to the collective actions of a host of independent *petty* influences in various combinations . . . Now the processes of heredity . . . are not petty influences, but very important ones . . . The conclusion is . . . that the processes of heredity must work harmoniously with the law of deviation, and be themselves in some sense conformable to it. (Galton, 1877, p. 512)

Guided by his data on sweet peas, Galton sought to account for this phenomenon in two respects: first, by showing that in at least one case his 1875 insight did provide an explanation for the means by which a heterogeneous set of parent seeds produced a second generation that both separately and in aggregate was in line with the law of error; second, by explaining how the stability in population variability could be reconciled with the first observation.

Galton found that if he looked separately at the seven groups of progeny seeds, groups classified by weight of parent seed from heaviest (Group K) to lightest (Group Q), each group separately followed a normal curve and, in addition, the curves, although centered at different weights, were equally disperse. He wrote, "I was certainly astonished to find the family variability of the produce of the little seeds to be equal to that of the big ones, but so it was, and I thankfully accept the fact, for if it had been otherwise I cannot imagine, from theoretical considerations, how the problem could be solved" (Galton, 1877, p. 513). Thus far, all was well. The groups of seeds had behaved just like the shot from the different compartments at the AB level of the quincunx. Each had produced a little normal curve (as in Figure 8.6), and the curves were equally disperse. Galton would not have been able to put the curves together to produce a combined normal curve because he had started with equal numbers in the groups of parent seeds (akin to a uniform distribution among the AB level compartments). He apparently felt secure in this respect, however, because of his knowledge that the sweet pea population was stable and the weights normally distributed. At any rate he did not allude to any difficulty on this point.

But how could one account for the seeds' stability? Why did the dispersion among the progeny seeds not lead to populations that were increasingly variable from generation to generation? Galton's answer—his second major finding—was *reversion*. The seven groups of progeny were normally distributed, but not about their parents' weight. Rather they were in every case distributed about a value that was closer to the average population weight than was that of the parent. Furthermore, this reversion followed "the simplest possible law"; that is, it was linear. The average deviation of the progeny from the population average was in the same direction as that of the parent, but only a third as great. The mean progeny reverted to type, and the increased variation was just sufficient to maintain the population variability.

Galton combined both effects in Figure 8.6: The sloping channels produced the reversion, and the quincunx below produced the dispersion that led to a population at the bottom, reinflated to be as disperse as that at the top. As an appendix Galton expressed these effects in algebraic terms:[3] Let the population distribution about the average be $(1/c\sqrt{\pi})\exp(-x^2/c^2)$, and let r represent the "fractional coefficient" of reversion. Then the reverted distribution (scaled down by r; the distribution immediately

3. Galton actually described a situation where the parents could be unequally productive, where two parents were permitted (unlike the situation of his sweet peas) and where a form of natural selection could operate. Consequently, my notation differs slightly from his.

below the sloping channels) is

$$\frac{1}{cr\sqrt{\pi}}\,e^{-x^2/r^2c^2},$$

and if the distribution of each group of progeny about its (reverted) mean is

$$\frac{1}{v\sqrt{\pi}}\,e^{-x^2/v^2},$$

then the distribution of the entire second generation will be

$$\frac{1}{d\sqrt{\pi}}\,e^{-x^2/d^2},$$

where $d = \sqrt{v^2 + r^2c^2}$. If the population is to remain stable, we must have $c = d$, so $c = \sqrt{v^2 + r^2c^2}$, or $c^2 = v^2/(1 - r^2)$.

The analysis was thus apparently complete: Reversion and family variability had been demonstrated empirically and Galton had shown how they combined to produce a stable, normally distributed population, reconciling the diversity of parental characteristics with the appearance of a normal population outline at the second generation. But there was one apparent gap in this analysis. Reversion had been demonstrated empirically, but Galton had given no reason *why* it should take place in such a neat linear manner. The proportionally sloping channels were an artificial construct, not a theoretical deduction. And there was no apparent reason why the relation $c = \sqrt{v^2 + r^2c^2}$ needed for stability should necessarily hold. A final step was needed, and it was to be 1885 before Galton took that step.

Symmetric Studies of Stature

Galton took the occasion of his 10 September 1885 presidential address to the anthropology section of the British Association for the Advancement of Science at Aberdeen to announce his next theoretical advance. By that time he had augmented his data on heredity with extensive data on a human population, data obtained through public appeal (with the promise of cash prizes as an incentive). The most useful part of the data was that on heights—or stature, as Galton referred to it. Galton's discussion of this topic was spread over several articles and was given in fullest form in his 1889 *Natural Inheritance*, along with much additional material.

The use of human data presented Galton with one new complication: Each individual had two parents, whereas with sweet peas it was thought that cross-fertilization did not take place. This led to three potential difficulties: First, the parents' heights were not directly comparable because the average height of women is less than that of men. Second, there was the possibility that the preferences of men and women for mates of like or contrasting height (sexual selection) might confound the results and make the analysis more difficult. And third, it raised the specter that it might be necessary to try to isolate the parents' contributions to inherited height separately. Galton overcame the first problem by the simple expedient of multiplying each woman's height by 1.08, a value he had determined to be the correct adjustment based on extensive data. The second of these difficulties Galton had actually come to grips with earlier, when he considered a different data set. In *English Men of Science* (1874b, p. 31) he had argued for the lack of sexual selection with regard to height by finding the probable error for husbands' heights to be 1.9″, and for wives' heights to be 1.7″. He wrote, "If there had been no sexual selection in respect of height, the sum of the heights of the two parents would also conform to the law of frequency of error, and that the probable error of the series would be $\sqrt{1.7^2 + 1.9^2} = 2.5''$. I find . . . 2.3″." Galton thought this agreement was close enough.[4] This comparison, which presupposes an implicit understanding of the effect of correlation upon the variability of sums, was incidentally the same test that Ebbinghaus used in 1885.[5] Galton gave essentially the same argument based upon his new data in *Natural Inheritance,* concluding, "marriage selection does not pay such regard to Stature, as deserves being taken into account in the cases with which we are concerned" (1889, p. 87).

The third difficulty was potentially more troublesome. Galton wanted to ignore the two parents' individual heights and replace them instead by the "mid-parent"—the average of the father's and mother's heights, the latter scaled up by 1.08. But would this do? Were the offspring heights only dependent upon the parental average, or would they also depend in some way upon how separated the two heights were? Galton made a partial test of this, by grouping 525 children according to the difference of the heights of their parents. He found "that they were no more diverse in the one case than in the other" (Galton, 1889, p. 90). The test was only partial in that it only guarded against effects of parental difference that led to a change in offspring dispersion. Still, it was sufficient to comfort Galton, and he went on to the main question, How was height passed from midparent to child?

In 1885 Galton was in much the same position as in 1877, when he had quantified what he now called regression through his analysis of the sweet

4. Curiously, it corresponds to a *negative* correlation of -0.15.

5. Galton gives no source for the test, but it may be based upon his reading of Airy (1861, §12), in which the effect of "entanglement" upon observations and their probable errors is discussed. Galton cited Airy as his standard source for error theory (for example, in Galton, 1877, p. 533).

pea data. But there was one major difference between his sweet pea data and his data on human stature, and that difference turned out to be quite important. The crucial distinction was that the human stature data involved a complete sample whereas the sweet pea data was only a collection of stratified samples. Galton had taken an equal number of sweet pea seeds of each of seven sizes and studied their progeny. He had learned a great deal about the conditional distributions of the progeny seed sizes, given the parent seed sizes: Their means "reverted," or regressed, toward mediocrity and their variances were equal. But the only theoretical basis he had for putting together these conditional distributions — his quincunx — was also unidirectional (what would later be called a Markovian structure, itself based solely upon conditional distributions). It is understandable that in 1877 he did not come upon an enlarged conceptual structure that would permit him to take one more step, one that would permit him to see regression in a broader perspective than the hereditarian one of 1877, which was bound to the generational time sequence. As the Galton of 1885 described his earlier self, "I was then blind to what I now perceive to be the simple explanation of the phenomenon" (p. 507). With the luck of the extraordinarily persistent, Galton's 1885 data provided a cure for his blindness.

The fact that Galton had a full sample rather than a set of conditional samples invited him to prepare a table in what was to be a most suggestive form. Table 8.1, from his 1885 address, is a cross-tabulation of 928 adult children, by their height and their midparent's height. We can relate this table to Galton's 1877 approach as follows: Each row might be viewed as a conditional sample, and the distributions of the counts within rows are like the little normal curves produced by the separate compartments in Figure 8.6, which mix together (in the Totals row) to form a large normal curve, the population distribution. But the table shows much more than that. For one thing, it suggests a symmetric treatment of the data; One can view the *columns* as little normal curves — distributions of midparents' heights given children's heights. And for another, it exhibits a striking internal "shape." Galton was quick to seize upon both of these.

Galton's own account is worth quoting:

> I found it hard at first to catch the full significance of the entries in the table, which had curious relations that were very interesting to investigate. They came out distinctly when I "smoothed" the entries by writing at each intersection of a horizontal column with a vertical one, the sum of the entries in the four adjacent squares, and using these to work upon. I then noticed (see . . . [Figure 8.7]) that lines drawn through entries of the same value formed a series of concentric and similar ellipses. Their common centre lay at the intersection of the vertical and horizontal lines, that corresponded to $68\frac{1}{4}$ inches. Their axes were similarly inclined. The points where each ellipse in

Table 8.1. Galton's 1885 cross-tabulation of 928 adult children born of 205 midparents, by their height and their midparent's height.

Height of the mid-parent in inches	Height of the adult child														Total no. of adult children	Total no. of mid-parents	Medians
	<61.7	62.2	63.2	64.2	65.2	66.2	67.2	68.2	69.2	70.2	71.2	72.2	73.2	>73.7			
>73.0	—	—	—	—	—	—	—	—	—	—	—	1	3	—	4	5	—
72.5	—	—	—	—	—	—	—	1	2	1	2	7	2	4	19	6	72.2
71.5	—	—	—	—	1	3	4	3	5	10	4	9	2	2	43	11	69.9
70.5	1	—	1	—	1	1	3	12	18	14	7	4	3	3	68	22	69.5
69.5	—	—	1	16	4	17	27	20	33	25	20	11	4	5	183	41	68.9
68.5	1	—	7	11	16	25	31	34	48	21	18	4	3	—	219	49	68.2
67.5	—	3	5	14	15	36	38	28	38	19	11	4	—	—	211	33	67.6
66.5	—	3	3	5	2	17	17	14	13	4	—	—	—	—	78	20	67.2
65.5	1	—	9	5	7	11	11	7	7	5	2	1	—	—	66	12	66.7
64.5	1	1	4	4	1	5	5	—	2	—	—	—	—	—	23	5	65.8
<64.0	1	—	2	4	1	2	2	1	1	—	—	—	—	—	14	1	—
Totals	5	7	32	59	48	117	138	120	167	99	64	41	17	14	928	205	—
Medians	—	—	66.3	67.8	67.9	67.7	67.9	68.3	68.5	69.0	69.0	70.0	—	—	—	—	—

Source: Galton (1886a).

Note: All female heights were multiplied by 1.08 before tabulation. Galton added an explanatory footnote to the table: "In calculating the Medians, the entries have been taken as referring to the middle of the squares in which they stand. The reason why the headings run 62.2, 63.2, &c., instead of 62.5, 63.5, &c., is that the observations are unequally distributed between 62 and 63, 63 and 64, &c., there being a strong bias in favour of integral inches. After careful consideration, I concluded that the headings, as adopted, best satisfied the conditions. This inequality was not apparent in the case of the Mid-parents." Galton republished these data in 1889, where they are referred to as the R.F.F. Data (Record of Family Faculties); he then noted that the first row must be in error (four children cannot have five sets of parents), but he claimed that "the bottom line, which looks suspicious, is correct" (p. 208).

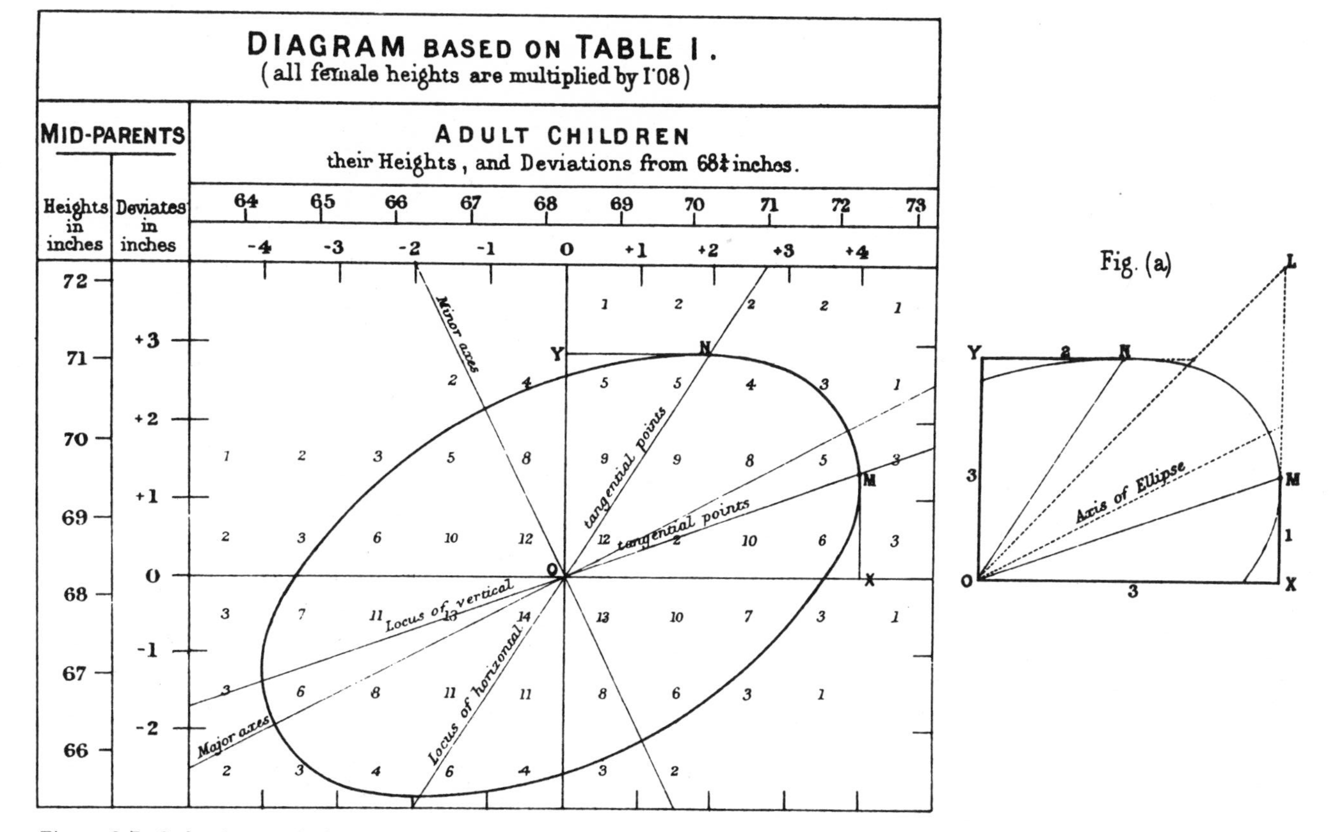

Figure 8.7. Galton's smoothed rendition of Table 8.1, with one of the "concentric and similar ellipses" drawn in. The geometric relationship of the two regression lines to the ellipse is also shown. (From Galton, 1886a.)

succession was touched by a horizontal tangent, lay in a straight line inclined to the vertical in the ratio of $\frac{2}{3}$; those where they were touched by a vertical tangent lay in a straight line inclined to the horizontal in the ratio of $\frac{1}{3}$. These ratios confirm the values of average regression already obtained by a different method, of $\frac{2}{3}$ from mid-parent to offspring, and of $\frac{1}{3}$ from offspring to mid-parent, because it will be obvious on studying . . . [Figure 8.7] that the point where each horizontal line in succession is touched by an ellipse, the greatest value in that line must occur at the point of contact. The same is true in respect to the vertical lines. These and other relations were evidently a subject for mathematical analysis and verification. They were all clearly dependent on three elementary data, supposing the law of frequency of error to be applicable throughout; these data being (1) the measure of racial variability, whence that of the mid-parentages may be inferred as has already been explained, (2) that of co-family variability (counting the offspring of like mid-parentages as members of the same co-family), and (3) the average ratio of regression. I noted these values, and phrased the problem in abstract terms such as a competent mathematician could deal with, disentangled from all reference to heredity, and in that shape submitted it to Mr. J. Hamilton Dickson, of St. Peter's College, Cambridge. I asked him kindly to investigate for me the surface of frequency of error that would result from these three data, and the various particulars of its sections, one of which would form the ellipses to which I have alluded. (Galton, 1886a, pp. 254–255)

What Galton had essentially done was to solve the whole problem by inspection (and minor adjustment of) this one table. He noted the concentric ellipses of constant frequency, located the two regression lines (and estimated their slopes as $\frac{2}{3}$ and $\frac{1}{3}$),[6] identified these lines of the means (and largest frequency) with the lines of tangency to the ellipses, and abstracted a simple mathematical hypothesis that could characterize the phenomenon. In modern terminology we would say that he assumed normal distributions and required three numbers. These were (1) the population variance, from which the variance of midparent height P followed, (2) the conditional variance of child's height p given midparent's height P, and (3) the slope of the regression line, r. Thus, in modern notation, Dickson received the problem in this form (heights are expressed in terms of deviations from the population mean of $68\frac{1}{4}''$): If P has a $N(0,\sigma^2)$ distribution, and p (given P) has a $N(rP,\tau^2)$ distribution, [alternatively, $p - rP$ has a $N(0,\tau^2)$ distribution] what are the contours of the bivariate distribution of (P,p), and what are the lines made by the points of tangency to these contours?

6. The regression of midparent upon child is half that of child upon midparent, because the midparent is the average of two uncorrelated heights, each with the same (population) variance. If $p =$ child's height and $P =$ the height of the midparent, then the slopes are $\mathrm{cov}(p,P)/\mathrm{Var}(p)$ and $\mathrm{cov}(p,P)/\mathrm{Var}(P)$, but $\mathrm{Var}(p) = 2\ \mathrm{Var}(P)$.

Given this statement, any moderately competent mathematician could carry through the development. Dickson filled the bill and made short work of it. Galton wrote:

> I may be permitted to say that I never felt such a glow of loyalty and respect towards the sovereignty and magnificent sway of mathematical analysis as when his answer reached me, confirming, by purely mathematical reasoning, my various and laborious statistical conclusions with far more minuteness than I had dared to hope, for the original data ran somewhat roughly, and I had to smooth them with tender caution. His calculation corrected my observed value of midparental regression from $\frac{1}{3}$ to $\frac{6}{17.6}$, the relation between the major and minor axis of the ellipses was changed 3 per cent. (it should be as $\sqrt{7}:\sqrt{2}$), their inclination was changed less than 2° (it should be to an angle whose tangent is $\frac{1}{2}$). It is obvious, then, that the law of error holds throughout the investigation with sufficient precision to be of real service, and that the various results of my statistics are not casual and disconnected determinations, but strictly interdependent. (Galton, 1886a, p. 255)

Galton arranged for Dickson's rather severely geometrical solution to be printed (Dickson, 1886; Galton, 1889, pp. 221–224).

The first key point that Galton had hit upon was the strict interdependence of the quantities involved. The regression coefficient, conditional variance, and population were bound together. The differing quantities he was considering were but different aspects of the same phenomenon. In particular, the linearity of regression and the relation $c = \sqrt{v^2 + r^2c^2}$ that he had found for sweet peas[7] were not a lucky circumstance that just happened to produce stability, they were a necessary *consequence* of stability (and, we might add, the normality assumptions). "We ought to expect filial regression, and . . . it should amount to some constant fractional part of the value of the mid-parental deviation" (Galton, 1885, p. 508; 1886a, p. 252). In this respect the "blindness" he had experienced in 1877 had been cured and his empirical finding of regression given a simple explanation.

The second key point—the consequences of the symmetry of the situation regarding child and midparent that was so apparent in Table 8.1—did not register immediately. Galton was not slow to spot the first implication, namely, that there was a second regression line, that of midparent on child. "The converse of this law is very far from being its numerical opposite. Because the most probable deviate of the son is only two-thirds that of his mid-parentage, it does not in the least follow that the most probable deviate of the mid-parentage is $\frac{3}{2}$, or $1\frac{1}{2}$ that of the son . . . It appears from the very same table of observations by which the value of the

7. Of course, for stature the relation is $c = \sqrt{v^2 + r^2c^2/2}$ because the regression is upon midparent size.

filial regression was determined when it is read in a different way, namely, in vertical columns instead of in horizontal lines, that the most probable mid-parentage of a man is one that deviates only one-third as much as the man does" (Galton, 1886a, pp. 253–254).

But a fuller appreciation of the consequence of symmetry was yet to come. The printed version of the 1885 address hints at completed work he could not discuss: "Still less can I enter upon the subject of fraternal characteristics, which I have also worked out." In early 1886 Galton was ready to report on part of this work (Galton, 1886b); and by the time *Natural Inheritance* was sent to the printer in late 1888, Galton had gone even further. This fraternal analysis was based on a new data set on the heights of brothers that he had obtained through personal contact with "trusted correspondents." The point was that once the relationship between child and midparent was seen as statistically symmetric, the lessons learned could be extended to truly symmetric relationships, such as among brothers. The phenomenon of regression was freed from its status as temporal or intergenerational in origin and, as we shall see later, this symmetry eventually led to the notion of "correlation."

Data on Brothers

Galton's consideration of brothers during the preparation of *Natural Inheritance* inspired further theoretical development, which he again expressed in the mechanical terms he felt most comfortable with. As part of the theoretical background for that book, he presented five "Problems in the Law of Error" (Galton, 1889, pp. 66–70). The first three problems were essentially restatements of his 1875 insight, again presented in terms of the quincunx. We might rephrase them as saying that if a normal deviation Y (with probable error b) is added to a normal deviation X (with probable error c), the result is a normal deviation (with probable error $q = \sqrt{c^2 + b^2}$). Here X represents the position of a shot after it has fallen to the AB level of the quincunx, Y the net change in its position between AB and the bottom, and $Z = X + Y$ the final position on the bottom.

The fourth and fifth problems were new, however, and were built upon the added insight of 1885. Galton presented them in terms of a target-shooting analogy, but they might as well have been phrased in terms of the quincunx. In those terms they can be stated as follows: (4) Given the deviation $Z = a$ of the final shot position on the bottom, what is the most probable value of the deviation X on the AB level? It is $[c^2/(c^2 + b^2)]\, a$. (5) Given $Z = a$, what is the probable error of X? It is $bc/\sqrt{b^2 + c^2}$. In this 1889 formulation it is interesting to note that the results amount to a simple application of Bayes's theorem: From the normal distribution of X and the normal conditional distribution of Z given X, the normal conditional distribution of X given Z is derived.

Actually there was one minor error in the 1889 treatment, which, although typographical, is illuminating because it shows how strong a role Galton's geometric intuition played in his work and how little he leaned on his weak algebra. Galton had written, "It is always well to retain a clear geometric view of the facts when we are dealing with statistical problems, which abound with dangerous pitfalls, easily overlooked by the unwary, while they are cantering gaily along upon their arithmetic" (Galton, 1889, pp. 66–67). Ironically, three pages later (p. 70) Galton fell into one of those pitfalls, as he gave the expected value of X given $Z = a$ as $\sqrt{c^2/(c^2 + b^2)}$ (which is actually the correlation coefficient of X and Z) instead of the correct $[c^2/(c^2 + b^2)]\, a$. The expression he gave did not even include the deviation a! That it was not a simple typographical error is shown by the fact that it was repeated on p. 127. That Galton did not lean on the expression is shown by the fact that all numerical work that should have followed from it was done correctly. But his early readers must have been bewildered. On 23 May 1889 John Venn wrote to Galton pointing out this error and another (and giving the correct expression); on 24 May Galton replied, "The first error is gross, and basely perhaps, I assign it either to a misprint *in part,* or in part or wholly to a blind & careless writing out of a formula from memory" (Galton Archives, folder 334). He quickly had an errata sheet prepared for insertion in unsold copies.

Galton's problems 4 and 5 show on a mathematical level how his appreciation of the symmetry of the situation permitted him to change his point of view and look back up the quincunx from the bottom level, as it were. But of real interest is the way he used these problems, essentially as a variance-components model for fraternal relationships.

Galton's examination of fraternal relationships was based upon the data summarized in Table 8.2, although some of his methods required more detail about the measurements than he provided. Galton based these data on returns from 295 families, including 783 brothers in all (p. 79). He explains that he included "each possible pair of brothers in each family: thus if there were three brothers, A, B, and C, in a particular family, I entered the differences of stature between A and B, A and C, and B and C; four brothers gave rise to 6 entries, and five brothers to 10 entries" (p. 92). A close inspection of the table, however, suggests that this description is inaccurate. The approximation to perfect symmetry in this table (and in one other similar table he generated in the same manner) is such that Galton must have included each pair of brothers in both possible orders — and made a few errors in tabulation. Thus three brothers gave rise to six entries, four to twelve, and so on. Galton evidently did not realize that this

Table 8.2. Galton's "Special Data" on brothers, giving the "Relative Number of Brothers of Various Heights to Men of Various Heights, Families of Five Brothers and Upwards Being Excluded."

Heights of the men in inches	Heights of their brothers in inches													Total	
	<63	63.5	64.5	65.5	66.5	67.5	68.5	69.5	70.5	71.5	72.5	73.5	>74	cases	Medians
74 and above	1	1	—	—	—	—	—	1	1	—	5	3	12	24	—
73.5	—	—	—	—	—	1	3	4	8	3	3	2	3	27	—
72.5	—	—	—	—	1	1	6	5	9	9	8	3	5	47	71.1
71.5	—	1	—	1	2	8	11	18	14	20	9	4	—	88	70.2
70.5	—	—	1	1	7	19	30	45	36	14	9	8	1	171	69.6
69.5	—	1	2	1	11	20	36	55	44	17	5	4	2	198	69.5
68.5	—	1	5	9	18	38	46	36	30	11	6	3	—	203	68.7
67.5	2	4	8	26	35	38	38	20	18	8	1	1	—	199	67.7
66.5	4	3	10	33	28	35	20	12	7	2	1	—	—	155	67.0
65.5	3	3	15	18	33	36	8	2	1	1	—	—	—	110	66.5
64.5	3	8	12	15	10	8	5	2	1	—	—	—	—	64	65.6
63.5	5	2	8	3	3	4	1	1	—	1	—	—	1	20	—
Below 63	5	5	3	3	4	2	—	—	—	—	—	—	1	23	—
Totals	23	29	64	110	152	200	204	201	169	86	47	28	25	1,329	

Source: Galton (1886b; 1889, p. 210).
Note: The near-perfect symmetry suggests that every pair of brothers was counted twice, contrary to Galton's own account.

would produce perfect symmetry, or he might have corrected the few discrepancies that occur. The net effect was to greatly exaggerate the symmetry already present in the data and to bias his subsequent estimates of variance components.

Estimating Variance Components

Galton's work abounds in informally derived estimates. His regression coefficients came from rough calculations based upon graphs. He used medians rather than means and median deviations from medians rather than traditional estimates based on squared deviations, not because of articulated theoretical reasons, but for ease of calculation and because they meshed well with Galton's intuitive understanding of the problem. His discussion of the estimation of fraternal variability is one of the very few places in which Galton put forth even vaguely explained formal methods.

Galton's aim was to estimate the variability within families of brothers' heights. He felt that the individual families were too small to provide accurate individual estimates, so he concocted four ad hoc methods. His descriptions (Galton, 1886b; 1889, pp. 124–129) are somewhat opaque, even when read with historical hindsight. In modern terms and notation, we can summarize his methods as follows: Consider each individual's height as consisting of two components, the ith individual from family j having height $Z_{ij} = X_j + Y_{ij}$, where X is a family mean, Y the individual's within-family deviation, and $i = 1, \ldots, n_j$. X, Y, and Z are to be thought of in the same way as in problems 4 and 5: X_j is normal with mean 0, probable error c; and, given X_j, Y_{ij} is normal with mean 0, probable error b. The aim is to estimate b. Galton's four methods were as follows:

I. Let $\tilde{Y}_j$ = the median of the Y_{ij}. Then we have

$$Y_{ij} = \tilde{Y}_j + (Y_{ij} - \tilde{Y}_j).$$

Treating the two terms of this sum as if they were independent (or at least uncorrelated), letting d = the probable error of $Y_{ij} - \tilde{Y}_j$, and assuming (erroneously) that the probable error of $\tilde{Y}_j$ is the same as that of the mean, $b/\sqrt{n_j}$, Galton had $b^2 = b^2/n_j + d^2$, so $b = d\sqrt{n_j/(n_j - 1)}$. Galton found estimates of d by pooling all $|Y_{ij} - \tilde{Y}_j|$ with $n_j = 4$, pooling all those with $n_j = 5$, and similarly with $n_j = 6, 7$. He thus found four estimates of b, which he averaged.

II. Galton's usual approximate handling of the data in Table 8.2 gave a slope $\frac{2}{3}$ for the regression line; that is, if heights are measured in deviations from the mean,

$$\mathrm{E}(Z_{1j}|Z_{2j} = a) = \frac{2}{3}a.$$

Now problem 4 says $E(X_j|Z_{2j} = a) = [c^2/(b^2 + c^2)]a$; and so, because $E(Z_{1j}|Z_{2j} = a) = E(X_j|Z_{2j} = a)$, $[c^2/(b^2 + c^2)] = 2/3$. Galton's data for the whole population gave $b^2 + c^2 = (1.7)^2$; hence he could solve for b.

III. Examination of the rows of Table 8.2 shows that each has a probable error of 1.24; take this to be the conditional probable error of Z_{1j} given Z_{2j}. Now $Z_{1j} = X_j + Y_{1j}$. Problem 5 gives the conditional probable error of X_j given Z_{2j} to be $bc/\sqrt{b^2 + c^2}$. The deviation Y_{1j} is independent of Z_{2j}, with probable error b. Hence the conditional probable error of Z_{1j} given Z_{2j} is

$$\sqrt{b^2 + \frac{b^2c^2}{b^2 + c^2}}.$$

Setting this equal to 1.24 and using $b^2 + c^2 = (1.7)^2$ as in method II, Galton could solve for b.

IV. Divide the median of the set of all $|Z_{1j} - Z_{ij}|$ (for all $i > 1$, all j) by $\sqrt{2}$; the result is an estimate of b. (Apparently Galton did not use all such pairs, but only a "random" selection.)

Naturally the four methods gave four different (though similar) answers, which he averaged to get $b = 1.06$ (p. 129). The only one of Galton's contemporaries to comment on these methods was Edgeworth, in a review in *Nature.* He noted that two of the methods (II and III) involved the variance-component model. The others could be used more generally, although "neither of the latter coincides with the theoretically best possible method" (Edgeworth, 1889). What Edgeworth had in mind as a more general situation was the same model where the X's are treated as fixed effects. His own work (Edgeworth, 1885c; Stigler, 1978a) suggests that the "best possible method" he had in mind was a multiple of the square root of the pooled sum of squares $\Sigma_{ij}(Z_{ij} - Z_{\bullet j})^2$. We might add with the benefit of hindsight that Galton's method of including the same individuals several times in Table 8.2 would introduce a bias in methods II and III, although the magnitude of the bias is difficult to determine.

Galton's use of a variance-components model in this context may have been novel, although such models can be found at least implicitly in Airy (1861, §12) and explicitly in Edgeworth (1885b,c)—both sources were known to Galton, although neither is cited in *Natural Inheritance.*

Galton's Use of Regression

Galton's own use of regression in *Natural Inheritance* shows either great naiveté or great optimism. Having found that the slope of the regression of child's height upon a midparent was $\frac{2}{3}$ (Figure 8.8), he blithely supposed that the same value held for all other characteristics. For example, he based calculations for both artistic ability (pp. 159–162) and "consump-

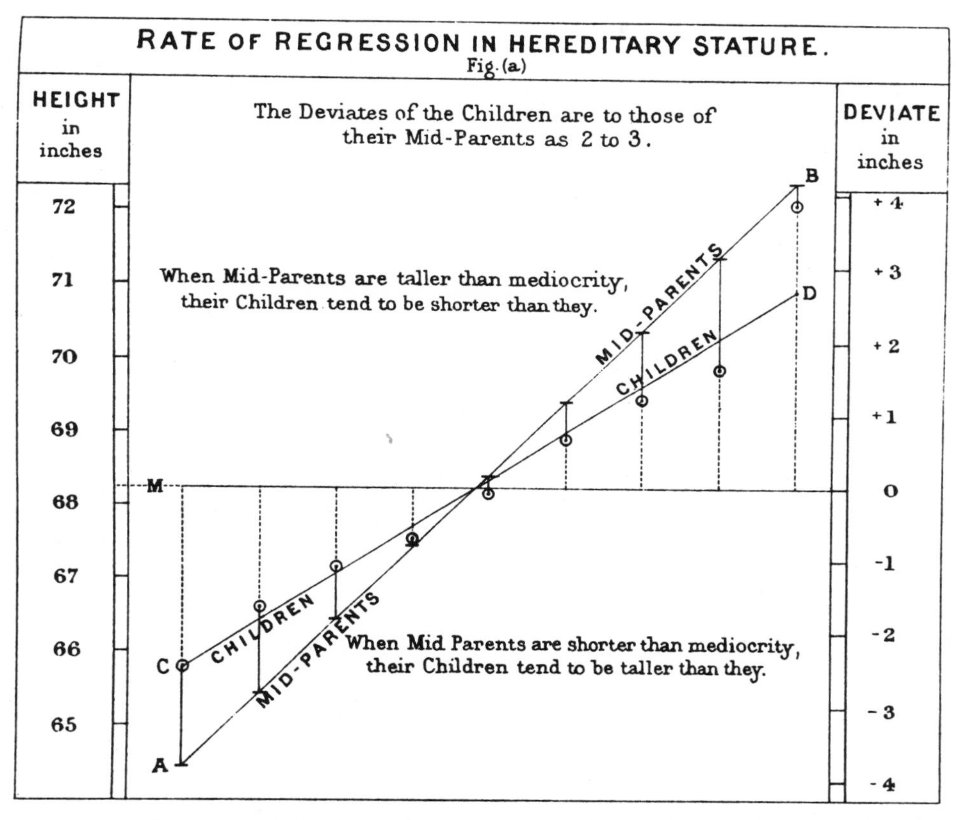

Figure 8.8. Galton's graphical illustration of regression; the circles give the average heights for groups of children whose midparental heights can be read from the line AB. The difference between the line CD (drawn by eye to approximate the circles) and AB represents regression toward mediocrity. (From Galton, 1886a.)

tivity" (or tendency to suffer from consumption or tuberculosis; pp. 183–184) upon this supposition, in both cases expressing data and conclusions in terms of percentiles and using his own earlier statistical scale. And when Galton wished to discuss the relationship between separated generations, or "collateral" relations such as cousins or nephews, he proceeded without scruple or hesitation to multiply regression coefficients. Thus the regression of nephew on uncle had slope $\frac{1}{3} \times \frac{2}{3} = \frac{2}{9}$ (p. 132). This practice, which was the only theoretical support to be given for the law of ancestral heredity (Swinburne, 1965; Provine, 1971, pp. 51–55, 179–187), was in accord with Galton's earlier data and views (for example, Figure 8.1) but is only valid in rather limited situations (for example, see Nesselroade, Stigler, and Baltes, 1980). Karl Pearson has also raised this point (Pearson, 1914–1930, vol. 3A, p. 24).

A subtler example of the gaps in Galton's understanding of his own invention can be found even earlier. When Galton had investigated "reversion" in sweet pea seeds, it is clear that he would have preferred to use seed diameter rather than seed weight as an analogy to human stature. But at least some of his work was done with weights, which he found easier to measure than diameter. In 1886 he published a "demonstration" of the equivalence of these two measures in an appendix (Galton, 1886a). He took a hundred seeds of each of seven selected weights and determined their average diameter by measuring them in one long row. The relationship between the weights and the average diameter was almost perfectly linear. He wrote that the results "show that I was justified in . . . accepting the weights as directly proportional to the mean diameters" (Galton, 1886a, p. 259). Now the problem is that Galton had fallen victim to what has since been named "the fallacy of ecological correlation." The tight relation between weight and *average* diameter did not imply a similarly tight relation for individual seeds. In fact, we should expect the contrary.

We cannot, of course, fault Galton for this, but it might shed light upon one question. Galton, we may recall, found a regression coefficient for sweet peas of $\frac{1}{3}$. Pearson (1920a, p. 194) felt that this value was too small and speculated that the sweet peas may not have in fact been self-fertilizing, as Galton had thought. Another intriguing possibility is that Galton had selected upon the basis of weight but had measured diameters of progeny in groups, thinking them equivalent. Galton's own account does little to clear up this point. Karl Pearson (1920a, p. 195) presented a diagram clearly labeled "diameters" that he claimed dated from the 1877 lecture; Galton (1885, p. 508) refers to the measurements as based on diameters. On the other hand, Galton (1877) refers to the measurements only as weights. The point is not of great importance, but it is amusing to speculate that it is at least possible that the very first regression coefficient was attenuated—biased toward zero—because Galton had used weights as an erroneous measurement of diameter.

On the more positive side it should be noted that Galton's intuitive insight could sometimes see quite subtle points that others were to miss, as when he remarked upon the potential of the regression phenomenon to mask actual effects and hence increase the difficulty in detecting any inherited acquired faculties (Galton, 1889, p. 197).

Correlation

The statistical concept Galton is best known for is correlation, yet correlation per se has played no role in the account so far. The truth is that while there is justice in associating correlation with Galton, the concept in its narrow interpretation as a single number measuring a bivariate relationship played only a minor role in Galton's own work; the word *correlation* does not appear in *Natural Inheritance.* Of course, we now recognize that two of the topics we have considered are inextricably tied to correlation: Galton's examination of the probable error of midparents' heights as a check for sexual selection, and the whole notion of regression. But the first of these really comes from error theory; and Galton's own conception of regression was most clearly enunciated as a unidirectional process. The more narrow idea of correlation was clearly just on the fringe of this work, however; and in late 1888 it briefly but effectively burst through.

In December 1888, after *Natural Inheritance* had gone to press, Galton read a short paper to the Royal Society, "Co-relations and their measurement, chiefly from anthropometric data." In it Galton carried the development of 1885 one step further. Not only could the regression problem be viewed symmetrically and not only were there two regression lines, but also, if the same statistical scale was used for both measurements, the two lines had the same slope! If both measurements — midparent's height and child's height, or, as in the 1888 article, forearm and head lengths — were expressed in units of their probable errors, then both regression lines had the same slope r. Hence this number could be taken unambiguously as an expression of the "closeness of co-relation." Galton illustrated this with a variety of measures of 350 men, measures made at his South Kensington laboratory, where he was embarked upon studies that involved identification and would eventually lead to an 1892 book, *Finger Prints.* Galton viewed his "index of co-relation" (Galton, 1888, p. 143) within the context of his earlier work on regression and emphasized its identity as a regression coefficient. Nevertheless, his intuitive understanding was clear: "It is easy to see that co-relation must be the consequence of the variations of the two organs being partly due to common causes."

The initial spelling of the term *co-relation* seems to have been a conscious attempt on Galton's part to distinguish his term from the word *correlation,* which was already in common use. In particular, the physicist W. R. Grove had published *The Correlation of Physical Forces* in 1846, a work that went through many editions. Grove wrote that two correlated ideas would be "inseparable even in mental conception," but he emphasized the asymmetric sense that "one . . . cannot take place without involving the other" (Grove, 1865, p. 183). In his 1874 *Principles of Science*, W. Stanley Jevons wrote: "Things are correlated (*con, relata*) when they are so related or bound to each other that *where one is the other is, and where one is not the other is not*" (vol. 2, p. 354). In Galton's personal copy of Jevons's book, this passage is marked, and Galton's marginal note proclaims: "Nice wd. never so with the common meaning (Groves's). Thus where motion is there heat may *not* be, but when motion is not then heat will [not] be" (Galton Archives). Within a year, however, the spelling had reverted to correlation.

Correlation was to play a much larger role in Karl Pearson's work; and in his mammoth biography of Galton, Pearson fastened upon a passage in Galton's memoirs as a description of the birth of the notion. Galton had written,

> As these lines are being written, the circumstances under which I first clearly grasped the important generalisation that the laws of Heredity were solely concerned with deviations expressed in statistical units, are vividly recalled to my memory. It was in the grounds of Naworth Castle, where an invitation had been given to ramble freely. A temporary shower drove me to seek refuge in a reddish recess in the rock by the side of the pathway. There the idea flashed across me, and I forgot everything else for a moment in my great delight. (Galton, 1908, p. 300)

Pearson wrote, "That 'recess' deserves a commemorative tablet as the birthplace of the true conception of correlation!" (1914–1930, vol. 2, p. 393), and variously dated the incident in 1889, which would have been after the paper on co-relation (1914–1930, vol. 2, p. 393; vol. 3A, p. 5), and in 1888 (vol. 3A, p. 50). Although such an association has romantic appeal, it is incorrect, as Hilts (1973, p. 233) has noted. The passage quoted appears in the section of Galton's memoirs where he was describing his statistical scale, not correlation, and there is no indication that Galton meant to associate the incident specifically with correlation. It seems much more plausible that the incident took place in the early or middle 1870s, and although it may have marked a key stage in Galton's development, that stage was not correlation. Indeed, Porter (1986) has found a hitherto unnoticed account that Galton wrote in 1890 for an American review (Galton, 1890) and in which Galton describes in some detail how he discovered correlation while *Natural Inheritance* was in press: While plotting forearm length versus height for his anthropometric data, he had noticed that the problem was intrinsically the same as that of kinship.

In the years immediately after 1889, Galton was preoccupied with other topics, notably the study of the classification of fingerprints, and he did not return to correlation, except in an advisory role to others. He had ended his 1888 article with a suggestive reference to a method involving "partial co-relation," and in 1890 he referred readers to that article, "to which it is likely I may be able to make additions before long" (Galton, 1890, p. 430). But he did not develop the topic further. He may not have had time; within three years it had been taken out of his hands by far more competent mathematicians. Galton could not have been surprised at that. He had opened up a new view of the intellectual landscape and he knew it: "I can only say that there is a vast field of topics that fall under the laws of correlation, which lies quite open to the research of any competent person who cares to investigate it" (Galton, 1890, p. 431).

9. The Next Generation: Edgeworth

Francis Ysidro Edgeworth (1845–1926), about 1892

GALTON'S WORK was widely known among his contemporaries, although its subtleties were not so widely appreciated and the full breadth of application of the work on regression was not apparent, except perhaps to Galton himself. His work was commonly viewed on two levels. On the one hand, the use of a statistical scale, of medians and percentiles as ways of characterizing distributions and reducing nonmetric data to a metric scale, was seen as a general technique of wide applicability. On the other hand, the idea of regression, when it was understood at all, tended to be viewed as specific to the studies of heredity that had inspired it.

The Critics' Reactions to Galton's Work

The promise of some of Galton's ideas was noted as far away as America, where John Dewey reviewed *Natural Inheritance* in September 1889. Dewey praised Galton's methodology and gave a brief tutorial on the use of the normal curve as a statistical scaling device. Dewey (1889, p. 333–334) wrote, "It is to be hoped that statisticians working in other fields, as the industrial and monetary, will acquaint themselves with Galton's development of new methods, and see how far they can be applied in their own fields." Dewey's description of regression was brief, however. He was skeptical about extending Galton's findings for the inheritance of stature to, say, the inheritance of wealth, unless the measurements were taken at widely separated time points. "The tendency of wealth to breed wealth, as illustrated by any interest table, and the tendency of extreme poverty to induce conditions which plunge children still deeper into poverty, would probably prevent the operation of the law of regression towards mediocrity" (Dewey, 1889, p. 334).

Other Americans were quick to notice Galton's work. Two years earlier, in 1887, a course had been introduced at Cornell University on "probabilities and least squares with sociologic applications, including some recent work of Galton." (This was possibly the first appearance of Galton's work in any university curriculum, although the actual extent of the coverage is not known.)[1] The American psychologist James McKeen Cattell, who spent the years 1886–1888 in England, was heavily influenced by Galton and brought that influence to bear upon his subsequent development of psychological laboratories at the University of Pennsylvania and Columbia University (Sokal, 1981, p. 221).

Galton's own countrymen were attentive to his work, but for the most part the immediate reception was not perceptive. His 1877 lecture on "reversion" had not inspired further work by others, although correspondence of the time does indicate that at least some scientists evinced an interest in it. His cousin George Darwin had offered criticism before the talk (Pearson, 1914–1930, vol. 3B, p. 465), and Stanley Jevons discussed reversion with him afterward (Black, 1977, vol. 4, pp. 200–202). Galton's 1885 address, in which "regression toward mediocrity" and the bivariate normal distribution were unveiled, also found a large but mostly uncomprehending audience. The address was given on 10 September 1885, and the next day the *Times* of London reported, "Another popular section was,

1. *Cornell Register* for 1887–88, p. 64. The course, Mathematics 24, was an outgrowth of one entitled "Probabilities and Insurance," which had been introduced in 1885 by Assistant Professor George W. Jones. From 1890 on, however, James E. Oliver was listed as instructor; and it seems plausible, given Oliver's familiarity with least squares (Newcomb, 1903, p. 72), that Oliver was responsible for this change in topics.

it need hardly be said, Anthropology. Mr. Galton, the president, prudently delayed the delivery of his address until 2 o'clock, so that those attending the other sectional addresses had an opportunity of flocking to the Long School, in which Mr. Galton discoursed on his rather novel studies." At the end of the lecture, Sir Lyon Playfair proposed a vote of thanks in words that could be taken as a subtle insult to Galton but more probably betrayed Playfair's lack of comprehension of the lecture: "He hoped that the regression and progression which they had been taught by those laws of Mr. Galton to believe in, would result in this, that those who were descended from him might be in a better position than himself. (laughter.)"

John Venn, the Cambridge logician, had a better understanding of Galton's work than most. His 1888 edition of *The Logic of Chance* (1st Edition, 1866; 2nd ed., 1876) included material on Galton's statistical scale and a brief discussion of regression. But even though Venn understood the relationship between regression and the maintenance of population variability, there is no indication in his book that he grasped the broader potential for the concepts Galton dealt with. Indeed, he and others can be excused by noting that Galton's own view of the potential of his work was not consistently clear. At times Galton eloquently evoked a general promise for his methods, as in the introduction to *Natural Inheritance:* "the road to be travelled over . . . is full of interest of its own. It familiarizes us with the measurement of variability, and with curious laws of chance that apply to a vast diversity of social subjects. This part of the inquiry may be said to run along a road on a high level, that affords wide views in unexpected directions, and from which easy descents may be made to totally different goals to those we have now to reach. I have a great subject to write upon" (1889, p. 3). But at other times Galton's view was much more narrowly focused upon heredity, encouraging his readers to believe that his methods were limited to that topic. In his 1885 address he had given as one important consequence of his work this definition of a racial "type": "The type is an ideal form towards which the children of those who deviate from it tend to regress" (Galton, 1885, p. 509). This was ironic as it seemed to bring Galton full circle, back to Quetelet's average man. If this was to be the payoff from Galton's work, then the work was a failure. But how was the work to be developed if no one other than its creator saw a broad potential, and he, as was the case, was both inconsistent in his interpretations and by 1890 drawn to the more immediately attractive (if less fruitful) pursuit of the study of fingerprints?

Pearson's Initial Response

Karl Pearson was eventually to play a key role in the development of Galton's work, but his initial reaction to a first exposure to the work was curious, not the least because it differed so much from his own and later accounts. Put bluntly, Pearson missed the boat in 1889; and he was not to begin his tour as a statistician until at least the latter half of 1892.

Karl Pearson was born in London on 27 March 1857 and graduated from King's College, Cambridge with mathematical honors in 1879 (Pearson, 1938; Eisenhart, 1981). His drive and curiosity soon outdistanced his mathematical ability. Throughout his life he did the work of three men, but before he turned to statistics they were in three different fields. After visiting Germany from 1879 to 1881 he immersed himself in German history, folklore, and philosophy and wrote extensively upon topics such as Martin Luther and Spinoza. As a physicist he made major contributions to the editing and completion of posthumous manuscripts by William Kingdon Clifford on the philosophy of science (published as *The Common Sense of the Exact Sciences* in 1885) and by Isaac Todhunter on the history of the theory of elasticity (vol. 1, 1886; vol. 2, 1893), and he also published several minor articles. He was an active force in intellectual politics, particularly regarding proposed changes in the structure of the University of London and current debates on socialism and the education of women. Pearson's bibliography was published in hardcover three years after his death (Morant, 1939), and it lists 648 items. Of these titles, exactly 100 appeared before 1893, the year Pearson finally started to publish on statistics; 9 of the 100 are classified as books!

In all these pursuits Pearson showed great zeal and sufficient ability to acquire a solid reputation as a man of science who could communicate with a general audience. He pursued his topics with such persistence that he apparently avoided any taint of dilettantism. In 1884 he became professor of applied mathematics and mechanics at University College London, where he developed into an excellent teacher; and in 1890 he added the duties (and honor) of the Gresham lectureship in geometry to an already immense burden.

During this period Pearson was teaching large numbers of engineering students at University College, but he still found time to participate in an evening series of lectures in a men's and women's club, where topics of socialism and sex were discussed and debated. It was in this context that he seems to have first encountered Galton's statistical ideas. Within weeks of the publication of *Natural Inheritance* Pearson had read the book and volunteered to lecture on it to the club. Pearson's lecture style was to prepare a fully written manuscript for reading, and the twenty-five-page manuscript of his lecture of 11 March 1889 survives in the Pearson archives. We might expect it to document a historical moment at which an extraordinarily energetic and ambitious man first sees the light that will guide him for the rest of his life. Instead it shows once again how difficult new concepts are to grasp and how resistant even an aspiring pioneer can be to new ideas.

Pearson's commentary on Galton was lucid, well-organized, mostly correct, and as perceptive as most, if not all, other commentators. Nevertheless, it was a view by a man wearing blinders. He saw only the application at hand; and if he explored the book's implications at all, it was as a contribution to questions related to sexual selection and to social policy (a topic on which Galton was notably silent). Pearson generally approved of Galton's work, although he did find some aspects to criticize: "Personally I do not feel quite satisfied with the application of Galton's results for stature to all forms of inheritance" (p. 21); and "There is no doubt that his general conception of regression is good; although the amounts he has settled for it may be open to criticism" (p. 24). Ironically, Pearson was most critical of Galton's application of his methods to disease, where Galton had dealt with the tendency of people to cluster into two classes by introducing the notion of a latent propensity to acquire disease and then applying his methods to the latent characteristic. "I think the data of hereditary disease are hardly sufficient to warrant his assumptions, and I am inclined to doubt whether his laws would hold for disease" (pp. 23–24). Pearson was later to make such latent propensity (or threshold or quantal response) models, which had appeared earlier in Fechner's work (see Chapter 7), a hallmark of his own work (see MacKenzie, 1981, chap. 7; Stigler, 1982b). Pearson also presented with evident approval Galton's naive multiplication of regression coefficients from, say, uncle to nephew, although later he criticized Galton on just this point (Pearson 1898, 1914–1930, vol. 2, p. 24).

Pearson thus had a first close look at Galton's statistical methods but was not able to see their promise. Far from viewing them as techniques for a general class of problems, he remained blinded by restrictions learned at Cambridge. If anything, Pearson was less enthusiastic than John Dewey had been, writing, "Personally I ought to say that there is, in my own opinion, considerable danger in applying the methods of exact science to problems in descriptive science, whether they be problems of heredity or of political economy; the grace and logical accuracy of the mathematical processes are apt to so fascinate the descriptive scientist that he seeks for sociological hypotheses which fit his mathematical reasoning and this without first ascertaining whether the basis of his hypotheses is as broad as that human life to which the theory is to be applied" (pp. 1–2).

Pearson's initial resistance to Galton's methods was so effective that nearly three years later in January 1892 when he sent his *Grammar of Science* to press, he appeared to have forgotten Galton. Pearson discussed heredity in that book, but only in terms of the work of Darwin and Weismann, whose book *Essays on Heredity and Kindred Biological Problems* also appeared in 1889. The word *correlation* is used only in the nonstatistical sense (see, for example, p. 34). The chapter on probability does not mention the possibility of its application to heredity, much less to social matters. John Venn's *Logic of Chance* is cited, but only in the 1866 edition. Galton is not to be found, even by indirect reference! Yet by November 1892 Pearson was beginning to lecture on the application of probability to inference, and by late 1893 he was contributing articles on what can be termed mathematical statistics to the Royal Society of London. What led to this transformation?

Pearson himself, writing in 1934, attributed his change in direction to his benefactor Galton and to his first encounter with *Natural Inheritance* in particular, writing, "It was Galton who first freed me from the prejudice that sound mathematics could only be applied to natural phenomena under the category of causation" (Pearson, 1938, p. 19), but this does not accord with the record noted earlier. He also frequently cited interaction with a friend, the biologist Weldon, as a major stimulus in his early work. Weldon clearly stimulated Pearson at a later time, but in one account Pearson stated that their friendship only began after the appearance of an 1892 Weldon paper (Pearson, 1906, p. 284). The earliest scientific correspondence with Weldon extant in the Pearson Archives is from November 1892, earlier exchanges being limited to university politics. It would seem that here, as is often the case, later recollections can be unreliable. Although it is of course impossible to reconstruct the actual course Pearson followed, strong evidence points to Edgeworth as the pivotal figure in this intellectual development.

Francis Ysidro Edgeworth

Even in the varied world of nineteenth-century statisticians, Francis Ysidro Edgeworth was an anomaly. Whereas most of those men we have discussed came from scientific backgrounds, Edgeworth's education was in classical literature. He was born 8 February 1845 in Edgeworthstown, Ireland, the family home of a creative and celebrated family (Keynes, 1933; Stigler, 1978a). In 1862 Edgeworth entered Trinity College, Dublin, and for several years concentrated on the classics, with marked success. In 1867 he entered Oxford University, where he graduated in 1869 with a first class in literae humaniores. Edgeworth next undertook the study of commercial law. There is little information available about Edgeworth's life from 1870 until he was called to the bar in 1877, but well before 1877 he must have realized that a career as a barrister could not satisfy either his intellectual curiosity or his ambition for recognition as a scholar.

There is no indication in existing records that Edgeworth's early training in mathematics, including that at Oxford, went beyond elementary algebra; but after leaving Oxford and while studying law, Edgeworth undertook a program of self-study in mathematics that was the equal of any university program of that time. The depth of Edgeworth's newly acquired mathematical talents is apparent in his earliest publications. His first substantial effort, *New and Old Methods of Ethics,* was published in 1877 shortly after he was called to the bar. It showed a confident and creative mastery of the calculus of variations, not to mention some knowledge of mathematical physics, and a thorough familiarity with the mathematical psychophysics of Fechner, Delboeuf, Helmholtz, and Wundt. By the early 1880s it is clear that his competence had expanded to the full spectrum of the mathematical sciences: Fourier's theory of heat; Poisson's mechanics; Cournot, Gossen, Jevons, and Walras on mathematical economics; Airy, Thomson and Tait, and Clerk Maxwell on physics; and, above all else, Laplace on the theory of probability.

In 1880 Edgeworth obtained a position as lecturer in logic at King's College, London, where he published the work that first brought him wide recognition. *Mathematical Psychics: An Essay on the Application of Mathematics to the Moral Sciences* was published in early 1881 and was a bold attempt to extend his earlier mathematical treatment of ethics to economics, in particular to incorporate utility theory with an analysis of economic contract and competition. The favorable notices the work received included reviews by the foremost economists of the day. Alfred Marshall, in the second of only two book reviews he ever published, wrote, "This book shows clear signs of genius, and is a promise of great things to come." Marshall wished the author had worked his theory through more fully before publication (advice Marshall himself followed to excess), "but, taking it at what it claims to be, 'a tentative study,' we can only admire its brilliancy, force, and originality." W. Stanley Jevons, whose *Theory of Political Economy* had been the subject of Marshall's only other review, was also generally favorable in a review in *Mind.* But although Jevons found the book a "very remarkable one" and saw in it "unquestionable power and originality," he also found the author's style puzzling and the book hard to read. "The book is one of the most difficult to read which we ever came across, certainly the most difficult of those purporting to treat of economic science."

The Jevons review led to early correspondence between Edgeworth and Francis Galton. Not long after the Jevons review appeared, Galton wrote to Edgeworth:

> 42 Rutland Gate SW
> Oct. 28/81
>
> Dear Sir:
>
> Permit me to express the very great interest with which I have been reading your powerful work of Math. Psychics, and especially those parts of it that claim the right of Mathematics to deal even with the loosest quantitative data. I write more especially, because I was led to a knowledge of your book by an article by Prof. Jevons in "Mind", in which he happens to speak of its being an unnecessarily difficult book to read. With that verdict I am totally at issue. It

strikes me that you have handled topics very difficult in themselves, with great lucidity and vivacity; and I do sincerely hope that you will not suffer yourself to be discouraged by that verdict and still less by those of such reviewers as the writer of the Saturday Review article which you quote. It is always the case with the best work, that it is misrepresented, and disparaged at first, for it takes a curiously long time for new ideas to become current, and the older men who ought to be capable of taking them in freely, will not do so through prejudice. It is a grand attempt that you are making, and by successive efforts more will assuredly be won. I trust you will continue to work at various branches of this wide subject . . .[2]

Believe me faithfully yours
Francis Galton

Edgeworth was in fact a distant cousin of Galton, and their contacts continued over the years. They never developed a close working relationship, however. It seems reasonable to surmise that conversations with Galton helped intensify Edgeworth's interest in statistics in the early 1880s. From 1883 through 1893 Edgeworth published nearly forty notes and articles in probability and statistics, one small book (*Metretike,* in 1887), and many reviews. As we shall see, much of this work bore the stamp of Galton's influence.

Edgeworth's Early Work in Statistics

After the appearance of *Mathematical Psychics* in mid-1881, Edgeworth published little before October 1883. What did appear (a few reviews and an outline of an article on rent) gave no indication of a shift in the direction of Edgeworth's thought, but internal evidence in later work suggests that he devoted considerable time and energy in this period to a study of the literature of probability and the method of least squares. It is evident from citations in *Mathematical Psychics* that Edgeworth had some knowledge of probability before 1881; he had read Laplace's *Essai philosophique sur les probabilités* and John Venn's *Logic of Chance* and was familiar with some aspects of the works of Galton and Quetelet, in particular with Galton (1875). By 1885, however, his knowledge of this literature had broadened to the point where it was unequaled in England.

Probability had played only a minor role in the *Psychics,* and statistics had played none at all, but in late 1883 Edgeworth began publication of a sequence of articles devoted exclusively to these topics. Viewed separately and out of the context of his work, they are puzzling, eccentric pieces. The

2. The omitted portion of this letter commented upon details; it is reprinted in full on the back cover of the *Journal of Political Economy,* vol. 85, no. 1 (February 1977). The letter is in the possession of David E. Butler of Nuffield College, Oxford.

titles of those published in the *Philosophical Magazine* and *Mind* from October 1883 through 1884 appear to describe their contents, but they give no indication of a common link: "The law of error," "The method of least squares," "The physical basis of probability," "The philosophy of chance," "On the reduction of observations," and "A priori probabilities." Edgeworth's plan, though, is clear both from the context of his work and from a closer study of these pieces themselves; it was to adapt the statistical methods of the theory of errors to the quantification of uncertainty in the social, particularly the economic, sciences. The type of questions Edgeworth sought to treat and the difficulties he saw in their treatment were described in an 1884 review of a posthumously published collection of Jevons's papers, *Investigations in Currency and Finance.* Edgeworth commented upon the beautiful diagrams in the book, which he thought would assist the reader

> to estimate the probability that the differences in the averages for different weeks and months are not accidental. The question which has been just indicated, one of the most delicate in statistics—namely, under what circumstances does a difference in figures correspond to a difference of fact—comes up often in these pages. Thus Mr. Jevons, comparing the amount of bills created in the different quarters of the year, speaks of a variation to the extent of about six per cent. as "no great difference." On the other hand, he regards it as noteworthy that, "out of 79,794 bankruptcies which were gazetted from the beginning of 1806 to the end of 1860, 28,391 occurred in the second month of the quarter, 26,427 in the third month, and only 24,976 in the first month." No doubt a similar disparity between "heads" and "tails" in the result of so many throws of a coin would prove a cause, a want of symmetry in the coin. But our knowledge of the behaviour of tossed coins rests at bottom upon observation and experiments such as those which Mr. Jevons once performed. That what is true of games of chance is true of bankruptcies is not to be assumed without examination. (Edgeworth, 1884, p. 38)

Edgeworth was to provide this examination.

Edgeworth's key work on this topic was contained in a series of four papers read in the year 1885. The first of these, "Observations and statistics: An essay on the theory of errors of observation and the first principles of statistics," was read on 25 May 1885 to the Cambridge Philosophical Society. It concentrated on statistical theory and summarized and extended his work of the previous two years. The second paper, "Methods of statistics," was read a month later, on June 23, to the international gathering to celebrate the jubilee of the [Royal] Statistical Society. It was concerned with methodology and presented, through an extensive series of examples taken from all manner of fields, an exposition of the application and interpretation of significance tests for the comparison of means. Much of the material in these two papers was presented at least in outline in

Edgeworth's evening classes in logic starting in April 1885 (see Appendix A). The third and fourth papers, "On methods of ascertaining variations in the rate of births, deaths, and marriages" and "Progressive means," were read at meetings of the British Association in September and October. The third presented a remarkable analysis for two-way classifications that anticipated many ideas of the analysis of variance. The fourth article, "Progressive means," was a brief discussion of the use of linear least squares for detrending time series, including the estimation of the coefficients' variabilities to permit significance tests for trend or comparisons of different series.

For Edgeworth the distinction between *observations* and *statistics* was an important one:

> Observations and statistics agree in being quantities grouped about a Mean; they differ, in that the Mean of observations is real, of statistics is fictitious. The mean of observations is a cause, as it were the source from which diverging errors emanate. The mean of statistics is a description, a representative quantity put for a whole group, the best representative of the group, that quantity which, if we must in practice put one quantity for many, minimizes the error unavoidably attending such practice. Thus measurements by the reduction of which we ascertain a real time, number, distance are observations. Returns of prices, exports and imports, legitimate and illegitimate marriages or births and so forth, the averages of which constitute the premises of practical reasoning, are statistics. In short observations are different copies of one original; statistics are different originals affording one "generic portrait." Different measurements of the same man are observations; but measurements of different men, grouped round l'homme moyen, are primâ facie at least statistics. (Edgeworth, 1885a, pp. 139–140)

Edgeworth's aim was to apply the tools developed in the previous century for *observations* in astronomy and geodesy, where a more or less objectively defined goal made it possible to quantify and perhaps remove nonmeasurement error and meaningful to talk of the remaining variation as random error, to social and economic *statistics* where the goal was defined in terms of the measurements themselves and the size and character of the variation depended upon the classifications and subdivisions employed. To do this he first disassembled these tools to determine carefully what lay beneath them: What conditions, what assumptions, what interpretations lay behind their successful use?

"Observations and statistics" focused on one theoretical aspect of this problem, the choice of a best estimate of a mean. Although the problem was presented as one of estimation, the underlying motivation, the implicit use to which the mean would be put, was the comparison of different sets of statistics. Edgeworth examined the philosophical interpretations of probability as applied to inference. He gave careful consideration to the objective frequentist or sampling theory view (he called it ordinary induction, assumed inversion, or the indirect method), but in the end he adopted the inverse probability, or Bayesian, view after giving cogent refutations to criticisms of this view by Cournot, Boole, and Venn. Edgeworth gave qualified endorsement to the use of uniform prior distributions, but only when the prior was based on experience: "We have a rough general experience that one value of the measurable occurs as often as another at any rate between the limits with which we are practically concerned" (Edgeworth, 1885a, p. 146). Like Laplace, Edgeworth embraced inverse probability for neither metaphysical (his priors were grounded on experience) nor dogmatic reasons. Like Laplace, Edgeworth was inconsistent in his application of inverse probability, reverting to sampling distributions when comparing means by significance tests.

In the selection of a best mean Edgeworth preferred choosing, with a general loss function, that value that minimized the posterior expected loss. He extensively discussed the relationship between the population distribution (the "facility-curve") and the best mean, his examples including the family of stable laws and others where averaging was detrimental, and densities where the median was to be preferred. He focused special attention upon the normal distribution (the "law of error" or "probability-curve"), which he parameterized as

$$y = (c^2\pi)^{-1/2} \exp\,[-x^2/c^2],$$

and he called c the *modulus* of the curve. He defended $2\Sigma(x_i - \bar{x})^2/n$ as an estimate of c^2 on the grounds that it maximized the posterior density, whereas $2\Sigma(x_i - \bar{x})^2/(n-1)$ "does not appear to correspond to the maximum of anything in particular." Edgeworth emphasized that he was appealing to the asymptotic normal approximation to the sampling distributions of means, not to an assumption that the populations were themselves normal. "What is required for the elimination of chance is not that the raw material of our observations should fulfill the law of error; but that they should be constant to any law" (Edgeworth, 1885b, p. 187).

The basic significance test employed in the "Methods of statistics" was extremely simple. Given two "means" (which could be medians or other estimates), first estimate their "fluctuations," a term Edgeworth invented to mean the modulus-squared or, in modern terminology, twice the variance. If one "mean" was a sample mean, $\bar{x}$, he recommended $2\Sigma(x_i - \bar{x})^2/n^2$ as an estimate of its fluctuation. If one was the median, he recommended $(2ny^2)^{-1}$, where y was the value of the population density (if known) at the median. Then, if c_1^2 and c_2^2 represented estimates of the fluctuations of the two "means", $(c_1^2 + c_2^2)^{1/2}$ estimated the modulus of the difference of the "means" (in modern terminology, $\sqrt{2}$ times the estimated standard deviation), and the test was performed by comparing the difference of the "means" with this modulus. If the "means" differed by more than twice this modulus, it was judged "exceedingly improbable" that the difference was accidental and some other cause was indicated. This was a rather exacting test (corresponding to a two-sided test at level 0.005); and Edgeworth considered smaller differences (say, 1.5 times the modulus) as worthy of notice, although he admitted the evidence was then weaker.

The main portion of "Methods of statistics" was given over to examples (often worked in several alternative ways) showing how this simple test, borrowed from the calculus of observations, could be used to assess the significance of statistical data on anthropometry, rates of births, deaths, and marriages, economic statistics, and such exotica as the flow to and from a wasps' nest, the attendance at club dinners, and the meter of Virgil's poetry. Edgeworth had encountered the work of Wilhelm Lexis in mid-1885 (Edgeworth, 1885a, p. 143), and in "Methods of statistics" he explained and criticized Lexis's methods for dealing with population statistics. One of the examples he gave was the data on bankruptcies that Jevons had presented and that Edgeworth had commented on in his 1884 review of Jevons's collected papers. (Jevons's offhand assessment of significance was found by Edgeworth to be "abundantly justified.")

Not all of Edgeworth's comparisons showed significant differences, and one of the more amusing examples concerned a wasps' nest. Edgeworth found the traffic rates at 8 A.M. and at noon on 4 September 1884 of a nest in Edgeworthstown, Ireland. Edgeworth's motivation for considering these data, which he gathered himself, was to show by analogy how import and export statistics might be handled. The rate at noon was slightly less than that at 8 A.M., and he commented: "If in an insect republic there existed theorizers about trade as well as an industrial class, I could imagine some Protectionist drone expressing his views about 12 o'clock that 4th day of September, and pointing triumphantly to the decline in trade of 2½ per cent. as indicated by the latest returns" (Edgeworth, 1885b, p. 209). Edgeworth found the difference, half the modulus, to be "insignificant."

The Link with Galton

The technical apparatus behind the test Edgeworth used was not new; it dated from Laplace. What *was* novel was the conceptual setting in which the apparatus was employed. Edgeworth, continuing along the path Galton had marked out in 1875, was subdividing populations that might have been considered homogeneous by Quetelet's test of fitting a normal curve and was then testing for differences between subpopulations using estimates of variability internal to the subpopulations. The tests might be viewed as akin to modern *t*-tests (but, of course, for large samples) or to special cases of the analysis of variance. Others had employed such tests, even with social data, but such usage was exceedingly rare and never made with the conscious, critical analysis of assumptions that Edgeworth brought to bear. For example, Edgeworth cited an 1852 *Handbuch der Statistik* by a Viennese administrator, Joseph Hain. Hain had reviewed earlier work (citing Quetelet, in particular) and had outlined in a few cases the application of the theory of errors to some examples involving social statistics, including the use of least squares to fit a trend to time series. Hain provided no conceptual rationale for such applications, however; and perhaps for this reason he was ignored and unknown to everyone except Edgeworth.

The conceptual rationale that Edgeworth provided was explicitly based upon Galton's 1875 insight. In "Methods of statistics," Edgeworth not only cited Galton's paper but also borrowed Galton's illustrative example regarding the sizes of fruit from different aspects of a garden. He included one figure (Figure 9.1) that might even have been taken from Galton's correspondence of the time: The smaller curves were said to represent the distributions of sizes of fruit from near particular walls, say, and the large curve the distribution for the garden as a whole. Edgeworth, in the context of this example, meant to provide a test "to ascertain whether the walls from which two baskets of pears have been gathered have important differences of aspect" (Edgeworth, 1885b, p. 204). Of course, this necessitated determining the dispersion of the appropriate smaller curves because that for the garden as a whole would be too large to serve as a basis for comparison.

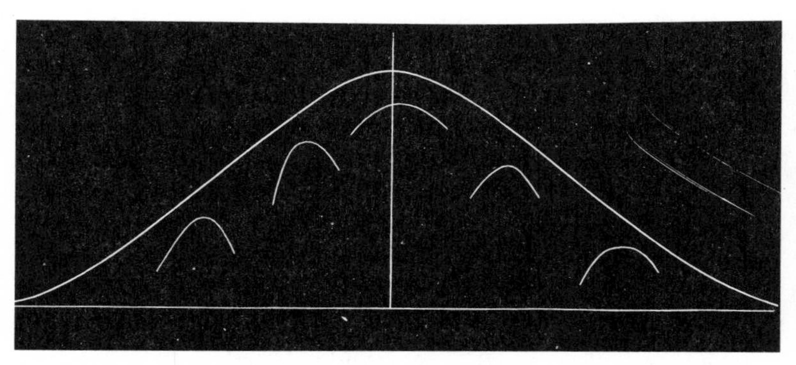

Figure 9.1. Edgeworth's 1885 illustration of Galton's 1875 insight. The smaller curves represent the distributions of the sizes of pears from specific parts of a garden (e.g., near a particular wall), and the large curve gives the distribution for the whole garden, thus showing the normal curve as a mixture of normal components. (From Edgeworth, 1885b.)

Edgeworth, then, used Galton's conceptual breakdown of large populations into a mixture of smaller populations as a basis for his test, which applied procedures adapted from the theory of errors to assess the meaningfulness of differences between the smaller populations. Galton's structure provided the needed theoretical anchor in the random world of social science. Classical error theorists had dealt with *observations,* measurements preclassified by theory, such as observations of the position of Saturn; Edgeworth used Galton's quincunx-derived conceptual apparatus to extend the classical apparatus to *statistics,* measurements of quantities that were in effect randomly determined themselves, for example, the position on the midlevel of a quincunx or the average size of fruit at a random location in a garden. The classical theory provided a rationale for the appearance of probability distributions (in particular, normal curves), where the goal was determining a fixed, objectively defined quantity from observations that were subject to measurement errors. Now these distributions could be rationalized for social data at arbitrary levels of classification, just as a quincunx could be cut at arbitrary levels. And the distributions, once rationalized, themselves served as objects for comparison. The conception was not yet fully realized in 1885, but it was a major advance from Quetelet, for whom a measurement error analogy had been necessary.

In 1885 Edgeworth did not go far beyond simple comparisons, but he did expend considerable energy on describing and developing techniques for dealing with special structures that we now call variance component models, a problem Edgeworth referred to as disentangling "entangled moduli." The work amounted to showing how to estimate dispersion in cross-classified additive models so that comparisons could be made between rows, between columns, or between cells. The main development was contained in the third paper, "On methods of ascertaining variations in the rate of births, deaths, and marriages," which was read on 12 September 1885 at the meeting of the British Association for the Advancement of Science at Aberdeen. This meeting is already famous in the history of statistics as that in which Francis Galton presented his address "Regression towards mediocrity in heredity stature."

Edgeworth's paper began by briefly reviewing some of the methods and concepts he had presented in "Methods of statistics," including the role of the modulus in statistical comparisons and the fact that, even though there is a strong tendency for data to be homoscedastic ("paradoxical as it sounds, fluctuation is one of the least flux things in the world"; Edgeworth, 1885c, p. 632), estimates of the fluctuation based on small samples are quite unreliable. ("The 'probable error' of a value based on sixteen observations is as much as an *eighth part* of the real thing observed. *A quarter* [of $2\sigma^2$] would not be a very improbable error"; Edgeworth, 1885c, p. 633.)

Edgeworth had introduced his problem, without solution, in "Methods of statistics" by noting as an example that if an investigator wished to use the registrar-general's records of marriages to determine whether one class had a higher propensity to marry than did another (to "test a theory that any particular class was more addicted to marriage, had a higher 'Matrimoniality' . . . than some other class"; Edgeworth, 1885b, p. 202), then there might be some difficulty in determining the modulus of the class's marriage rate. The registrar-general's records would be classified by place and year, and the fluctuation of a randomly selected return could be thought of as $C_t^2 + C_p^2 + C^2$, where C_t^2 was a "time-fluctuation," C_p^2 a "place-fluctuation," and C^2 a fluctuation independent of year and place. Then, if the data on social classes were from a single county, $C_t^2 + C^2$ would be the relevant fluctuation for making comparisons, whereas if it were from a single year, $C_p^2 + C^2$ would be the relevant quantity. In either case, if an investigator wished to use the records of the registrar-general to enlarge the data base beyond that available for social classes separately (thus improving the accuracy of the estimate of fluctuation used for comparisons), then it would be necessary to extract separate estimates of C^2, C_t^2, and C_p^2 from the registrar-general's two-way classification. The situation is similar in some respects to the variance-component model considered by Galton four years later, and it is plausible that Edgeworth's work here had at least a slight reciprocal influence upon Galton.

Edgeworth's solution was explained indirectly in examples involving the meter in Virgil's poetry and data on English death rates by year and county. The solution was insightful and foreshadowed much of twentieth-century work on the analysis of variance, but it was (and remains) extremely hard to decipher and had little discernible impact upon subsequent development (Stigler, 1978a, §5). One reason for its difficulty was that it presented complex material without recourse to mathematical notation. No doubt Edgeworth's aim was to render the methods accessible to the majority of the members of the [Royal] Statistical Society (in whose journal it appeared), but the effort was not successful. In some respects it was like passages in Laplace's *Essai philosophique sur les probabilités* (1814), in which Laplace attempted to communicate deep ideas to the common man, with mixed results. (For example, Laplace had attempted to convey the exact *formula* for the normal density to laymen using words alone.) Edgeworth's explanations were occasionally eloquent, as in the following passage explaining an additive effects model:

> The site of a city consists of several terraces, produced it may be by the gentler geological agencies. The terraces lie parallel to each other, east and west. They are intersected perpendicularly by ridges which have been produced by igneous displacement. We might suppose the volcanic agency to travel at a

> uniform rate from west to east, producing each year a ridge of the same breadth. It is not known prior to observation whether the displacement of one year resembles (in any, or all, the terraces) the displacements in the proximate years. Nor is it known whether the displacement in one terrace is apt to be identical with that in the neighbouring terraces. Upon the ground thus intersected and escarped are built miscellaneous houses. The height of each housetop above the level of the sea is ascertainable barometrically or otherwise. The mean height above the sea of the housetops for each acre is registered. It is required from these data to elicit the fluctuation of the height of houses above the ground. (Edgeworth, 1885c, pp. 639–640)

In one short paragraph Edgeworth tells the reader more without mathematical symbols than many recent texts convey in pages of algebra. The rows and columns of a two-way table are given physical meaning, and the fluctuation sought is clearly described, regardless of whether or not the effects are conceived of as fixed or random. The end result was unsuccessful, however, although it seems to have had a later influence upon Bortkiewicz (1895, 1896; also Edgeworth, 1896a,b). Bortkiewicz did not adopt Edgeworth's approach, but after encountering this work he developed Lexian theory along similar lines. For us the main importance of this work of Edgeworth is in the way the central question was framed in relationship to Galton's work and in the light this throws upon subsequent work of Edgeworth that did influence Pearson in fundamental ways.

Edgeworth, Regression, and Correlation

Much of Edgeworth's work through 1885 can be interpreted as using Galton's 1875 insight as a rationale for bringing the techniques of the classical theory to bear on problems in the social sciences. His appreciation of the potential of Galton's further development of regression and correlation came, for the most part, much later. Edgeworth's fourth 1885 paper, "Progressive means," can be viewed as a slight movement in that direction, however. In that work the focus was quite narrow. He mainly considered the use of classical least squares as a way to test for the statistical significance of a trend over time, as for example in English death rates from 1838 to 1878. The only overt link to Galton seems to be the choice of the word *progressive,* apparently because it was antithetical to *regressive* and more appropriate to Edgeworth's declining death rates. If Edgeworth saw a connection between Galton's regression and this work, he did not betray it in his writing of that period. By 1892, however, his view had changed.

Like Pearson, Edgeworth came upon *Natural Inheritance* soon after publication. Edgeworth's notice of the book took the form of a review in *Nature,* published on 25 April 1889. The review was highly favorable; but although it showed that Edgeworth had a better awareness of a broad potential than did Pearson, it also showed more than a hint of cautious skepticism. Edgeworth praised the mathematical development, noted that "the author has restated the law of error in a form adapted to sociological investigations" (p. 603), and stated that even physicists and astronomers could learn from Galton's methods of estimating variation. Nevertheless, Edgeworth's descriptions of regression and the more novel aspects of Galton's work were carefully qualified and restrained. Galton's reasoning was "anthropometrical," and Edgeworth was reluctant to admit its validity beyond the data on stature, for which related studies could provide at least partial corroboration. As for other areas, such as eye color, where "exact methods of measurement" were not available, he commented, "How far Mr. Galton has triumphed over this imperfection of his data, it must be for specialists to decide" (p. 604). And, as if to mirror Pearson's doubts, he added, "Our misgivings increase when we go on to apply the calculus to the returns as to disease which are obtained from the family records" (p. 604). Here also a natural resistance to Galton's bold use of new ideas took its toll. Was Edgeworth, like Pearson, to miss the boat?

In the latter half of 1890 Edgeworth wrote on the statistics of examinations; but, although his work of that period showed a growing sense of the broad scope of applicability of statistical methods, even to examiners' evaluations of Latin prose, there is no hint that he was thinking about regression. This was no longer true on 11 May 1892, when Edgeworth delivered the first of his series of Newmarch Lectures at University College London. The syllabus for these lectures was distributed in advance (correspondence from Edgeworth in the Pearson archives shows that Pearson had received one by April 25); it is reproduced in Appendix B. The sixth lecture was to be on correlation, and the material outlined was to form the subject of a series of short papers Edgeworth published over the next year.

The precise genesis of Edgeworth's interest is unclear. He cited Galton's 1888 paper on "co-relations," but the more immediate catalyst seems to have been the biologist Weldon, who sent to Edgeworth on 11 February 1892 a paper applying Galton's ideas on correlation to measurements of shrimp. Galton was in contact with both Weldon and Edgeworth at the time. Indeed, Edgeworth borrowed Galton's quincunx for display at his Newmarch Lecture. Thus it seems plausible that it was Galton who stimulated Edgeworth to consider the opportunities open to a statistical theorist to deal with problems involving simultaneous measurements of several organs.

The first of Edgeworth's papers appeared in August in the *Philosophical Magazine* as "Correlated averages." Now, the multivariate normal distribution was far from unknown in 1892. One appearance, as a limit to the multinomial distribution, could be found in work published over a century earlier by Lagrange (1776). Laplace (1812, p. 324) had found a multivariate normal joint limiting distribution for least squares estimates in 1812; Bravais had examined the geometry of the distribution in two and three dimensions in 1846, and various continental and American men, from Andräe to Helmert to Czuber to De Forest, from 1858 to 1885, had considered the distribution, usually in the context of errors in the plane or in target shooting (see Plackett, 1983, for a recent account).[3] Not all of these works were well known (quite the contrary), but it remains true that by the late 1880s the distribution was sufficiently common that Edgeworth felt justified in giving a name (Probabiloid) to the contours of the bivariate normal posterior distribution of two linear model coefficients (Edgeworth, 1887b, p. 284). In fact, Edgeworth's 1892 "Correlated averages" began with a rather baldly stated *assumption* that a collection of multivariate measurements were multivariate normal:

> The "correlation" between the members of a system such as the limbs or other measurable attributes of an organism may in general be expressed by the formula
>
> $$\Pi = Je^{-\mathrm{R}}dx_1,\ dx_2,\ dx_3,\ \&\text{c.};$$
>
> where
>
> $$\begin{aligned}\mathrm{R} &= p_1(\mathbf{x}_1 - x_1)^2 + p_2(\mathbf{x}_2 - x_2)^2 + \&\text{c.},\\ &+ 2q_{12}(\mathbf{x}_1 - x_1)(\mathbf{x}_2 - x_2) + 2q_{13}(\mathbf{x}_1 - x_1)(\mathbf{x}_3 - x_3) + \&\text{c.};\end{aligned}$$
>
> $\mathbf{x}_1, \mathbf{x}_2, \mathbf{x}_3$ &c. are the average values of the respective organs; x_1, x_2, &c. are particular values of the same; $p_1, p_2, \ldots, q_{12}, q_{13}$ are constants to be obtained from observation; J is a constant deduced from the condition that the integral of Π between extreme limits should be unity. The expression Π represents the probability that any particular values of x_1, x_2, &c. should concur. It enables us to answer the questions: What is the *most probable* value of one deviation x_r corresponding to *assigned* values x_1', x_2', &c. of the other variables? and What is the *dispersion* of the values of x_r about its mean (the other variables being assigned)? (Edgeworth, 1892a, p. 190)

3. No doubt there is much technical literature that is associated with practical problems in ballistics and could be added to the literature on bivariate normal distributions cited by Seal (1967) and Plackett (1983). Didion (1858) gave not only the density for the uncorrelated case (p. 43), but also a contour plot that was based upon the positions of 1300 shells and that he said he had made in 1823 (p. 36). An official report, dated 31 March 1880, of the Ordnance Office of the U.S. War Department gives some minor mathematical development; it was published as "Small-Arm Firing" by Henry Metcalfe, with mathematical appendix by O. E. Michaelis, in *Ordnance Notes No. 199* (Washington, 5 June 1882), part of vol. 6 of *Ordnance Notes from the Ordnance Office, War Department, Washington, D.C.* A later entry in this same series (*Ordnance Notes No. 342*, 8 April 1884) presented a translation of a chapter by E. Jouffret under the title "Probability of Fire," from the book *Les Projectiles* (Fontainbleau, 1881). No doubt a large and separate literature continued.

Edgeworth rather weakly justified the assumption of multivariate normality by the claim that each measurement ("each organ") was itself separately normally distributed (a fact "proved by theory and observation") and by the further claim that this was sufficient to imply multivariate normality. He promised a proof of this in a subsequent paper. The second claim, however plausible in 1892, is of course strictly speaking false. We should therefore not be surprised that when the proof did come (Edgeworth, 1892b), it was not fully satisfactory. Insofar as it was a proof at all, it was a vaguely stated argument for a multivariate central limit theorem. But the key point was not the source of the multivariate normal distribution; rather the key point was the use to which it was to be put.

We could restate Edgeworth's announced goal as follows, recalling that, within the normal framework that Edgeworth assumed, the "most probable" and "expected" values were one and the same: What is the conditional expected value of x_r, one of a set of multivariate normal variables, given the values $x_1', x_2', \ldots$ of the other variables? And what is the conditional dispersion of x_r given $x_1', x_2' \ldots$? The answers to these questions, determining the conditional relationship of x_r to the other x's and the uncertainty of that relationship, are central goals of modern multiple regression analysis. A modern analyst might expect Edgeworth to proceed rather directly to the method of least squares as the appropriate way of estimating the relationship, using as data a collection of sets of values (x_1, $x_2, \ldots, x_r$) of all variables. Edgeworth did not do that; indeed it was another five years before the problem came to be viewed in that manner. Instead, Edgeworth severed the problem of estimating the constants that would determine the distribution Π from the problem of determining the conditional distribution of x_r given $x_1', x_2', \ldots$. He separately estimated averages, dispersions, and correlations, and then, supposing them known, moved on to further questions. With such an approach it is not surprising that he was not led to treat the question as an application of least squares.

Edgeworth's approach was a natural one to take. He was not dealing with his own data; what empirical work he did present was in terms of data of Galton. In fact, the data do not appear to have been available to Edgeworth in raw form; for the most part he dealt with summary tables or with summary statistics. Even if the raw data had been available, it is doubtful that it would have been in a form conducive to a unified analysis. Galton had been focusing upon pairwise relationships, for which complete vectors of measurements were not essential. Missing values were common. As Galton wrote, his study had been based upon data on 350 males, but "the exact number of 350 is not preserved throughout, as injury to some limb or other reduced the available number by 1, 2, or 3 in different cases" (Galton, 1888, p. 137).

The strategy Edgeworth adopted could be summarized as follows: First,

Table 9.1. The data from which Galton in 1888 found the first published correlation coefficient.

	Length of left cubit in inches, 348 adult males								
Stature in inches	Under 16.5	16.5 and under 17.0	17.0 and under 17.5	17.5 and under 18.0	18.0 and under 18.5	18.5 and under 19.0	19.0 and under 19.5	19.5 and above	Total cases
71 and above	—	—	—	1	3	4	15	7	30
70	—	—	—	1	5	13	11	—	30
69	—	1	1	2	25	15	6	—	50
68	—	1	3	7	14	7	4	2	48
67	—	1	7	15	28	8	2	—	61
66	—	1	7	18	15	6	—	—	48
65	—	4	10	12	8	2	—	—	36
64	—	5	11	2	3	—	—	—	21
Below 64	9	12	10	3	1	—	—	—	34
Totals	9	25	49	61	102	55	38	9	348

Source: Galton (1888).

Note: Galton's ad hoc semigraphical approach gave the value $r = 0.8$. His totals for the fourth, sixth, and ninth rows are incorrect; they should be 38, 47, and 35.

find a method of determining the "coefficients of correlation"[4] from the data at hand for each possible pair of variables. Second, show how the constants of the multivariate distribution Π (that is, $p_1, p_2, \ldots, q_{12}, q_{13}, \ldots$) could be determined from the pairwise correlation coefficients, supposing those to be known. And third, show how to determine the conditional expectations and dispersions he wanted in terms of the p's and q's.

Estimating Correlation Coefficients

Edgeworth's contributions on the first step included some of his least successful work in statistics. Galton had attacked the problem of determining a correlation coefficient in a sensible, if ad hoc, way. Starting with a table showing cross-classified, grouped data (such as Table 9.1), Galton had first found the median value within each row (for example, median cubit length for each stature) and plotted it against the row value (for example, stature), both measurements being expressed in terms of devia-

4. Edgeworth introduced this term here, although Galton used "index of correlation" in 1888.

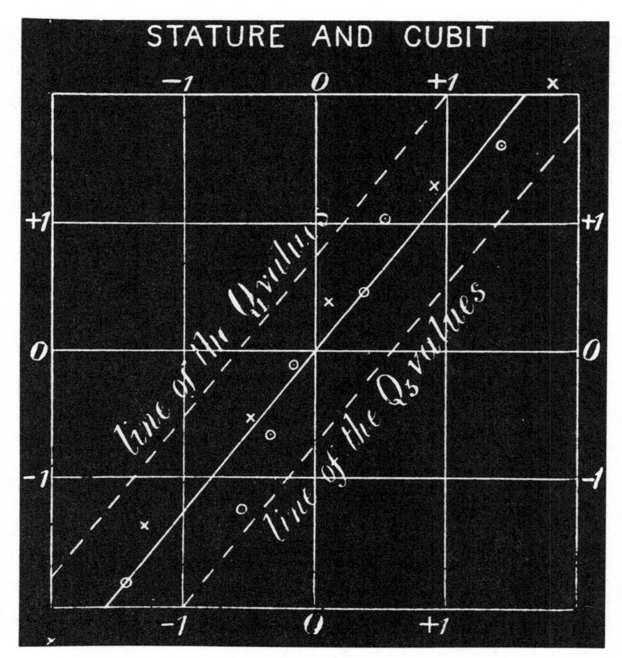

Figure 9.2. Galton's graphical determination of a correlation coefficient, based upon the data in Table 9.1. Galton converted both stature and cubit measures to standard units (number of probable errors from the median), and then for each row plotted median cubit versus stature (the circles), and for each column plotted median stature versus cubit (the crosses). In both cases the medians were plotted along the x *axis. He found the solid line by eye, and took the correlation coefficient as the inclination of that line to the vertical, which he judged to be* r = *0.8. Dotted lines approximating within row (and within column) quartiles are also shown. (From Galton, 1888.)*

tions in his standard scale. Galton had then repeated the procedure with rows and columns interchanged. The result was two sets of points, both approximating the same straight line (Figure 9.2). He had determined the slope of that line by eye, to the nearest 0.05. That slope was his correlation r. Edgeworth later gave what he implied was Galton's method as "arranging the observations in small groups, taking the quotient [$\Sigma y_i/\Sigma x_i$] for each group, and the arithmetic mean of all these quotients as the value for ρ" (Edgeworth, 1893b, p. 101). But this misses the flavor of Galton's more intuitive and graphic approach. Edgeworth's own initial suggestion for estimating the correlation coefficient ρ (to use his symbol), was computationally simple but unsatisfactory. Rather arbitrarily he focused on the individuals with above-average values of what he considered the independent variable but excluded individuals with extreme values "with respect to which the law of error is liable to break down." For example, for the data of Table 9.1 he suggested focusing upon the 55 + 38 = 93 individuals with cubit lengths between 18.5 and 19.5. He then (1) found the average cubit length for those ninety-three individuals, (2) found the average stature for those ninety-three individuals, (3) converted both figures to deviations from the population averages in standard units, (4) estimated the correlation coefficient as the ratio of standardized average stature divided by standardized average cubit length. As a method this is not absurd—it amounts to considering the correlation coefficient as a regression slope on a standardized scale (just as Galton had) and estimating it crudely, after some rather coarse grouping, on the basis of a portion of the data. The grouping in the independent variable (cubit) tended to bias the result toward zero, however. But Edgeworth's presentation was marred by a series of blunders in his first example. He forgot to subtract population averages in standardizing, and he took the ratio in the wrong order. In a climax to this comedy of errors he arrived at the same answer, $\rho = 0.8$, that Galton had found! Had Edgeworth done his work correctly, he would have found $\rho = 0.68$, which, as we would expect from the bias due to grouping on the independent variable, is too small.

In a subsequent paper Edgeworth partially redeemed himself from this fiasco with a second method for estimating ρ. The method was buried in a more general paper, and he did not emphasize it elsewhere or apply it. In a paper on the calculation of the moduli of estimates (recall that modulus equals $\sqrt{2}$ standard deviation) Edgeworth reconsidered the estimation of a correlation based upon a simple pair of variables, x and y. He supposed that he was dealing with standardized variables, "each measured from the corresponding average, in units of the proper modulus" (Edgeworth, 1893b, p. 98). In a brief paragraph he considered the variable y with x "assigned," that is, conditional on x. This gave him, since the conditional expected value of y/x was ρ, a collection of estimates of ρ that should be combined with weights inversely proportional to each one's modulus-squared. Now, the modulus of y with x assigned was $\sqrt{1-\rho^2}$ (assuming the variables were standardized to have modulus one); so that of y/x was inversely proportional to x. Thus the y/x should be combined with weights x^2. The "best value" for ρ was then (using S for summation, as Edgeworth did):

$$\frac{S(x^2 \cdot y/x)}{S(x^2)} = \frac{S(xy)}{S(x^2)}.$$

This amounted to finding the least squares estimate of ρ, regressing y on x, though Edgeworth did not describe it in those terms. Furthermore, he added, the error for this estimate had modulus

$$\frac{\sqrt{S[x^2(1-\rho^2)]}}{S(x^2)} = \frac{\sqrt{1-\rho^2}}{\sqrt{S(x^2)}} = \frac{\sqrt{1-\rho^2}}{\sqrt{\frac{1}{2}n}};$$

the last equality was due to "the modulus of x being unity by hypothesis," $(2/n)S(x^2) = 1$. This corresponds to a standard deviation of $\sqrt{1-\rho^2}/\sqrt{n}$ (Edgeworth, 1893b, pp. 100–101).

Thus Edgeworth had derived something similar to what has become known as "Pearson's" product moment estimate of the correlation coefficient. There was one doubtful point, namely, the standardization used: In stating that the estimate should be applied to measurements expressed in "units of the proper modulus," did Edgeworth mean the empirical modulus $\sqrt{(2/n)S(x^2)}$ or the theoretical modulus $\sqrt{2}\sigma_x$? If the latter, his estimate differed from Pearson's estimate, although the difference would tend to be small when n is large. On the other hand, if Edgeworth intended the empirical modulus (and his use of $S(x^2) = n/2$ in the calculation of the estimate's modulus would support this interpretation), then his "best value" reduces to $2S(xy)/n$, which is exactly Pearson's estimate, allowance being made for Edgeworth's use of moduli instead of standard deviations for scaling. But why then did Edgeworth not himself give the reduction $S(x^2) = n/2$ in the formula for his estimate instead of only in the formula for its modulus? Also, if the empirical modulus was intended as the unit of measurement, then the modulus of y with x assigned is no longer $\sqrt{1-\rho^2}$, and Edgeworth's modulus for his estimate is in error. The point must remain doubtful, although, as we shall see, Pearson's own initial analysis is equally doubtful on the same point. Edgeworth (1908–1909, p. 395) later stated that the empirical modulus had been intended, but this view was probably colored by later developments. Indeed, Edgeworth, like Pearson later, probably did not at first understand the importance of the distinction. In any event Edgeworth did not pursue the matter further or emphasize this method. He described it as involving "laborious multiplication" and preferred a version of his earlier method because it was "more convenient" and, based upon a comparison of moduli, only "worse than the best method in the degree $\sqrt{\pi}:\sqrt{2}$ or 1.25 times" (Edgeworth, 1893b, p. 101). Edgeworth's second method apparently passed unnoticed until after Pearson's work, at which time Edgeworth called it to Pearson's attention (see Chapter 10).

Edgeworth's Theorem

The remainder of Edgeworth's program involved working directly with the multivariate normal distribution. Edgeworth had written the distribution as

$$\Pi = Je^{-\mathrm{R}}\, dx_1 dx_2 dx_3 \ . \ . \ . \ ,$$

where $\mathrm{R} = p_1(\mathbf{x}_1 - x_1)^2 + p_2(\mathbf{x}_2 - x_2)^2 + \&\text{c.} + 2q_{12}(\mathbf{x}_1 - x_1)(\mathbf{x}_2 - x_2) + 2q_{13}(\mathbf{x}_1 - x_1)(\mathbf{x}_3 - x_3) + \&\text{c.}$, with $\mathbf{x}_1, \mathbf{x}_2, \ldots, \mathbf{x}_r$ being the expected values of the variables $x_1, x_2, \ldots, x_r$, which were supposed to be standardized to have unit moduli (that is, variances $\frac{1}{2}$), and J being a normalizing constant. Edgeworth described R as a "quantic of the second order" (1892b, p. 523); we now call it a quadratic form. Edgeworth's analysis was presented in the determinantal notation and terminology of the time, but we shall use a modern matrix notation to clarify issues for a modern reader. Arranging the constants $p_1, p_2, \ldots, q_{12}, q_{13}, \ldots$ in matrix form as

$$\Delta = \begin{bmatrix} p_1 & q_{12} & q_{13} & \cdots & q_{1r} \\ q_{12} & p_2 & q_{23} & \cdots & q_{2r} \\ \cdot & & & & \\ \cdot & & & & \\ \cdot & & & & \\ q_{1r} & q_{2r} & q_{3r} & \cdots & p_r \end{bmatrix}$$

and letting

$$\mathbf{x} = \begin{bmatrix} \mathbf{x}_1 \\ \mathbf{x}_2 \\ \cdot \\ \cdot \\ \cdot \\ \mathbf{x}_r \end{bmatrix}, \qquad x = \begin{bmatrix} x_1 \\ x_2 \\ \cdot \\ \cdot \\ \cdot \\ x_r \end{bmatrix},$$

we have

$$\mathrm{R} = (\mathbf{x} - x)^T \Delta (\mathbf{x} - x) = (x - \mathbf{x})^T \Delta (x - \mathbf{x}).$$

Edgeworth introduced the symbols $\rho_{12}, \rho_{13}, \ldots$ for the correlation coefficients of x_1 and x_2, x_1 and x_3, and so on.

Now, if the entries of Δ were available, the original questions (find the conditional expectation and dispersion of x_r given the values $x_1', \ldots, x_{r-1}'$ for the other variables) could be answered easily. The conditional distribution in question was simply the univariate normal distribution proportional to $e^{-\mathrm{R}}$, where the given values $x_1', \ldots, x_{r-1}'$ are substituted for $x_1, \ldots, x_{r-1}$ in R. Then the conditional modulus of x_r was simply the reciprocal of the coefficient of x_r^2 in R, namely, $1/p_r$. And the "most probable" or expected value could be found as the value corresponding to the maximum of the density, by solving $dR/dx_r = 0$; this gives, taking the $\mathbf{x}_i = 0$,

$$2p_r x_r + 2q_{1r}x_1' + 2q_{2r}x_2' + \ldots + 2q_{r-1,r}x_{r-1}' = 0$$

or

$$x_r = \frac{1}{p_r}(q_{1r}x_1' + q_{2r}x_2' + \ldots + q_{r-1,r}x_{r-1}').$$

The difficult part was to find the elements of $\mathbf{\Delta}$ (that is, $p_1, \ldots, q_{12}, \ldots$) in terms of the pairwise correlation coefficients, the ρ's. A modern statistician knows that the answer can be expressed elegantly: $\mathbf{\Delta}$ is simply the inverse of the correlation matrix

$$\mathbf{\Sigma} = \begin{bmatrix} 1 & \rho_{12} & \rho_{13} & \cdots & \rho_{1r} \\ \rho_{12} & 1 & \rho_{23} & \cdots & \rho_{2r} \\ \cdot & & & & \\ \cdot & & & & \\ \cdot & & & & \\ \rho_{1r} & \rho_{2r} & \rho_{3r} & \cdots & 1 \end{bmatrix},$$

and so the (i, j)th entry in $\mathbf{\Delta}$ is just the (j, i)th cofactor of $\mathbf{\Sigma}$ divided by the determinant $|\mathbf{\Sigma}|$. (This is true in Edgeworth's case because he was working on a scale where all variances were $\frac{1}{2}$; with a modern standardization to unit variances, an additional factor of 2 would be needed.) Edgeworth in fact arrived at just this solution (in Edgeworth, 1893c), writing ${}_1\Delta$ for our $|\mathbf{\Sigma}|$ and Δ for our $|\mathbf{\Delta}|$), but only after a piecemeal development involving several articles.

In the 1892 "Correlated Averages" Edgeworth had noted that the solution for $r = 2$ was essentially contained in Galton's work: In that case the exponent could be written in terms of the correlation as

$$R = \frac{x_1^2}{1-\rho^2} - \frac{2\rho x_1 x_2}{1-\rho^2} + \frac{x_2^2}{1-\rho^2}.$$

He went on to derive the solution for $r = 3$ and $r = 4$, essentially using an inductive argument starting with the case $r = 2$ and applying that result to all two-dimensional marginal distributions for $r = 3$ and $r = 4$. As an application he found the constants for the joint distribution of the first four values of a random walk. This work contained several misprinted subscripts, but the logic was clear. He simplified the argument somewhat in another article (Edgeworth, 1893a) before he finally realized that not only were the entries of $\mathbf{\Delta}$ proportional to cofactors of $\mathbf{\Sigma}$ but also the constant of proportionality was $|\mathbf{\Sigma}|^{-1}$. This last result was given for general r in yet another article (Edgeworth, 1893c). Karl Pearson (1896) was to name this result "Edgeworth's Theorem." (Seal, 1967, and Plackett, 1983, have discussed portions of this work.)

Edgeworth did little else on this topic at the time. He made some cumbersome attempts to calculate probable errors for his various estimates of correlations using propagation of error ideas (Edgeworth, 1893b). And he argued that Cournot's criticism of Quetelet's average man (that an individual of all average parts might be an unviable monster) could be dismissed on correlational grounds; the most probable or expected value of a multivariate normal distribution did indeed correspond to the separate averages of the separate organs (in modern terms, the expected value of a random vector is the vector of expected values) (Edgeworth, 1893d). The next major step, however, was to be up to Karl Pearson.

10. Pearson and Yule

Karl Pearson (1857–1936), in 1890

George Udny Yule (1871–1951)

I hope that you flourish in Probabilities.

—From a letter from Edgeworth to Pearson, 11 September 1893

KARL PEARSON's appointment to the Gresham lectureship in geometry in December 1890 obligated him to give regular public lectures. In his application for the post he had listed as possible topics "lectures on the elements of the exact sciences, on the geometry of motion, on graphical statistics, on the theory of probability and insurance" (Pearson, 1938, p. 21). In the first two series of lectures, in March and April of 1891, he presented the first two of these topics. The syllabi (Pearson, 1938, pp. 132–141) of these lectures are essentially an outline of *The Grammar of Science,* which Pearson sent to press in January 1892. It was only with the third series of lectures, in November 1891, that a hint of statistical interest became publicly apparent.

Pearson's topic on that occasion and in the two series of lectures given in January and May 1892 was the "Geometry of Statistics." The syllabi (Pearson, 1938, pp. 142–145) indicate that the lectures were exclusively concerned with graphical display. Pearson must have felt that the treatment of statistical graphics presented an opportunity for him to honor his titular commitment to geometry, develop his interest in social issues, and attract a much broader audience than a narrow technical lecture series would have. The syllabi show an abundance of fancy terminology: stigmograms, euthygrams, epipedograms, histograms, chartograms, hormograms, topograms, stereograms, radiograms, and isodemotic lines. He did not treat statistical inference, however. Nor did he mention Galton by name, although a reference to "Meteorological chartograms, cyclone and anti-cyclone, construction of contour lines for parts of the ocean where there are no meteorological stations" (Pearson, 1938, p. 145) suggests that he may have been familiar with Galton's *Meteorographica.*

Pearson the Statistician

In November 1892, in the first series of lectures Pearson gave after Edgeworth's Newmarch Lectures, there was a noticeable shift in focus. Pearson chose his topic to be "Laws of Chance: Being the Elements of the Theory of Probability in its Relation to Thought and Conduct." Here at last Pearson concentrated his energies upon the use of probability in the assessment of statistical evidence. He broke no new ground, although he covered much old ground. The only contemporary references he gave were to Venn, Edgeworth, and Westergaard (a book he had, it turns out, borrowed from Edgeworth). He stated that the course was "to deal with the discovery, not the application of laws of chance" (Pearson, 1938, p. 147), and he emphasized questions of logic and controlled experiments, such as Monte Carlo roulette.[1]

Pearson had encountered Edgeworth's work on a priori probabilities by 1888 (Pearson, 1888, p. 35) and had cited it approvingly then and in *The Grammar of Science* (p. 175). (C. S. Peirce, 1892, in a review of the latter in *The Nation* remarked, "He adheres to Laplace's doctrine of indirect probabilities in its least acceptable form, relying here upon Mr. F. Y. Edgeworth's cobwebs.") The Pearson Archives show the two to have been in correspondence at least from April 1891, when Edgeworth tried unsuccessfully to get Pearson to submit an article to the new *Economic Journal*

1. The first of these lectures was posthumously published in *Biometrika* 32, part 2 (1941): 89–100.

(see Kendall, 1968); and by April 1892 they were corresponding on probability. Apparently at that time Pearson had access to Weldon's data, possibly as material for the illustration of graphical methods;[2] and on April 25 Edgeworth wrote:

> My dear Pearson,
> With respect to Mr. Weldon's paper what I want is to construct from his data a table which would show the mean value of one attribute say "postspinous portion" corresponding to—not every observed value of *one* other alternative as in his tables—but every observed *pair* of two other attributes, say total carapace length and sixth abdominal tergum.
> The work would not be so laborious as it seems in virtue of a certain abridged method which is referred to in my syllabus. (Pearson Archives)

Edgeworth went on to ask for access to the original data and for Pearson's advice on which triplet was best, and he offered to lend Pearson Westergaard when next they met. The reference to the problem Edgeworth was to consider in "Correlated averages" was clear, although the absence of an example based on Weldon's data suggests that Edgeworth was unsuccessful in obtaining those data in the form he wanted.

Pearson's letters to Edgeworth have not survived, but Edgeworth's half of the correspondence (which extended over the next decade) and Pearson's manuscript notes on works of Venn, Edgeworth, and Westergaard make it clear that at this period Pearson was greatly expanding his intellectual horizons and that Edgeworth was his major intellectual colleague in the endeavor. During 1893 Pearson made progress and by October of that year his first original work was revealed to the public. His lectures and correspondence, however, showed signs of progress somewhat earlier.

In a series of lectures from 31 January through 3 February 1893 he put much emphasis upon Edgeworth's significance tests. But he showed his individuality by measuring differences in terms of "standard deviations" or "S.D.'s," a name he coined at that time,[3] rather than in terms of Edgeworth's modulus (1 standard deviation = modulus $\div \sqrt{2}$). It was in this series that Pearson first unleashed an attack upon the randomness of Monte Carlo roulette, claiming that the published evidence was incompatible with a fair wheel. Pearson later developed this attack into a marvelous polemic that he published in the *Fortnightly Review* in Spring 1894 (reprinted with minor changes in his 1897 volumes, *The Chances of Death*). His 1894 conclusion was dramatic:

> Clearly, since the Casino does not serve the valuable end of a huge laboratory for the preparation of probability statistics, it has no *scientific raison d'être.* Men of science cannot have their most refined theories disregarded in this shameless manner! The French Government must be urged by the hierarchy

2. The syllabi of May 1892 and before mention no related application, however.
3. Manuscript notes show Pearson using "standard divergence" in late 1892.

of science to close the gaming-saloons; it would be, of course, a graceful act to hand over the remaining resources of the Casino to the Académie des Sciences for the endowment of a laboratory of orthodox probability; in particular, of the new branch of that study, the application of the theory of chance to the biological problem of evolution, which is likely to occupy so much of men's thoughts in the near future. (Pearson, 1894b, p. 193; 1897, vol. 1, p. 62)

The tests of the roulette wheel he performed in the article were naive—separate tests of separate, selected sets of outcomes. The February 1893 syllabus (Pearson, 1938, p. 148) shows a bit more awareness of the subtleties of the problem, stating: "Question as to the 'correlation' of the 37 numbers, difficulty solved by a theorem of Professor Edgeworth's." Actually, the question at issue was a question of testing goodness of fit for a multinomial distribution, and it was to lead in 1900 to Pearson's introduction of the chi-square test. Pearson had been corresponding about this question with Edgeworth in January 1893.[4] Edgeworth had emphasized that different frequency counts were correlated and had suggested that, working with a normal approximation to the distribution of deviations from equal frequency, this would involve a thirty-six dimensional distribution. Somewhat sarcastically he wrote, "The illustration in *space of thirty-six dimensions* will be admirably suited for a popular audience. I shall like to attend on the occasion—a few years hence, I presume, to allow time for the arithmetic being done." On the back of the envelope he added, "I forgot to say that the arithmetic involves among other operations the evaluation of a determinant of the 36th order."[5] Although he suggested some ways of transforming the multivariate integral, Edgeworth was unable to see the simplifying step Pearson was later to discover, either at this time or a year later when the matter arose again. (Pearson, 1965, §4, reproduces much of the relevant correspondence; Plackett, 1983, also discusses the discovery of chi-square.)

Skew Curves

Roulette may have given Karl Pearson his first introduction to a nontrivial problem in mathematical statistics. But his first published work—and his initial fame—came from his contributions to the analysis of skew curves.

4. Edgeworth's letters in the Pearson Archives are dated January 6, January 7, and January 10 but do not give a year. The references to roulette and the work being for "a popular audience" point to 1893 because the syllabus for February 1893 is the only one to stress that topic. This is also consistent with the mention of the "theorem of Professor Edgeworth's" in the syllabus.

5. Edgeworth explained the dimension in a postscript to one letter: "Why [do] I say 36 rather than 37? Apply the method in case the original data related to three cases 0, 1, and 2. You will see you have only *two* independent observations." Thus he was correcting for the loss of one degree of freedom through the restriction on the sum of the counts.

The nineteenth century is often spoken of as a period when all distributions were assumed normal. That is not true. It is true that throughout most of the century it was assumed that observations of a single phenomenon, homogeneous with respect to all but random, individually insignificant factors, would follow the normal curve. Indeed, as we have seen, Quetelet and others took the appearance of that curve as a validation of those hypotheses. But the curve was not to be generally assumed without some check. Poisson (1829, pp. 20–22) had proposed one check (based upon a comparison of even and odd sample moments), and Cournot (1843, pp. 227–228) had suggested that means and medians be compared, a suggestion later endorsed by Edgeworth (1887a, p. 333).[6] Quetelet's test was less formal; essentially it was visual inspection of his two columns of observed and calculated frequencies. However, as we have also seen, these checks were relatively insensitive to all but gross departures from normality; and the result was an appearance of a generally normal world, even when Laplace's hypotheses did not hold. Indeed, Galton's pathbreaking work was directed at the separation of these normal worlds into separate normal colonies on the basis of correlated measurements.

As empirical material accumulated, however, statisticians became more conscious of situations where ostensibly homogeneous observations betrayed nonnormal, particularly skewed distributions. Quetelet himself had encountered such data (Quetelet, 1846, p. 180; 1852); and he suggested that they be represented by an asymmetric binomial distribution. In 1879 Galton had both introduced a powerful way of dealing with asymmetry and enlarged the scope of applications for the normal curve with his suggestion that the logarithms of the observations be analyzed. Drawing upon ideas of Fechner, Galton reasoned that many vital phenomena were multiplicative in effect. Jevons (1863; Stigler, 1982c) had earlier advanced this argument as a justification for using the geometric mean as an index number. Galton went further in noting that multiplicative effects and the Laplacian hypotheses would imply that the logarithms of observations are normal. At Galton's instigation a Cambridge mathematician (Donald McAlister) carried out a mathematical development of what we now call the lognormal distribution (McAlister, 1879). In *Natural Inheritance* Galton mentions in several places (for example, pp. 95, 118) that he had applied these ideas to his data, which had shown some asymmetry, but found the improvement "insignificant."

In 1887 a latent interest in this topic was rekindled by a letter to *Nature* from John Venn. Venn protested against the theoretical assumption that

6. Edgeworth's formal development of this test erroneously treated the sample mean and median as independent, despite the fact that Laplace had found the joint distribution of the two in 1818, in work Edgeworth referred to elsewhere (for example, Edgeworth, 1887b). See Stigler (1973a) for a discussion of that work of Laplace.

there was only one law of error; and in two charts he presented examples based on large numbers of meteorological measurements (of pressure and temperature) that he claimed showed marked departure from normality. Actually, the figures were sufficiently bell-shaped that they might have passed as normal in a less demanding age, but Venn used them as a challenge: "What I trust is that these results may be the means of calling forth some discussion by practiced experts in this branch of statistical inquiry, which may serve to confirm or correct my results, and in the former case to offer some explanation of the causes of the phenomena" (Venn, 1887, p. 412).

A year earlier Edgeworth had considered the question of fitting asymmetrical binomial distributions to asymmetrical frequency data in a paper, "The Law of Error and the Elimination of Chance," in the *Philosophical Magazine* (Edgeworth, 1886b). The paper included a suggestion that Poisson's higher order approximation to the distribution of sums could be useful for this purpose. In response to Venn Edgeworth wrote to *Nature* (briefly listing the ways Laplace's hypotheses could have failed with Venn's data) and also prepared a longer paper, "The Empirical Proof of the Law of Error," for the *Philosophical Magazine* (Edgeworth, 1887a). In that paper he examined two other data sets for nonnormality with a simple goodness-of-fit test. He broke the range of data into two classes and found the odds strongly in favor of the normal distribution over some simple alternatives (see Stigler, 1978a, including Eisenhart's discussion).

Karl Pearson's first statistical publication, a 26 October 1893 letter to *Nature,* was in a direct line of descent from these works. Citing Venn's letter, Pearson wrote, "I have recently obtained a generalised form of the probability curve which fits with great accuracy such curves" (Pearson, 1893a, p. 615). The general treatment would have to wait; in the meantime Pearson wanted to announce that he could do a better job of fitting asymmetrical point binomials to such data than Edgeworth had: "Accordingly I proceed *not* by the method suggested in Prof. Edgeworth's 'Law of Error and the Elimination of Chance' (*Phil. Mag.* p. 318, April 1886), but by a method of higher moments." Pearson considered a system of binomial probabilities spaced apart by c, based on n trials with probability p on each, and adjusted c, n, p and the location so that the binomial agreed with the frequency curve in its first four moments about the mean.

Pearson's letter hinted at "a generalised form," and indeed he was at that time at work on a project that would run to a total of over a hundred printed pages and lead to his election to the Royal Society. The work was presented to the Royal Society on 18 October 1893, a week before the letter to *Nature.* An abstract was published in the *Journal of the Royal Statistical Society* in December 1893, and the account finally appeared in full in 1894 and 1895 in two monstrous memoirs in the *Philosophical*

Transactions of the Royal Society. The memoirs were the first two of a series of "Contributions to the Mathematical Theory of Evolution." The work grew from Pearson's interest in statistical graphics and seems to have been specifically developed to analyze data Weldon had brought him in 1892 purporting to show a "double-humped" frequency curve for crab measurements. The mathematical tools were borrowed from mechanics, with a heavy emphasis upon moments. In some respects, however, the statistical approach was rooted in a much earlier period.

The first portion of the 1893 work was concerned with the dissection of an asymmetrical frequency curve into a mixture of two normal curves, namely, determining $c_1, c_2, b_1, b_2, \sigma_1, \sigma_2$ so that the curve would be given as

$$\frac{c_1}{\sigma_1\sqrt{2\pi}} \exp\left[\frac{-(x-b_1)^2}{2\sigma_1^2}\right] + \frac{c_2}{\sigma_2\sqrt{2\pi}} \exp\left[\frac{-(x-b_2)^2}{2\sigma_2^2}\right].$$

The possibility of such a dissection was of course implicit in Quetelet's work and was mentioned explicitly by Galton in 1869 (see Chapter 8), but Pearson seems to have been the first to give an explicit mathematical method for accomplishing it. The algebra involved was frightening (including the solution of ninth-degree equations) because Pearson sought to accomplish his dissection by matching five moments of the mixture to those of the frequency curve (Pearson, 1894a). The referee's report prepared by Galton for the Royal Society was favorable but reserved:

> I regret that I am not competent to give a useful opinion on the soundness of Prof. Karl Pearson's paper.
>
> But assuming its mathematics to be correct (as is most reasonable to expect) I should recommend its publication in the Transactions as a vigorous and original attempt to supply a statistical want.
>
> Such misgivings as I may have as to its practical utility being as great as they [sic] seem to appear to the author, I quite put aside in making this recommendation; for, if the paper be sound & solid, without doubt others will occupy themselves with different aspects of the same problem & achieve greater simplicity in the end. This páper will in any case be the first step.
>
> Francis Galton
> Nov. 22/93
>
> (Royal Society Archives, RR.12.12)

In a separate letter to Pearson, Galton went further in expressing doubts, writing on 25 November 1893 that

> My misgivings, rightly or wrongly based, about the practical application of your method are (1) if there be really 3 or more components and if the given curve be dissected into only 2, then neither of the 2 calculated components can be right. (2) I do not see how you can get rid of the large bulk of *blended* cases. (3) As Prof. [George] Darwin said [when the paper was read to the

Royal Society], it seems to me that observed curves of frequency are never so exact in contour as to lend themselves to exact & minute treatment. I once amused myself by mixing together a large number of measures of male & female adults and was painfully surprised by finding that the result did not deviate anything like so sensibly as I had expected from a normal curve. (Galton Archives)

The Pearson Family of Curves

The dissection of a curve into two normal components was but part of Pearson's 1893 work. His abstract shows that he also presented the first of what was to be a whole family of skew curves. We now call the curve the gamma distribution. It was derived as an approximation to an asymmetric binomial, in this form:

$$y = \frac{a}{\sqrt{2\pi\mu_2}}\left[\frac{\sqrt{(2\pi\beta)}\beta^{\beta}e^{-\beta}}{\Gamma(\beta+1)}\right]\left(1+\frac{\mu_3}{2\mu_2^2}x\right)^{\beta-1} e^{-(2\mu_2/\mu_3)x}$$

μ_2 and μ_3 being moments about the mean of the binomial, $\beta = 4\mu_2^3/\mu_3^2$, Γ the gamma function, and a the area under the curve. Pearson initially called this a "generalised form of the normal curve of an asymmetrical character"; later it became known as a Type III curve. Unknown to Pearson at the time, just this curve was derived and presented as a useful skew curve by the American Erastus De Forest (De Forest, 1882–83; Stigler, 1978c). When Edgeworth called De Forest's work to Pearson's attention in the summer of 1895, Pearson acknowledged De Forest's priority in a letter to *Nature* (Pearson, 1895b).[7]

Pearson's analysis of the symmetric binomial probability function had led him to a simple relationship for each section of a polygon constructed through the graphed binomial probabilities:

$$\frac{\text{slope of polygon}}{\text{mean ordinate}} = -\frac{2\text{ mean abscissa}}{2\sigma^2},$$

which he noted was the same in form as that for the normal curve:

$$\frac{\text{slope of curve}}{\text{ordinate}} = -\frac{2\text{ abscissa}}{2\sigma^2}.$$

He italicized this discovery in the published version (1895a, p. 54): "Hence: *this binomial polygon and the normal curve of frequency have a very close relation to each other, of a geometrical nature, which is independent of the*

7. Pearson wrote to Yule on 2 August 1895: "I saw Edgeworth and he told me with some glee that an American had in 1884 reached my skew curve of type III! So he has and quite nicely: see *Nature* this week" (Pearson Archives).

magnitude of n." In terms of a differential equation Pearson (1895a, p. 79) gave this as

$$\frac{1}{y}\frac{dy}{dx} = -\frac{x}{c_1}.$$

By working with the asymmetric binomial he was led to similarly corresponding difference and differential equations. The differential equation, which he solved to find his "generalised curve," was in this form:

$$\frac{1}{y}\frac{dy}{dx} = \frac{-x}{c_1 + c_2 x}.$$

(Interestingly, essentially the same analysis had led De Forest to the same curve.) By the time this work was published in 1895 Pearson had added a similar analysis of the hypergeometric distribution and had arrived at a third differential equation (1895a, p. 79):

$$\frac{1}{y}\frac{dy}{dx} = \frac{-x}{c_1 + c_2 x + c_3 x^2}.$$

The remainder of the finished paper presented a detailed categorization of the solutions of these differential equations into five groups, or "Types," and descriptions of how the moments of the data could be used to distinguish among the types and to choose a curve to fit the data. Type I was an asymmetric beta density, Type II a symmetric beta density, Type III the aforementioned gamma, and Type V the normal. Type IV was the family of asymmetric curves

$$y = \frac{y_0}{(1 + x^2/a^2)^m} e^{-\nu \tan^{-1}(x/a)}.$$

Pearson assessed the fit by what he called "a rough test of the goodness of fit," a calculation of "mean percentage errors" (Pearson, 1895a, p. 50). The description of exactly what he meant was unclear. Yule (1896a, p. 332) described the criterion as

$$\frac{\text{area between curve and polygon}}{\text{area of whole curve or polygon}},$$

where by polygon he meant the connected frequency histogram. Somewhat later (11 November 1898) Edgeworth wrote to Pearson asking for clarification. Letting Y_i and y_i be the heights of the frequency histogram and the corresponding heights of Pearson's fitted curve, respectively,

$i = 1, \ldots, n$, Edgeworth asked Pearson whether he meant

$$\frac{1}{n}\sum_{i=1}^{n}\frac{|Y_i - y_i|}{Y_i} \quad \text{or} \quad \frac{\sum_{i=1}^{n}|Y_i - y_i|}{\sum_{i=1}^{n} Y_i}.$$

From Yule's 1896 description it is clear that the second was meant, although Edgeworth's first guess is interestingly close to Pearson's later (1900) chi-square statistic, namely,

$$\sum_{i=1}^{n}\frac{(Y_i - y_i)^2}{y_i}.$$

Pearson's actual statistic can be written in modern notation as $\int|f-g|d\mu$, where f is the fitted density, g the empirical one, and μ a "counting measure." In this form it is recognizable as a distance measure now arising in statistical theory.

If this were all Pearson had done, it is conceivable that his work would have been ignored—the fate of De Forest's work of 1882–83 and Thiele's work of 1889 (see Hald, 1981). But in thirty pages of detailed examples he demonstrated the successful flexibility and practicality of the system with a force that bludgeoned any potential skeptic into submission. The examples ranged from Venn's barometric pressures to the heights of St. Louis schoolgirls to Weldon's crabs to statistics on pauperism. The publication of the manuscript had been delayed for over a year while Pearson assembled this evidence, and the result showed him to have been wise in this precaution. Not only was his work more general than others, it was practical and came to public view with a record of proven accomplishment.

In some respects, though, Pearson's statistical attitude in this work was distinctly that of Quetelet: The appearance of the curve betrays its origins. "When a series of measurements gives rise to a normal curve, we may probably assume something approaching a stable condition; there is production and destruction impartially round the mean. In the case of certain biological, sociological, and economic measurements there is, however, a well-marked deviation from this normal shape, and it becomes important to determine the direction and amount of such deviation" (Pearson, 1894a, p. 2). In his initial work, Pearson took the genesis of his curves quite literally, even speaking of the underlying asymmetric point binomial as a "dissection" of the curve. He wrote of his Type III curve, "The importance of this first dissection of asymmetrical frequency curves lies in the fact that it measures the theoretical number n of contributory 'causes' and the odds $p:q$ that an element of deviation will be *positive*" (Pearson, 1893c, p. 677). By 1895 this literal view had softened, and the emphasis was upon the curves as evidence of homogeneity, notwithstanding their nonnormality, because the closeness of fit showed the phenomenon in question to be explainable in a simple way as the result of a homogeneous set of elementary random disturbances.

At first glance this was Queteletian imperialism run rampant. Not only were normal distributions evidence of homogeneity, but a great array of nonnormal distributions were as well. This spreading of the domain of homogeneity was antithetical to the thrust of Galton's most innovative work, and we might expect Galton and Weldon to have been displeased. In his official reaction, Galton was open-minded as ever. His referee's report of 1 January 1895 recommended that the second paper also be printed: "It is a most original paper, and a decided advance in Statistical Theory" (Royal Society Archives, RR.12.369). But some signs of skepticism can be found in letters. In earlier private correspondence Galton had reminded Pearson of Galton's own lognormal distribution, which he felt would often be adequate. And Galton had put in an oblique word for his own program, writing on 18 November 1893, "The law of frequency is based on the assumption of perfect ignorance of causes, but we rarely *are* perfectly ignorant, & where we have any knowledge it ought of course to be taken into account" (Galton Archives). Somewhat later he tried again to steer Pearson into incorporating other variables into his approach, writing on 17 June 1894, presumably in reference to a draft of Pearson, 1895a:

> About the charts—
>
> They are very interesting, but have you got the *rationale* quite clear, of your formulae, and the justification for applying them? Or are they mostly mere approximative coincidences?
>
> The estimated mean for tints [an example in Pearson (1895a, pp. 90–92)] must fall into line with some other physiological processes, but which? & what is the common feature in their causation?
>
> Wherever the normal law is found inapplicable, it is most probable that some big dominant causes must come in, each of which might be isolated & separately discussed and the law of its frequency ascertained. (Galton Archives)

Weldon, in letters to Galton, was more frank in his criticism, but in essential agreement:

> [23 January 1895; evidently referring to the example in Pearson, 1895a, p. 84]
>
> I fail to see how a method which treats American school children—the most heterogeneous material on earth of its kind—as homogeneous can be reliable.

[27 January 1895]

Of course, I see the power of Pearson's methods, so far as one can from the mere popular account which he gave on Thursday. Also, I see that there are many cases of "skew" variation: but all cases which he has given, of variation with an unmistakably skew frequency, are taken from phenomena which are changing with a rapidity much greater than that of any organs in crabs, or such creatures.

Pauperism, divorces, and the like, have only been invented, in their present form, for a short time, and as he himself shows, the maximum frequency changes its position at least in ten years.

The point I am anxious about is the theoretical value of *slight* skewness, when observed in such measures as mine.

[11 February 1895; referring to a draft magazine article by K.P. attacking a Weldon Committee report on variation in crabs]

Here, as always when he emerges from his cloud of mathematical symbols, Pearson seems to me to reason loosely, and not to take any care to understand his data . . . Can we not get some mathematician on our Committee?—The position now seems to be that Pearson, whom I do not trust as a clear thinker when he writes without symbols, has to be trusted implicitly when he hides in a table of Γ-functions, whatever they may be.

[3 March 1895]

The more I think over Pearson's theory of skew variation, the less evidence I see that it is a real thing. (Galton Archives)

Weldon's principal reservation, which he argued forcefully with both Galton and Pearson, was that the evident asymmetry and other nonnormality could be, and in Weldon's view often would be, simply a masking of heterogeneity due, for example, to growth and unequal age in the population under study. Pearson insisted on treating all measurements together, sometimes at a cost Weldon would not accept. On 6 March 1895 Weldon wrote:

About the mathematicians. I feel the force of what you say, naturally. But I am horribly afraid of pure mathematicians with no experimental training.

Consider Pearson. He speaks of the curve of frontal breadths, tabulated in the report, as being a disgraceful approximation to a normal curve. I point out to him that I know of a few great causes (breakage and regeneration) which will account for these few abnormal observations: I suggest that these observations, because of the existence of exceptional causes, are not of the same value as the great mass of the others, and may therefore justly be disregarded. He takes the view that the curve of frequency, representing the observations, must be treated as a purely geometrical figure, all the properties of which are of equal importance; so that if the two 'tails' of the curve, involving only a dozen observations, give a peculiarity to its properties, this peculiarity is of as much importance as any other property of the figure.

For this reason he has fitted a 'skew' curve to my 'frontal breadths'. This skew curve fits the dozen observations at the two ends better than a normal curve; it fits the rest of the curve, including more than 90% of the observations, *worse.* Now this sort of thing is always being done by Pearson, and by any 'pure' mathematician. (Pearson, 1965, p. 337; Galton Archives)

Weldon thought that Pearson's dogmatic insistence that the entire population of measurements be brought under one curve, even at the cost of a poorer fit, was mathematical myopia. Weldon was quite content to analyze a rationally selected subpopulation, by excluding a small fraction of extreme cases. Pearson distrusted this as rationalization and remained adamant on the point. He did come to grant Weldon's argument that growth (or the "moving mean") could produce nonnormality, but he would not budge on the matter of excluding extreme values from his analysis. In his obituary for Weldon, Pearson did state, though, that it was through discussions with Weldon that he came to work on the influence of selection upon variability (Pearson, 1906, pp. 291–292).

Pearson versus Edgeworth

Pearson's work on skew curves did claim one major casualty: Edgeworth. In 1883 and again in 1886 Edgeworth had discussed asymmetric corrections to the normal curve as an approximation to the distribution of sums. When Pearson was at work on his "generalised form of the probability curve" in the fall of 1893 he had been in regular communication with Edgeworth; and it is clear from his published comments that one incentive to his work on this problem had been to do better than Edgeworth had. Edgeworth's immediate reaction was to develop his own approach further, to meet the challenge in a collegial way. He sent Pearson a paper doing this even before Pearson's first presentation to the Royal Society. Pearson was angry that Edgeworth would try to scoop him; and Edgeworth, to mollify him, agreed to hold off publication. Much of the surviving relevant correspondence is reprinted in Freeman's discussion in Stigler (1978a, p. 319).

The following summer Edgeworth sent a long paper on this topic to the Royal Society, which he read on 21 June 1894. The major mathematical referee (James W. L. Glaisher) was very slow, only advising against publication the following May.[8] Despite favorable reports by Galton and George Darwin, the paper was rejected. It was now Edgeworth's turn to be angry. Although he did not act impolitely, he probably (and incorrectly) blamed Pearson for the rejection, as Pearson thought he did (Stigler, 1978a, p. 319). One sign of Edgeworth's annoyance was that even though he had discovered De Forest's work by December 1894, when he added an appendix to his Royal Society manuscript, he waited until July 1895 to

8. It is possible Glaisher was not the first mathematical referee tried.

inform Pearson about it—after it was too late for Pearson to add a citation to his own paper. Another is that Edgeworth published a vigorous critique of Pearson's work in September 1895. Pearson's equally vigorously annotated copy survives in the University College Library.

The essence of Edgeworth's argument was that the connection between Pearson's underlying binomials and the curves derived from them was too tenuous to serve as a rationale and that the curves could only be viewed as ad hoc empirical constructs. In particular, they could not be used to infer homogeneity. They fit data well, but "it is a nice question what weight should be attached to this correspondence by one who does not perceive any theoretical reason for the formulae" (Edgeworth, 1895, p. 511). Edgeworth's own favored approach, on the other hand, was derived rather explicitly from assumptions of homogeneity. Edgeworth found, for example, that his own curve did not fit Pearson's divorce data well, whereas Pearson's did. Pearson, in his copy, underlined this passage on p. 511: "But there is no certainty about the cause of these observations, no science, as there is in the case of the general law of error [Edgeworth's]. Perhaps the statistics of divorce do not form even a particular law of error." In the margin Pearson wrote, "In other words, if my curve fits & yours does not, the statistics are not 'chance' statistics in their distribution."

The point is that Edgeworth thought he saw in Pearson's paper what he had seen in Lexis's work, an unsupported appeal to binomial hypotheses when the data did not warrant it. Pearson's marginal comment had Edgeworth right—Edgeworth thought that Pearson's derivation from the binomial had been presented as the rationale for the curves and that Pearson had intended to conclude from a good fit that the binomial hypotheses were validated. Edgeworth felt that this conclusion was unsupported and unsupportable. In addition, Edgeworth believed that if *his* theoretically based general law of error did not fit, then the statistics were not "chance" in the sense that they were not homogeneously susceptible to a large number of random contributory causes. And Edgeworth had Pearson right, at least the Pearson of 1893. Pearson had entered into this work with the view that a close fit would validate the hypotheses used in the derivation. Indeed, Pearson never fully discarded this belief; but by 1895 he had changed his view slightly, without admitting it to Edgeworth and the world and possibly without realizing that he had done so. In one margin of Edgeworth's critique, next to a criticism of Pearson's logic in inferring simple binomial elemental causes from a fit, Pearson wrote, "The question is not whether the elements *are* these but whether their total *effect* can be thus simply described."

What Edgeworth had not reckoned with was that by 1895 Pearson had apparently moved toward viewing his curves in terms of his more general approach to the philosophy of science. That philosophy was explained in the *Grammar of Science* and was characterized by C. S. Peirce (1892) as "Kantian nominalism"—laws of nature were mere "products of the perceptive faculty," and "the logic man finds in the universe is but the reflection of his own reasoning faculty." That is, Pearson came to view his curves precisely as mental constructs that could concisely summarize empirical evidence, and the homogeneity he found was a product of this mental construction rather than of any actual underlying contributory causes, the existence of which Pearson may well have denied. From Pearson's evolving point of view he was perfectly justified in treating data that fit his curves as if they had arrived according to his binomial hypotheses because the hypotheses thus viewed were a kind of mental shorthand rather than a realistic depiction of the world.

Pearson's student G. Udny Yule tried to clarify the argument in a reply to Edgeworth that he included in an article on pauperism the following year. Yule's interpretation of Pearson's approach was, first, that with simple experiments (such as tosses of a die) we could consider each toss as influenced by a multitude of causes. But the number of tosses n could be small. Then a large number of contributory causes is effectively grouped in a small number of "component cause-groups." Yule went on:

> Now when we pass from fitting a point-binomial to a known discontinuous number of observations to fitting it to a series known to be really continuous, our numbers n, p, and q no longer admit of anything but a metaphorical interpretation, if indeed they can be considered as anything further than constants describing the form of the distribution . . . Just as n in the case of dice throwing is, as it were, the number of distinct channels through which the multitude of small contributory causes affect the form of the [distribution], so we assume as a tentative working hypothesis that it bears the same significance in the case of continuous variation. (Yule, 1896a, p. 331)

Yule's version of the argument might be considered as a form of a "factor model" in which he wanted to interpret the n from the fitted curve as the number of separate independent factors affecting the data. Later in the paper he succumbed to the temptation of advancing, however tentatively, the interpretation that an increase from $n = 3.3$ in 1870 to $n = 13.5$ in 1890 for one curve fit to data on pauperism indicated an increase in the complexity in the economic conditions of pauperism (Yule, 1896a, p. 340).

When Yule sent Pearson the draft of this paper in the spring of 1896, Pearson replied in an undated letter with unqualified approval:

> I took up your paper after dinner tonight and read it straight through. I am very delighted with it and am, perhaps, the only person (Miss Lee excepted) who can appreciate the vast amount of arithmetic work in it. I have no criticism to make and think you put the case for skew curves quite fairly and soberly . . . I particularly like your emphasis on the distinction between

> contributory cause and contributory cause group, and I think you have put the whole thing in a way which might be understanded of [sic] the statistician, if he can grasp any theory at all . . . I think the great service of the paper will be to show how to apply the new methods. (Pearson Archives)

Edgeworth's rejected paper remains in the Royal Society Archives; only an abstract was published (Edgeworth, 1894b). Most of the paper was subsumed in a later paper (Edgeworth, 1905). With the wisdom of hindsight we can see that the paper (the first full treatment of the Edgeworth series) merited publication, although we can sympathize with the referee, who mistook Edgeworth's generous citation of his predecessors as a sign of lack of novelty and let Edgeworth's presentation obscure the importance and scope of the paper. Even the favorable referee (Darwin) felt the paper "requires revision" (Royal Society Archives, RR.12.99)

Edgeworth's greatest misfortune was to allow himself to become fixated upon this idea. Over the next thirty years he published enough on this topic to fill more than two inches of file space. His relations with Pearson eventually regained cordiality; even his 1895 critique of Pearson ended on a positive note.[9] But he could never shake the idea that his was the proper way to deal with data; as a result his labors were immense and little rewarded. He was drawn from other, potentially more fruitful pursuits and was utterly unable to attract a following for this work, save only the estimable Arthur Bowley. Edgeworth was unable to appreciate either the force of Pearson's work or how, despite the fact that it was retrograde in some respects, it carried the seeds of a twentieth-century revolution. But even though he may be pitied, he should not be faulted for shortsightedness, for it was only with the work of R. A. Fisher after 1920 that those seeds found fertile soil.

Pearson's skew curves furnished a flexible and broadly useful tool for statistical workers, and for more than twenty years they were standard equipment for statisticians, not the least of the reasons for this being that the statistical laboratory Pearson built up was tireless in developing and proselytizing them. As devices for dissecting social and other data, for identifying and describing nontrivial contributory causes, however, they were failures. Pearson may have realized this, because he ceased to emphasize that claim and turned his attention to other work. That other work was to inspire a final nineteenth-century step in the extension of statistical methods to the social sciences.

9. The critique was not positive enough to assuage Pearson, however, who wrote a long complaining letter to Edgeworth. Pearson's letter does not survive, but Edgeworth's reply begins, "You bring so many charges against me—(1) misinterpretation, (2) mathematical error, (3) logical fallacy and (4) unjustifiable tone. My withers are not wrung equally by all these." 5 November 1895, Pearson Archives; quoted in Wishart's discussion in Stigler (1978a, p. 315).

Pearson and Correlation

Pearson's major early contributions to regression analysis, the proximate inspiration for the final stage in the nineteenth-century spread of statistical methods to the social sciences, were contained in the two long memoirs in his "Contributions to the Mathematical Theory of Evolution" series. The first of these, "Regression, heredity and panmixia," was sent to the Royal Society on 28 September 1895; the second, "On the probable errors of frequency constants and on the influence of random selection on variation and correlation," with L. N. G. Filon, was sent on 18 October 1897. During the 1894–95 term at University College, Pearson presented a long course on statistical theory. One of the two or three persons in attendance was G. Udny Yule, and a summary of Yule's lecture notes is given in Pearson (1938, pp. 154–159).[10] By the spring of 1895 Pearson had exhausted his material on skew curves and turned to correlation. As he worked through Galton, Weldon, and Edgeworth, he prepared the first memoir, which both synthesized and extended those works. "The investigation of correlation which will now be given does not profess, except at certain stated points, to reach novel results. It endeavors, however, to reach the necessary fundamental formulae with a clear statement of *what assumptions are really made,* and with special reference to what seems legitimate in the case of heredity" (Pearson, 1896, p. 121).

Two of Pearson's novel contributions in this first paper involved the correlation coefficient. Here he derived what he called the "best" value for a correlation coefficient. He considered n pairs of organs, each pair with the bivariate normal distribution

$$\frac{1}{2\pi\sigma_1\sigma_2} \cdot \frac{1}{\sqrt{1-r^2}} \exp\left\{-\frac{1}{2}\left[\frac{x^2}{\sigma_1^2(1-r^2)} - \frac{2xyr}{\sigma_1\sigma_2(1-r^2)} + \frac{y^2}{\sigma_2^2(1-r^2)}\right]\right\},$$

where r stood for the correlation coefficient and σ_1 and σ_2 for the respective standard deviations. By blurring the distinction between the actual standard deviations and their values as given by data ("or, S denoting summation, since $\sigma_1^2 = S(x^2)/n$, $\sigma_2^2 = S(y^2)/n$" Pearson, 1896, p. 125). he was able to write the joint probability of all the n organs' deviations as proportional to

$$(1) \qquad \frac{1}{(1-r^2)^{n/2}} \exp\left\{-n\left[\frac{1-\lambda r}{1-r^2}\right]\right\},$$

where $\lambda = S(xy)/n\sigma_1\sigma_2$. He went on to argue that taking $r = \lambda$ maximized this joint probability; hence this value was the "most probable" value for r.

10. The estimate of "two or three" students is from a 1920 report on the history of the biometric laboratory that Pearson prepared. In addition to Yule, Alice Lee attended the lectures; I do not know the identities of any other students.

Pearson was effectively treating (1) as a posterior distribution, although he omitted any statement of the nature of the assumptions implicit in this, notwithstanding the italicized statement I quoted in the previous paragraph. But this approach did allow him to find an approximate standard deviation for his value λ of r, namely,

$$\frac{1 - r^2}{\sqrt{n}(1 + r^2)}.$$

By the time he prepared the second paper with Filon, Pearson had realized that this last result was in error and that by taking $\sigma_1^2 = S(x^2)/n$ and $\sigma_2^2 = S(y^2)/n$ he had effectively found a conditional (given the empirical standard deviations) posterior standard deviation. Pearson presented $(1 - r^2)/\sqrt{n}$ as the correct result (Pearson and Filon, 1898, p. 192).

A modern reader can find much to criticize in Pearson's loose mathematical and logical reasoning here, but the effect was unmistakable. A basic formula for estimating the correlation coefficient had been finally presented and a means of assessing its accuracy given in terms that left no room for doubt about the practical usefulness of the formula. Edgeworth had given an equivalent formula for estimating a correlation coefficient, but it had been buried. It also had been subject to the same conceptual vagueness on standardization as Pearson's formula, but Edgeworth had never cleared up the point and had never corrected his estimate of dispersion. In fact, Edgeworth had de-emphasized it.

Edgeworth noticed Pearson's work when it appeared, writing to Pearson on 19 June 1896, "I notice with interest that you obtain my formula for (the most accurate) determination of coeff. of correlation by a different and I think more accurate method than that which I employed." On 21 June 1896 he wrote, "The formula to which I alluded as having been given by me is the expression for the correlation between two attributes or organs x and y viz. $g = Sxy$ divided by a certain denominator—I gave it in my paper *exercises on calc of errors* in Phil. Mag. July 1893 (near the beginning of the paper). I haven't yet had the leisure to consider afresh how far the method which I adopted was legitimate" (Pearson Archives). The topic was dropped in later correspondence. Pearson's reply is lost, but he seems never to have granted Edgeworth any claim to priority, as he never referenced this work in print. Edgeworth did not forget his earlier work, however. He called specific attention to it in 1896 and took an opportunity in 1908 to reprint most of the relevant paragraph (Edgeworth, 1896c, p. 534; 1908–1909, p. 395; Bowley, 1928, pp. 115–116).

In contrast to Edgeworth, Pearson knew what he had, and he made the most of it. He went on to illustrate the use of his formula, including a reworking of some of Galton's data. Pearson found much more variation in correlation coefficients than Galton's ad hoc approach had revealed. By finding diversity of strength of association where Galton and Weldon had found uniformity, the power of the technique was amply demonstrated.

Beyond this innovation the most important features of Pearson's 1895 paper were his synthesis of all that had gone before in multiple correlation theory and his detailed working out of how that theory could throw useful light upon problems in heredity. Pearson had clearly internalized "Edgeworth's theorem" (which he presented under that name), and he went to some length to show how its consequences could resolve many questions left hanging or glossed over by Galton in *Natural Inheritance*.[11] It may indeed be true, as Pearson had admitted, that most of the material presented was not, strictly speaking, novel; but the unified presentation carried great force nonetheless. Only the limitation to data on heredity gave it an appearance of narrow scope. It is ironic that the referees were lukewarm. Galton wrote that it should be printed, as "It wd be too dull to *read*." George Darwin wrote, "I think he is often lacking in clearness of exposition" (Royal Society Archives, RR.12.94 and RR.12.95).

While Pearson's paper was in press, Udny Yule began his own investigations into correlation; and Yule communicated one of his results to Pearson in January 1896. Pearson had shown (following Galton) that when a bivariate distribution is bivariate normal then the regression of one variable upon the other is linear. This statement can be written in modern notation as follows: Bivariate normality implies that $E(Y|X) = \rho(\sigma_1/\sigma_2)X$, when both variables are centered at expectations, where ρ is the correlation coefficient and σ_1, σ_2 the standard deviations of Y and X. Yule discovered a generalization: If the assumption of normality was replaced by the assumption that the regression was linear [that is, $E(Y|X) = \beta X$ for some β], then the coefficient β was still given by $\beta = \rho(\sigma_1/\sigma_2)$, regardless of the nature of the bivariate distribution. Thus if one accepted (as it was natural for the developer of the method of moments to do) that ρ, σ_1, and σ_2 could be estimated by the moment estimators r, s_1, and s_2 in general, the scope of Pearson's analysis was broadened. The assumption of normality could be weakened to the assumption of linearity of regression. Pearson was aware that this did not fully solve the problem of dealing with skew correlation, because he did not think that linear regression would often hold in nonnormal situations. He wrote to Yule on 26 January 1896, "this seems to me a very nice generalisation, but I fear it won't help us in skew correlation, for the fundamental problem in that case is, it appears to me, to find the law

11. When Pearson came to write his own history of correlation (Pearson, 1920a), he went to great lengths to retract the credit he had earlier given Edgeworth in this regard. I agree with Seal (1967) and Plackett (1983) that Pearson's hindsight failed him in this matter, and the commentary reflects well neither upon Pearson nor upon the general trustworthiness of the later recollections of great scientists.

connecting the regression of each array and type . . . Still the theorem gives one more confidence in using the expression $r(\sigma_1/\sigma_2)$ for results which are not normal, but for which the skewness does not considerably upset the colinearity of the means" (Pearson Archives). With Yule's consent Pearson added footnotes to his paper announcing (with credit to Yule) this slightly enlarged scope of application (Pearson, 1896, pp. 128, 147).

Pearson's long paper with Filon two years later may be considered a completion of this portion of his statistical program as well as a launching point for much of R. A. Fisher's later work. Pearson and Filon, working (again implicitly) in a Bayesian framework, considered the general question of determining the standard deviations of Pearson's "best" values of the coefficients of a bivariate normal distribution (including a correction of Pearson's earlier error regarding the standard deviation of the sample correlation coefficient) and of the coefficients he had found for several of his skew curves. Fisher later (1922, p. 329) was to call attention pointedly to the fact that the application of their analysis to this latter case was wrong because Pearson did not employ the same criterion of "best" (that is, "most probable" or maximum likelihood) in finding coefficients for his skew curves that he had used for the correlation coefficient. But at the time, the work, however faulty, filled the gap that permitted Pearson to charge ahead, applying his scheme to all manner of statistics and giving at least plausible descriptions of the accuracy of his results. Only one important step remained to be taken in 1895, a step crucial to the completion of the larger program that Galton had launched. Before two years had passed G. Udny Yule was to take that step.

Yule, the Poor Law, and Least Squares: The Second Synthesis

George Udny Yule (born 18 February 1871, died 26 June 1951) had studied engineering at University College and physics under Hertz in Bonn before accepting a demonstratorship under Karl Pearson in 1893 (Kendall, 1952). When he came to Pearson's laboratory he may have expected to study applied mathematics, but by 1895 he was a statistician. His duties included assisting Pearson in a drawing class and preparing specimen diagrams for a treatise on graphics that Pearson planned in June 1893 but never completed. The first published notice of Yule's statistical work was an acknowledgment by Pearson of his assistance in calculation (Pearson, 1894a, p. 10), and already by 1895 Yule was showing signs of moving in independent directions.

One important example in Pearson's long paper on skew curves was due to Yule. Yule had come upon a book by Charles Booth—*The Aged Poor*—part of which Booth had read to the Royal Statistical Society in March 1894; and Yule had discovered in it material on pauperism that could serve to illustrate the techniques he was learning from Pearson. To Pearson's paper Yule contributed one example of the fitting of skew curves to the frequency of paupers per poor law union[12] in England and Wales in 1891 (Pearson, 1895a, pp. 102–104 and plate 7, fig. 17). In April 1896 Yule read to the Royal Statistical Society a long paper based on this material (Yule, 1896a). Half of the presentation was a marvelously clear tutorial on Pearson's skew curves, and the other half was an application of those curves to data on pauperism. The paper did not break new ground in statistical theory, but it provided instruction in the use of Pearson's techniques in a way Pearson had not, opening them to a much larger audience and clarifying several points Pearson had left obscure in his own account.

Galton rose in the discussion of Yule's paper not only to express his admiration but also to remind the audience that shorter methods (in particular, Galton's own method of percentiles) were available. (These remarks were amplified and replied to by Yule at a later time; Galton, 1896; Yule, 1896c.) The recorded discussion did not indicate whether or not Galton went further and told Yule that six years earlier he had suggested applying correlation to a different aspect of data on pauperism when he wrote:

> There seems to be a wide field for the application of [correlation] to social problems. To take a possible example of such problems, I would mention the relation between pauperism and crime. I have not tried it myself; but it is easy to see that here, as in every case of relation, success would largely depend on finding quasi-normal series to deal with. Both pauperism and crime admitting of many definitions, it would be necessary to restrict the meanings of those words for the purpose of the inquiry, so that the cases to be dealt with shall be fairly homogeneous in respect to all important circumstances. To do this is the business of the statistician, who becomes assured of the soundness of his judgment in devising his restrictions when he finds that his statistics are of a quasi-normal character. (Galton, 1890, p. 431)

Yule may not have needed such a hint, although his subsequent actions were entirely consistent with it. Indeed, he went on to make another use of the data on pauperism, one that is much more important to our present purposes.

In *The Aged Poor* Charles Booth had claimed that "the proportion of relief given out of doors bears no general relation to the total percentage of pauperism" (Booth, 1894, p. 423; quoted in Yule, 1895, p. 603). At that time relief to the poor was of two types. "Out of doors" relief, or out-relief, was assistance granted in their own home to persons judged wholly unable

12. Under nineteenth-century British law, a union (or poor-law union) consisted of two or more parishes combined for administrative purposes. It was thus akin to an administrative district.

to work, for example, to the aged or the infirm. It was in contrast to "indoors" relief given to the able-bodied, who would be taken into workhouses where they would perform various labors in return for their subsistence. In-relief carried with it a social stigma that out-relief did not. A high incidence of in-relief was considered to be associated with strict enforcement of legal criteria by the union, and a high incidence of out-relief was considered to be associated with lax or lenient enforcement. The relationship in question was that between the proportion of total relief (indoor plus outdoor) given "out of doors" in a union and the percentage of the population in that union classified as paupers. Thus Booth claimed there was no relation between these; in modern terms, he claimed that there is no relation between the ratio of direct welfare to work-relief and the incidence of poverty, essentially by district.

In the summer of 1895 Yule had just finished hearing Pearson's lectures on correlation. He constructed correlation tables of Booth's data to test the statement and found quite the contrary. For example, for the data given in Table 10.1 he found, after computing means for each row, that

Table 10.1. Yule's "correlation table" showing the relationship between pauperism and prevalence of out-relief, using data from 1891.

	Percentage of population in receipt of relief								
Ratio	0–1	1–2	2–3	3–4	4–5	5–6	6–7	7–8	8–9
0–1	1	9	13	3	3	1	—	1	1
1–2	—	18	17	7	—	1	—	—	—
2–3	2	19.5	23.5	18	5	1	—	—	—
3–4	1	24	30	31	10	4	—	—	—
4–5	—	11	34	26	19	7	—	—	—
5–6	—	9	14	27	20	11	1	—	—
6–7	—	1	20	17	15	6	—	1	—
7–8	—	6	4	13	12	7	—	—	—
8–9	—	—	5	7	6	1	—	—	—
9–10	—	—	2	1	3	7	4	—	—
10–11	—	—	1	3	5	—	1	—	—
11–12	—	—	2	—	1	—	—	—	—
12–13	—	—	1	—	—	—	—	—	—
13–14	—	—	—	—	1	—	—	—	—
14–15	—	—	—	—	1	—	—	—	—
15–16	—	—	—	—	—	—	1	—	—
16–17	—	—	1	—	—	1	—	—	—

Source: Yule (1895, p. 609).

Note: "Ratio" is the number of out-paupers to one in-pauper; the data are for 580 poor law unions. Based on these data Yule computed the correlation coefficient as 0.388, with a probable error of 0.022.

"the rise in the mean percentage pauperism, as the proportion of out-relief is increased, is as marked as could be desired" (Yule, 1895, p. 604). He also computed the correlation coefficient and its probable error (0.388 and 0.022, respectively) to support the argument; the probable error was based upon Pearson's first (and erroneous) formula.

Now Yule was sensible of one aspect of these data that rendered the strict application of Pearson's work on correlation coefficients questionable, namely, an evident skewness in the bivariate distribution. In his first publication on this in Edgeworth's *Economic Journal* he carefully qualified his statement: "Though, as we have said, no great stress can be laid on the value of the correlation coefficient (the surfaces not being normal), its magnitude may at least be suggestive" (Yule, 1895, pp. 604–605). In a short paper sent to the Royal Society in December 1896 (Yule 1897a), Yule went on to attack the problem of skew correlation from a theoretical point of view. Yule's 1896 work was not the first attack on this problem; he cited Edgeworth (1894a), who had alluded to an extension of the generalized law of error to two dimensions. Edgeworth's short paper was inconsequential, though; it was little more than a notice that in skew cases regression lines need not be straight and that means and modes would no longer coincide (so caution was advised). In addition Pearson had discussed possible models for skew surfaces in lectures in the winter of 1895, apparently without reaching any firm conclusion (Pearson, 1938, p. 158). Yule did much more.

Yule's work on correlation reached full expression in a paper in the December 1897 *Journal of the Royal Statistical Society.* The paper, "On the theory of correlation," was not primarily one of new results — Yule himself stated that "few of the results given in the sequel are entirely new" and referred nonspecifically to the works of Edgeworth, Galton, and Pearson (Yule, 1897b, p. 812). Rather the paper had a new and broader outlook that at once put the developing theory of correlation in a perspective from which it could deal with the problems of the social sciences and reconciled it formally with the traditional method of least squares from the theory of errors. As such Yule's work marked the completion of a final stage in the development of what could be called Galton's program and formed a cap on nineteenth-century work on statistics for the social sciences.

Yule set the stage for his presentation as follows:

> The investigation of causal relations between economic phenomena presents many problems of peculiar difficulty, and offers many opportunities for fallacious conclusions. Since the statistician can seldom or never make experiments for himself, he has to accept the data of daily experience, and discuss as best he can the relations of a whole group of changes; he cannot, like the physicist, narrow down the issue to the effect of one variation at a time. The problems of statistics are in this sense far more complex than the problems of physics. (Yule, 1897b, p. 812)

Yule's encounter with skewed surfaces had helped free him from two of the strictures of his predecessors' approach. First, he no longer felt confined by the multivariate normal distribution, a confinement that had led Edgeworth and Pearson to focus upon the coefficients of the exponent in that distribution as the targets of their analyses rather than upon the lines or planes they determined. Second, being freed of the normal density, he could take a fresh look at just what a regression line was doing, a look that led him back to Legendre's method of least squares.

One way in which Yule developed the problem was the approach he had communicated to Pearson in January 1896, namely, his observation that for some purposes the assumption of normality could be weakened to that of linearity of regression. A more important insight came from another route, however. Yule began with the case of two variables, x and y, measured as deviations from their respective means. Figure 10.1 shows his schematic version of a correlation table of such measurements (for example, of the type given in Table 10.1). He called the rows of the table "x-arrays" and the columns "y-arrays"; Figure 10.1 shows the x-arrays as horizontal strips. Thus an x-array is the set of all data pairs with a given y value (rounded off); in Table 10.1 the thirty-two unions in the first row form an x-array. Yule's figure represented the means of the x-arrays by "x" and called their locus the "curve of regression of x on y" (Yule, 1897b, p. 814). He noted that in many cases it would not differ much from a straight line, which he would call the "line of regression." In the case of "normal correlation" it would ideally *be* a straight line.

Yule's idea, even in the case where the regression curve was not exactly a straight line, was to take as his objective the determination of that straight line which came closest to the curve of regression in a certain sense (the line RR in Figure 10.1). He then reasoned as follows (I have changed Yule's notation only by the addition of subscripts and the use of Σ instead of S): Consider the data of the ith x-array as pairs $(x_{ij}, y_i), j = 1, \ldots, n_i$, and let $X = a + bY$ be the equation of the line RR. Let d_i be the distance from the array mean $\bar{x}_i = \Sigma_j x_{ij}/n_i$ to the line,

$$d_i = \bar{x}_i - (a + by_i);$$

and let σ_i^2 be the array standard deviation,

$$\sigma_i^2 = \frac{1}{n_i} \sum_{j=1}^{n_i} (x_{ij} - \bar{x}_i)^2.$$

Then Yule derived what we now, following Fisher, call the analysis of variance breakdown: For the ith array,

$$\sum_{j=1}^{n_i} [x_{ij} - (a + by_i)]^2 = n_i\sigma_i^2 + n_i d_i^2,$$

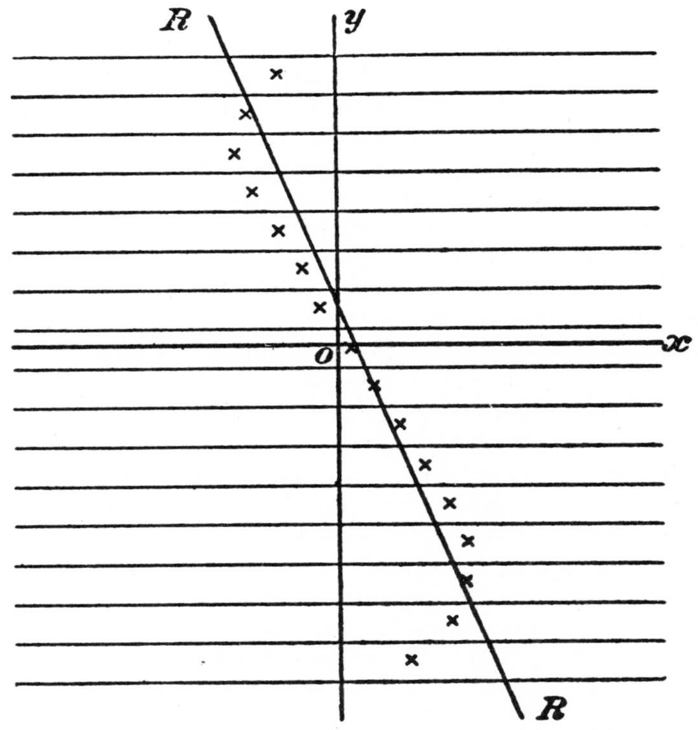

Figure 10.1. Yule's schematic drawing based on a correlation table. The x*'s represent means of the values within* x*-arrays (horizontal strips corresponding to rows of the table), and the line RR is the straight line closest (in a least squares sense) to the "curve of regression" given by the* x*'s. (From Yule, 1897a,b.)*

and summing over arrays gives

$$\sum_i (n_i d_i^2) = \sum \sum [x_{ij} - (a + by_i)]^2 - \sum_i n_i \sigma_i^2 .$$

Because Yule would choose as his best-fitting line the one that minimized $\Sigma(n_i d_i^2)$ and because $\Sigma(n_i \sigma_i^2)$ did not depend on a and b, he was led to choose a and b to minimize

$$\sum \sum [x_{ij} - (a + by_i)]^2,$$

that is, to the least squares estimates (Yule, 1897b, p. 815). He recognized that this reasoning carried over to higher dimensions and that this meant that his coefficients a and b could be thus determined from the classical normal equations of the method of least squares—so named by Gauss, as if to predict that Yule would reconcile them with "normal correlation" nearly ninety years later!

Yule's focus upon arrays was a direct echo of Galton, and his insistence upon simple linear relations was carried over from Edgeworth's and Pearson's development of Galton's work with normally distributed schemes. Yet although Pearson had been willing to accept Yule's previous approach (at least when the regression was linear), this new approach bothered him. In September 1896, before Yule's paper was submitted, Pearson wrote to Yule that "I feel still unconvinced by your arguments . . . I cannot see that your initial hypothesis is a satisfactory one to start a theory of correlation from—not nearly so satisfactory *to me* as the hypothesis of linear regression from which you first started" (Pearson Archives). The gist of Pearson's objection in this and a second letter dating from the same month had two components: his suspicion of Yule's goal of estimating a single equation, and his skepticism about the ad hoc choices of a straight line for that equation and the sum of squared errors as a criterion.

First, Pearson was reluctant to accept what he thought he saw as Yule's formulation of the general problem as one of recovering a "single-valued" relation between x and y and, in particular, in forcing a complex, highly multivariate relation into one between a small number of variables, say, two. Pearson could accept Yule's approach for some problems in physics, where he could really see the goal as recovering $y = \psi(x)$ amid noise (although even there he could see no a priori reason for the relation to be linear). But Pearson had difficulty bringing himself to that point of view for problems in biology:

> In physics you know by experience that the finer your methods of observation and your powers of observation, the more nearly you get your two variables related by a single valued equation and you are justified in trying to find the value of its constants . . . The key to your method is, such a relation between the two variables actually exists in nature, it is the axiom from which you start. In biology you start with the exact opposite—no such single valued relation exists, but I understand by correlation the theory which endeavors to supply its place. (Pearson Archives)

Pearson had already written of his resistance to the idea that a biological relationship could be characterized in a mathematically exact and deterministic way, saying, "The causes in any individual case of inheritance are far too complex to admit of exact treatment" (Pearson, 1896, p. 115). It appeared to him that Yule was tending in the wrong direction. He wrote to Yule, "In Biology there might possibly be a single-valued function relation, if we could take account of thousands of variables, but we never attempt in our observations to make any approach to such a relation" (Pearson Archives). Pearson argued that the proper approach started with a frequency surface, not with a simple relation.

The second component of Pearson's objection was directed at Yule's choices of a linear relation and squared error, choices Yule admitted were due to convenience, not to logical necessity. This point was related to Pearson's first: Pearson wanted to start with a frequency surface and, if a regression line was sought, find that line appropriate to the surface. If the surface followed the normal law, then he could accept the route to straight lines fit by least squares. But, "why should not another law even of symmetrical frequency lead to the *p*th powers of the residuals being a minimum?" (Pearson Archives).

Yule's replies to these two letters do not survive. But from the fact that both of Pearson's letters reiterate the same points and from the character of the published paper (Yule, 1897b), we can infer that Yule did little more than shade his wording in minor ways. He did make two apparent concessions to Pearson, emphasizing that the single-valued function that was his goal was an average relationship and that the choices of a linear form for it and a squares-of-error criterion were somewhat arbitrary:

> In actual statistics we know that we will find not one value of x_1 but a whole series, fluctuating round some *mean.* It is this *mean* value that is given by the estimated x_1, to a greater or less degree of approximation, according as the actual regression relation approximates more or less closely to the linear form . . .
>
> It must be understood that we only take a *linear* characteristic relation because it is the simplest possible form to calculate. We could on precisely the same principles solve for some more general equation such as—
>
> $$\begin{aligned} x_1 = a_{11} & \\ & + a_{12}x_2 + b_{12}x_2^2 + c_{12}x_2^3 + \dots \\ & + a_{13}x_3 + b_{13}x_3^2 + c_{13}x_3^3 + \dots \\ & + a_{14}x_4 + b_{14}x_4^2 + c_{14}x_4^3 + \dots \\ & + \dots\dots\dots\dots + \dots \end{aligned}$$
>
> which would doubtless give the mean of the x_1-array with a greater degree of accuracy." (Yule, 1897b, p. 816)

> We shall use the condition of least squares. This is done solely for convenience of analysis. Using S to denote "the sum of all quantities like" we might make S(nd) a minimum, counting d always as positive, or S(nd^4), but the first condition is the easiest to use, gives good results, and is a well known method. (Yule, 1897b, p. 814)

In some respects Yule and Pearson were not far apart. Both were building on concepts that had originated with Galton, but their quite different areas of application—biology and the social sciences—led them in different directions. Pearson was comfortable with frequency surfaces. They were a natural extension of the frequency curves that had been the key to his first statistical success, and they were a natural way of conceptualizing relationships such as those he studied between different organs. In such problems potential causation was not a major issue—a large foot would not "cause" a large head and a single-valued relation between them was not particularly useful in capturing the essence of a weak relationship. More to the point was the degree of association, that is, correlation, and the way in which it was related to the measured individuals' relationships (Were the head and foot on the same body? Or were they from father and son, or first cousins?). Pearson was interested in the way family structure affected the association of two measurements, not (with the exception of his work on selection) on the way one measurement might affect the other.

Indeed, whereas we may now view the Pearson of 1895 as a statistician attacking problems in inheritance and give prime attention to the statistical methods that in retrospect were his major contribution, Pearson himself was more concerned with the opportunity to contribute to the substantive scientific problem he faced. His first major statistical papers were given the heading, "Contributions to the Mathematical Theory of Evolution," a phrase that changed in 1896 to "Mathematical Contributions to the Theory of Evolution," perhaps signifying a subtly increasing emphasis on method. Even on 2 August 1895 when he wrote to Yule about his work it was to say, "I am now going to finish my paper on Panmixia" (that is, Pearson, 1896), not "I am now going to finish my paper on correlation." And on 16 September 1895, again to Yule, Pearson wrote, "I had a most kindly and encouraging letter from Francis Galton about my Heredity paper. He is really a fine old fellow to take my modifications of his views so well." Pearson's reactions to Yule's work were thus strongly affected by his somewhat limited view of the techniques as tools in biology. From this perspective his concerns seem sensible (even insightful); from a broader perspective they may not.

Yule, on the other hand, was attempting to go beyond biological problems into realms where a different type of relationship was very much the issue. And from this different perspective Yule saw things differently. He looked upon a regression line as a surrogate for a causal relation rather than as a mere characteristic of a frequency surface. The role of normality became less important. Indeed, in his first paper on this, Yule was emphatic, with the italicized statement: *"In any case, then, where the regression appears to be linear, Bravais'*[13] *formulae may be used at once without troubling to investigate the normality of the distribution. The exponential character of the surface appears to have nothing whatever to do with the result"* (1897a, p. 481).

13. Yule's terminology here reflects Pearson's, which as Pearson (1920a) later noted was overgenerous to Bravais. Bravais's (1846) investigations of spatial laws of error were advances within the Laplace–Gauss tradition but contained no hint of an idea of correlation or regression. His work in Pearson's hands, however, proved useful in framing formal approaches in those areas.

We now know that this was a somewhat naive overstatement, but in 1897 it had a liberating effect. It appeared no longer necessary to make detailed checks of normality (something that was—and still is—exceedingly difficult in higher dimensions); an analyst could charge on in and, with some checks of approximate linearity and approximately equal standard deviations, make full use of the newly evolving technology.

Regression was no longer a law of inheritance and no longer tied to stringent joint distributional assumptions, but it was at last seen as essentially what the method of least squares had, at least superficially, been all along—a fitting of straight lines. This new view led Yule briefly to use a new name for a regression equation—characteristic relation. He wrote:

> We may say that the present method consists in forming a linear equation between any one variable x_1 of a group, and the other variables x_2, x_3, x_4, &c.; this equation being so formed that the sum of the squares of the errors made in estimating x_1 from its associated variables x_2, x_3, x_4, &c., is the least possible. This relation is termed the characteristic relation. It is evident that if there are n variables, n characteristics can be formed between them, expressing each one in turn in terms of the others. The magnitude and sign of the coefficients of the x's on the right of such an equation show in what direction and to what extent the average of x_1 will be altered when x_2, x_3, x_4, &c., undergo alterations of any given magnitude and sign. (Yule, 1897b, p. 817)

One consequence of this newly discovered relation between regression and least squares was to be that a great array of algorithms that astronomers and geodesists had been developing to simplify the solution of normal equations and the calculation of the probable errors of coefficients was available. Yule does not seem to have realized that at first, however, and he proceeded to show how the solutions of the normal equations could be expressed in terms of pairwise correlation coefficients. Thus even though the problem of the estimation of correlation coefficients was no longer formally severed from that of estimating the regression equation (as in Edgeworth), the practical effect was the same. All correlation coefficients would be calculated, and then through Yule's relations (adapted from Edgeworth and Pearson) the regression coefficients would be found.

Yule gave the names "net or partial regression" for the coefficients in multiple regression equations; and what we now call a partial correlation coefficient he termed "*net* coefficient of correlation" (Yule, 1897b, p. 833). He also introduced the multiple correlation coefficient: "R_1 may in fact be regarded as a coefficient of correlation between x_1 and (x_2, x_3)." In this case he proposed the name "coefficient of double correlation" (Yule, 1897b, p. 833). His analysis included a proof that adding a variable to a regression equation could not increase the standard deviation around the fitted relationship and would actually decrease it unless the partial correlations were zero (Yule, 1897b, pp. 834–835, 837). Yule gave instructions on how to calculate the needed means, standard deviations, and correlations. He discussed how second or higher order surfaces could be used in place of lines or planes, even though he thought the arithmetic for that prohibitive. All in all it was a tour de force—a nearly complete tutorial on correlational analysis as well as a striking reformulation of existing work.

Yule included an application to some of his pauperism data in Yule (1897b); and a year and a quarter later he read an extended treatment of that topic to the Royal Statistical Society (Yule, 1899). This paper shows that Yule had realized the power of his methods; the paper was in its way a masterpiece, a careful, full-scale, applied regression analysis of social science data. Yule was concerned with the relationship between pauperism and the out-relief to in-relief ratio; in modern terms we might put the questions this way: How could the welfare policy of the day be most effective in dealing with poverty? Did strict enforcement of criteria increase or decrease poverty? Actually, Yule showed an insight that not all of his descendants have shared in asking the question in terms of *changes* in the variables: Was a change in pauperism over a decade to be ascribed to a change in proportion of out-relief?

Yule sought to answer the question with a regression equation, but he worried about the possible effects of other variables upon his conclusions. In his introduction he put it this way:

> Suppose a characteristic or regression equation to be formed from these data, in the way described in my previous paper, first between the changes in pauperism and changes in proportion of out-relief only. This equation would be of the form—
>
> change in pauperism =
> A + B × (change in proportion of out-relief)
>
> where A and B are constants (numbers)
>
> This equation would suffer from the disadvantage of the possibility of a double interpretation, as mentioned above: the association of the changes of pauperism with changes in proportion of out-relief might be ascribed *either* to a direct action of the latter on the former, *or* to a common association of both with economic and social changes. But now let all the other variables tabulated be brought into the equation, it will then be of the form—
>
> change in pauperism =
> $a + b \times$ (change in proportion of out-relief)
> $+ c \times$ (change in age distribution)
> $+ d \times$
> $+ e \times$ } changes in other economic, social, and moral factors
> $+ f \times$
>
> Any double interpretation is now—very largely at all events—excluded. It cannot be argued that the changes in pauperism and out-relief are both due to the changes in age distribution, for that has been separately allowed for in

> the third term on the right; b × (change in proportion of out-relief) gives the change due to this factor *when all the others are kept constant.* There is still a certain chance of error depending on the number of factors correlated both with pauperism and with proportion of out-relief which have been omitted, but obviously this chance of error will be much smaller than before. (Yule, 1899, p. 251)

Thus Yule used the regression equation as a device both to uncover the relationship he sought and to allow for potentially influential changes in the other variables he had at hand. He expressed his variables in terms of percentage changes, and when he was finished he had an equation such as this, for the decade 1871–1881 (1899, p. 258):

change per cent. in pauperism =
− 27.07 per cent.
+ 0.299 (change per cent. in out-relief ratio),
+ 0.271 (change per cent. in proportion of old),
+ 0.064 (change per cent. in population)

Yule's analysis of pauperism was well known at the time, but it did not have a pronounced or immediate effect upon either social policy or statistical practice. Over the period 1905–1909 a Royal Commission on the poor law met to consider reforms; and one might expect that Yule's analysis would have been introduced as part of the evidence, perhaps even that Yule would have been called to testify. But apparently the only time Yule's work was mentioned was indirectly, by A. C. Pigou, in a form that guaranteed it would be given little attention. In 1908, shortly before his appointment as Alfred Marshall's successor as professor of political economy at Cambridge, Pigou submitted written testimony that included a critique of Yule's work, a chapter entitled "The Limitations of Statistical Reasoning." Pigou's judgment was harsh: "I submit to the Commission that the practical importance of reasoning of this order has been exaggerated, and that, by it, little that is significant can be proved." After describing Yule's approach as "the method of triple correlation" and quoting from his conclusions, Pigou wrote:

> It is sometimes supposed that investigations of this character have great practical importance, because they go to prove that a more extended use of out-door relief, such as is at present often advocated, would have injurious results. In my opinion statistical reasoning cannot be rightly used for this purpose.
>
> The fundamental objection to it is that some of the most important influences at work are of a kind which cannot be measured quantitatively, and cannot, therefore, be brought within the jurisdiction of statistical machinery, however elaborate. It is not merely that other forces beyond those taken into account *may* be operating in a way that would tend to reverse the conclusion

prima facie suggested by the statistics. If that were all, this conclusion would merely be rendered less certain; it would remain a probable conclusion. But, among the non-quantitative facts of the situation is one which obviously accounts, at all events for a part of the correlation on which the conclusion depends. It is well known that, during recent years, those unions in which out-relief has been restricted have, on the whole, enjoyed a general administration much superior to that of other unions. (Pigou, 1910, p. 986)

Pigou had noted, and expressed in forceful terms, what is still the most frequently encountered criticism of the use of regression analysis for causal reasoning, namely, the possibility that an unmeasured "lurking variable" (to use a phrase coined by G. E. P. Box more than a half-century later) was producing the appearance of a causal effect. In general terms Yule had anticipated this criticism. Yule had specifically discussed the potential effect of omitted variables in his introduction of his main equation (as quoted earlier: "There is still a certain chance of error depending on the number of factors correlated both with pauperism and with proportion of out-relief which have been omitted"). Indeed, over the next several decades Yule was second to no one in warning of the pitfalls of spurious or nonsense correlation. The only novel point in Pigou's critique was his claim to have identified quality of administration as such a factor, for which his evidence was the question-begging "It is well known." One can imagine Yule replying that Pigou had neither demonstrated that "careful administration" existed as other than strict enforcement of criteria (and thus was captured by a low out-relief ratio) nor examined the possible policy consequences of a high correlation of the type he claimed existed, namely, that instead of simply restricting out-relief it would presumably be necessary to introduce a better administrative structure that would in turn be charged with restricting out-relief. Pigou's ad hoc speculation that if out-relief was increased and in some sense more carefully administered, then poverty would decline could not, of course, be disproved from the data Yule used. That these issues remain with us today is testimony to their inherent difficulty.

With more than eighty years' accumulated wisdom we might think Yule naively optimistic in his belief in the power of his methods. Yet in many ways the paper stands the test of time quite well. Indeed, an embarrassingly high percentage of modern work is not up to Yule's standard. The salient point, however, is that at last a quantitative step had been taken toward what I have called the de Keverberg dilemma, the problem of categorization that Baron de Keverberg had posed to Quetelet in 1827 (Chapter 5). Yule's formulation and use of regression analysis was not a full solution—we will never find a full solution—but it completed a masterly step toward one. In principle, by measuring all relevant variables one could have the variables themselves, through their appearance in the regression equation, define a multitude of categories. Yet through the constraints imposed by the model, it was possible to sift and evaluate and compare the effects of the variables. One or more factors could be "kept constant" (Yule, 1899, p. 262) by the statistical analysis, even when it was impossible to do that through experimental control, and the signs and magnitudes of the other variables' effects evaluated.

Yule developed his approach to correlation via regression a bit further in 1907 (Yule, 1907), introducing the notation that became the standard over the following half-century. By the 1920s Yule's approach came to predominate in applications in the social sciences, particularly in economics, but Karl Pearson maintained his preference for the use of correlation coefficients as estimates of characteristics of frequency surfaces. For some years Pearson kept working toward a skew surface that would do for multivariate problems what he felt his family of curves had done in one dimension, but he never found a satisfactory solution. In 1905 he gave a qualified endorsement of Yule's approach as a way of coping with multivariate skew surfaces until he, Pearson, overcame the analytic difficulties that kept him from a full theory of skew surfaces. Pearson did note of Yule's approach, "Its chief advantage is that it makes little or no assumption as to the distribution of frequency; its chief defect lies even in this advantage of generality: it does not enable us to predict the probability of an individual with a given combination of characters" (Pearson, 1905, p. 495). Nonetheless, Pearson's own work went increasingly in other directions, emphasizing the use of correlation coefficients even for qualitative data.

Edgeworth was more amenable to Yule's formulation. Although leery of the application of the methods in markedly nonnormal situations, he described Yule's regression equation as "the ideal method of eliminating chance" (Edgeworth, 1902, p. 285; 1911, p. 396).

The Situation in 1900

Yule's work provides a symmetric ending to this narrative: We began this history with the method of least squares and we shall end with the method of least squares. Yet the story is far from being circular. Over a two-century period there had been sporadic progress in the measurement of uncertainty, a sequence of developments we may think of as leading toward a completion of the logic of the quantification of science, by eventually permitting the formal evaluation and comparison of measurements. In this sequence the early achievements in astronomy and geodesy were in some respects like the later successes in the social sciences. In both cases a key barrier had been the lack of a conceptual structure that permitted the combination of observations; and in both cases the theory of probability had played a crucial role in overcoming the barrier, with the method of least squares supplying the means for completing the calculations. Beyond these broad similarities, however, lay a number of important differences.

In astronomy the combination of observations had required both the anchor of the mathematical theory of gravitation and the growing knowledge of the behavior of random sums. The theoretical structures of celestial mechanics not only defined the goals for the astronomers' empirical work but also helped the astronomers to reach those goals. By giving a link between different measurements of the same body, Newtonian theory provided a route by which the measurements could be combined, a way in which the relatively small numbers of major causes could be incorporated in one equation and related to the observations. Yet, as the example of Euler shows, that link was not enough by itself. A theory of errors was also needed, both the idea that combining measurements was beneficial, not harmful, and a means for turning the combination to inferential use. Mayer grasped this intuitively in 1750 and a few years later Simpson added a formal analysis for simple means, but it took over a half-century before the grand Gauss–Laplace synthesis of 1810 was achieved. The delay is ample testimony to the difficulties of formulating, much less solving, the problem. Nevertheless, by the time of Laplace's death in 1827 a major success had been recorded, and the use of probability to measure, compare, and interpret uncertainty was well on the way to becoming a commonplace in astronomy and geodesy.

In the social sciences the problem took on a different face. It was not that theory was lacking: By the middle of the nineteenth century several economists had given mathematical expression to the theories of Adam Smith and David Ricardo. But even though these theories might have captured the major causes that were of interest to the economist, they did not incorporate the myriad causes of little concern in economic theory that nonetheless had a major impact upon the data used to test that theory. Cournot could relate price and demand in theory, all other things being held constant, but he could not hold all other things constant in the real world. A new conceptual approach was needed, and it arose from an unexpected quarter—from studies of heredity.

Galton, Edgeworth, and Pearson assembled the structure; Yule completed it by finding a variation on their advances that finally provided a formulation and analysis for questions in the social sciences. It is ironic that in some respects the tool Yule used was an old tool, the method of least squares. Of course, the matter was much subtler than that; it was not Legendre's or even Gauss's or Laplace's least squares that Yule found, but a superficially similar tool that had been transformed by the concepts developed by Galton and others. Legendre's least squares had been around and freely available for ninety-five years. But when it had been tried before, in isolated instances, it had not answered the right questions. By 1900, though, the questions could be reformulated so that the astronomers' least squares could be used to new purpose. It was not merely an elegant variation of language that called for the term *regression analysis* for this conceptually new use of least squares. In 1805 the coefficients were constants, deriving their meaning from an exterior theory. In 1900 the coefficients were interior theoretical constructs, given their meaning in the context of Galton's variance component models as what we now characterize as conditional expectations of multivariate distributions. In this new framework least squares provided the homogeneous subclassifications for analysis Quetelet had sought while offering a sufficiently restricted setting that Cournot's worries about post-data selection could be addressed in at least limited ways.

The realization that the regression and correlation concepts that had emerged in Galton's studies of heredity were intimately related to the least squares of the beginning of that century was the second great synthesis of the history of statistics. In this second synthesis, probability had played a role rather different from that in the theory of errors. In the theory of errors, the probability calculus had revealed order in chaos with the central limit theorem, and that discovery had made the measurement of uncertainty in aggregate measurements possible. In the social sciences the same magnificent theorem had posed a problem of seemingly incredible difficulty: How could the known diversity of causes be reconciled with this always present order? How could the normal distribution Quetelet had found be disassembled to allow a study of causes? Galton's quincunx had led to the answers to these questions, by suggesting a new role for conditional probability. In the theory of errors, conditional probability had permitted inference about the constants of astronomers' theories. In regression analysis, conditional probability made possible the very definition of the quantities about which the statistician was interested in making inferences.

Looking back, we can see four distinctly different solutions to the problem of the combination of observations, four ways in which the variation in external factors could be allowed for in order to permit an attempt at aggregation: through external theory (as in astronomy); through experimental design (as in psychology); with internal regression analysis; and with large amounts of multiply classified data, through fine cross-classification. Actually, the last of these was not really feasible in the nineteenth century. When data were plentiful (for example, a census), it was in principle possible to look at all manner of categories; but in practice it was not. In regard to the U.S. census, Herman Hollerith wrote, "Until the census of 1890, we never even knew the proportion of our population that was single, married, and widowed . . . To have divided the native born into those of native parentage and those of foreign parentage, would have been practically impossible with the methods of 1880. To obtain the population classified according to age, sex, and birthplace of mother could not have been considered" (Hollerith, 1894, p. 678). This direct assault upon the de Keverberg dilemma would require modern tabulating equipment, even when the data were plentiful and at hand.

The conceptual triumphs of the nineteenth century had been the product of many minds working on many problems in many fields, and one of the most striking of their accomplishments was the creation of a new discipline. Before 1900 we see many scientists of different fields developing and using techniques we now recognize as belonging to modern statistics. After 1900 we begin to see identifiable statisticians developing such techniques into a unified logic of empirical science that goes far beyond its component parts. There was no sharp moment of birth; but with Pearson and Yule and the growing numbers of students in Pearson's laboratory, the infant discipline may be said to have arrived. And that infant was to find no shortage of challenges.

As modern statisticians know, the introduction of least squares for the analysis of social data, like all advances that are in any sense revolutionary, created as many or more questions than it answered, and immense efforts have gone into tackling those questions in the twentieth century. Many of the lines of development we have studied were to continue to flourish. The methods of fitting frequency curves that Edgeworth and Pearson had proposed were to be dramatically transformed by Ronald Fisher into a general theory of parametric inference, a theory that was in turn to lead to a sharper framing of testing problems in the hands of Jerzy Neyman and Karl Pearson's son, Egon. Yule's development of correlation and regression was to open the way to the whole of multivariate analysis in the works of Fisher and Harold Hotelling. Inverse probability was to excite many debates and produce sharper new concepts: Fisher's fiducial probability, Neyman's confidence intervals, and L. J. Savage's personal probability. The benefits that experimental design had bestowed upon psychology, coupled with multivariate analysis in the work of Fisher, were to create a new age in experimental agriculture. Yule himself was to go on to make major contributions to qualitatively categorized data analysis and effectively to invent modern time series analysis. But those efforts, and the many and significant successful solutions that have been found, have just served to confirm the importance of what had come before. From the doctrine of chances to the calculus of probabilities, from least squares to regression analysis, the advances in scientific logic that took place in statistics before 1900 were to be every bit as influential as those associated with the names of Newton and Darwin.

APPENDIX A

Syllabus for Edgeworth's 1885 Lectures, King's College, London

LOGIC OF STATISTICS

by F. Y. Edgeworth, M.A.

Statistics will be understood here principally in the third (occasionally in the first and second) of three senses, which seem to constitute respectively the most popular and the most philosophical definitions, and that which is a fair compromise between the conflicting attributes of a good definition. According to the first definition, Statistics is the arithmetical portion of social science; according to the second definition, the science of Means in general (including means of physical observations); according to the third, the science of those Means which relate to social phenomena. Of course "Means" imply the correlative conception of terms of a "series," or members of a class, whose mean is to be taken.

The science of Means may be summed up in two problems; (1) To find how far the difference between any proposed Means (*e.g.* the average mortalities in different occupations) is accidental, or indicative of a law; (2) To find what is the best Mean, whether for the purpose contemplated by the first problem, the Elimination of Chance, or other purposes.

LECTURE II

The Elimination of Chance is effected by the Calculus of Probabilities, a powerful but dangerous instrument, which requires not only skill in handling mathematical formulae but also judgment to discern how far mathematics are applicable to each class of phenomena. Formulae founded upon games of chance are to be transferred with caution to real life. The "ballot-box of nature" is not so simple as has been supposed. Often the calculus affords rather an appropriate general conception than definite numerical results; in this respect resembling the mathematical theory of Economics.

Even in cases most favourable to calculus there remains a large *rôle* for ordinary inductive logic. For example, in the case of some recent experiments in so-called *Psychical Research* the returns prove conclusively that some agency other than chance has been at work. But the calculus is silent about the nature of that agency. Whether it is more likely to be some form of illusion, or a less familiar cause, must be decided by common sense.

LECTURE III

The most important piece of mechanism in the apparatus for eliminating chance is the *Law of Error;* so called because in the case of physical observations it expresses the frequency with which each particular error, or deviation to a certain extent from the true point, usually and approximately occurs. The law is extensively fulfilled in social, as well as in physical, phenomena. But its employment in art far exceeds its fulfilment in nature. For example, let there be a large group of men whose statures do not range under the *law of error;* as in one of the French Departments where the different races have not been perfectly fused. Nevertheless, if a considerable set, say a thousand men, is taken at random from that large group, and the mean of the statures of that set is formed, and again the mean of another and another set of a thousand, the means thus formed do (tend to) range under the typical *law of error.* Accordingly the reasoning based upon that law may be employed to determine whether within that department there is a significant difference of stature between different classes, *e.g.* artisans and agriculturalists. What is required for the application of the law of error is not that the raw material of our statistical returns should be shaped according to that law; but that they should be *constant to any law* of divergence.

LECTURE IV

The principal quaesitum in any case to which the law of error is applicable is the numerical value of that constant (technically termed the *modulus*) which expresses the particular divergence, "*ecart,*" or fluctuation, of the quantities which vary according to that general law.

It is not to be assumed without examination that the numbers of births or deaths fluctuate about their Mean in exactly the same way as the numbers of balls of a certain colour drawn at random from an urn known to contain a certain proportion of balls of that colour. The ideal case of games of chance affords an *inferior limit* of fluctuation which is rarely reached in social statistics. In general in social statistics, just as in physical observations, the fluctuation must be inferred from specific experience, by examination of a great number of observations of the class under consideration.

The fluctuation being ascertained, we can assert confidently that the difference between two statistical figures is either not even *prima facie* significant, or corresponds to a real difference in fact, or at least affords an hypothesis deserving attention. Thus a theory that one profession is more favourable than another to the birth of male infants, based as it has been upon some thousand births in which the ratio of male to female was 107 to 100, instead of, as usual in England, 104 to 100, is not *prima facie* entitled to much consideration. On the other hand, the smaller difference between 104 in England and 105 on the Continent is (in view of the greater number of observations) certainly significant of a real difference. And the changes of the ratio from time to time are also proved to be important. The difference of a small fraction per cent. In the mortality of different occupations, though based only on some hundred deaths, is indicative of a real difference between the conditions of those occupations.

LECTURE V

In general the statement that the numbers of any phenomenon—it may be Births or Bankruptcies or Bills of Exchange—occurring at one time or place (or under any other category) have differed by so much per cent from the number occurring at another time or place is, *per se*, little significant. A knowledge of what extent of difference was to be expected, of the fluctuation incidental to each class of phenomenon, brings out the hidden value of our datum. The *law of error* is the grand touchstone of statistical arguments. An inferior test is to break up the sum totals into partial sums, and observe whether there is the same sort of difference between the parts belonging to different totals as between the totals. The evidence of consilient curves (*e.g.* those representing commercial fluctuations and solar phenomena) is of this character.

Problem II.—In selecting the best Mean, the purposes to be regarded appear to be twofold: first, the practical need of taking one particular quantity as the representative of a set of fluctuating quantities; and, secondly, the theoretical purpose of eliminating chance.

LECTURE VI

In the selection of a Mean, there are two subjects of choice: *A* the genus, and *B* the species, of the Mean, *e.g.;* we determine the genus as the (weighted) Arithmetical Mean when, if the returns to be meaned are x_1, x_2, x_3, &c., we determine that the Mean shall be of the form $h_1x_1 + h_2x_2 +$ &c., *divided by* $h_1 + h_2 +$ &c. We determine the species when we assign particular values to the constants, or *weights,* h_1 h_2, &c.

A—The genera may be placed in the three groups—(1) The Arithmetical Mean. (2) Two other Means which compete with the Arithmetical Mean in respect of convenience. (3) All other Means. The grounds of preference differ according as the quantities to be measured do, or do not, range under the typical *law of error*. In the former case there is a peculiar propriety in the use of the Arithmetic Mean. In the latter case the advantage of that Mean is reduced to mere convenience; an advantage which is disputed by the second group of genera.

B—*Weight* is of two kinds: that depending on mere mass, or number of observations *ceteris paribus;* and specific weight. The former presents little difficulty. The latter is determined either by the more exact method which is contemplated in Laplace's Method of Least Squares, or by a more rough-and-ready inference that one source of information is less accurate than another. The latter method is the one principally employed in Social Statistics. And accordingly specifically weighted observations do not seem to play a great part in the more exact portions of the science.

LECTURE VII

An important case of weighted Mean occurs when we balance different authorities; whether in general, or with reference to a statistical argument. For it often happens that even reasonings based on figures are not reliable, when accompanied with party-spirit personality, and other sure marks of a bad authority. On a question like that of vaccination, the layman is at a disadvantage; which he must minimise as far as possible by estimating the weight of his authorities.

The fallacies incident to imperfect information, and which amateurs can only guard against by the principle of authority, are twofold. First, erroneously assuming that the conditions of the Method of Difference are fulfilled. For example, it might be argued that, because the number of convictions at one time or place was greater than at another, therefore the criminality was greater; whereas the difference might really be due to a change in the criminal law, or to a difference between the legal terminology of two countries. The second fallacy is more peculiar to our subject: assuming that a certain extent of numerical difference is significant, when a knowledge of the fluctuation to be expected would show the contrary. On the nice question what extent of fluctuation is in any class of phenomena significant of real difference the "undemonstrated opinions" and acquired sight of the specially-educated man possess much authority.

The idea of attaching different weight to different sources of information is justified by its success in Physics. But to assign numerical weights is beyond the powers of the Logic of Statistics.

The Course will begin on Monday, April 20. *Fee,* 14*s. for the Term, payable in advance in the College Office.*

[The lectures were presented in Evening Classes during Easter Term, 1885.]

Syllabus for Edgeworth's 1892 Newmarch Lectures, University College, London

ON THE USES AND METHODS OF STATISTICS

by Professor F. Y. Edgeworth, M.A., D.C.L.

I. FIRST PRINCIPLES

The extent of the subject here treated is that which is denoted by two leading definitions of statistics, viz: the study of numerical statements relating to society, and the theory of means. The subject may be divided according as the element of induction is less or more prevalent. First come general directions as to the acquisition of data; *e.g.*, that figures should be accurate, and terms unambiguous. Examples of the violation of these rules; together with other precepts and cautions. Use of relative figures (per head, per cent, &c.) Analysis of the data.

References: Conférences sur la Statistique (Rozier Editeur), 1891; Pidgin, *Practical Statistics,* 1888; Giffen, *International Statistical Comparisons,* Economic Journal, June, 1892.

II. GRAPHICAL METHODS

The Cartesian system of co-ordinates. Integration and interpolation. Case where several dependent variables (*i.e.* diseases from different causes) are referred to one independent variable (*i.e.* the time). The case of one variable dependent on two independent variables is properly represented by a surface; but curves of level and variously coloured planes are more convenient. Methods of expressing variation of a quantity relative to its initial, or average, value. Miscellaneous devices for exhibiting numerical relations to the eye.

References: Marey, *La Méthode Graphique,* 1885; Favaro, *Leçons de Statique Graphique* (translated into French by Terrier), ch. v. with appendix by the translator. Levasseur, *La Statistique Graphique,* Journal of the Statistical Society, Jubilee vol., 1885; Marshall, *The Graphic Method of Statistics,* Ibid; Cheysson, *Les Cartogrammes à teintes graduées,* Journal de la Société de Statistique de Paris, 1887; Scribner's *Statistical Atlas of the United States;* Longstaff, *Studies in Statistics,* 1891.

III. THE DOCTRINE OF AVERAGES

The general idea of a *mean* comprehends innumerable species, of which the most important are, the Arithmetic Mean, the Median, the Greatest Ordinate (or centre of greatest condensation) and the Geometric Mean. A cross division is between simple and weighted means. Concrete instances of these varieties. Subtle distinction between so-called objective and subjective means. Peculiar prestige attaches to the means of which the constituents are grouped according to the Probability Curve, or *law of error. A priori* demonstration, and empirical verification, that this form arises under certain conditions.

References: Venn, *Logic of Chance,* Third Edition, 1888, chap. xviii., and xix.; Venn, *On . . . Averages,* Journal of the Statistical Society, 1891; Galton, *Statistics by inter-comparison,* Philosophical Magazine, 1875; Bertillon, *Moyenne,* Dictionnaire Encyclopédique des Science Médicales; Edgeworth, *On the choice of Means,* Phil. Mag., 1887, *On the empirical proof of the law of error,* Ib., 1887.

IV. TYPES AND CORRELATIONS

The 'mean man' has for stature, length of cubit, height of knee, &c, the respective means of the statures, lengths, &c., of a great number of men. Reply to the objection that such a combination of partial means may not form a possible whole. Relation between the deviation of one organ or attribute, *e.g.* stature, from its mean value, and the "correlated" deviation of another attribute, *e.g.* length of cubit, from its mean; as established by Mr. Galton, and illustrated by Mr. H. Dickson. Abridged method of ascertaining the co-efficient which expresses the correlation. Extension of the Galton–Dickson method to the correlation between three attributes, *e.g.* stature, length of cubit and height of knee. The formula for the most probable value of one attribute, *e.g.* stature, corresponding to assigned values of two other attributes, *e.g.* length of cubit and height of knee, may be ascertained either from the three simple correlations, between stature and cubit, stature and height of knee, cubit and height of knee; or by observations special to the case of three variables. Correlation between any number of attributes.

References: Quetelet, *Anthropométrie;* Galton, *Family Likeness in Stature,* Proceedings of the Royal Society, 1886; *Co-relations and their measurements* Ibid. 1888; Weldon, *Correlated Variations,* Ibid. 1892.

V. THE STATISTICAL PART OF INDUCTIVE LOGIC

Passing *Insurance* and other direct applications of statistics, we come to the investigation of causes. The inductive method to which statistics lends itself, *the Method of Agreement,* is liable to the fallacy *Post hoc propter hoc;* of which numerous examples occur. The *Method of Concomitant variations* is facilitated by the use of parallel curves. The *Method of Residues* is exemplified when in comparing the death rates of different classes, we make allowance for their different ages; and in similar cases.

References: Mill, *Logic;* Giffen, *Essays on Finance,* and Article in June No. of Economic Journal; Humphreys, *Value of death rates as a test of Sanitary conditions,* Journal of the Statistical Society, 1874, *Class Mortality Statistics,* Ibid. 1887.

VI. THE ELIMINATION OF CHANCE

One case of the Method of Residues, for which there exists a technical apparatus, is where the agency allowed for consists of those "fleeting causes" called chance. The simple method of eliminating chance, described by Mill (*Logic,* iii, xvii, 4), and the higher method derived from the theory of error. The latter method is particularly applicable where the deviation from the average value of a ratio—*e.g.* that between male and female births—follows the analogy of the simpler games of chance. In other cases the higher theory affords rather regulative ideas than exact conclusions; in this respect, comparable to the use of the mathematical theory of economics.

References: Westergaard, *Grundzüge der Theorie der Statistik,* 1891; Duesing, *Das geschlechtverhaltniss in Preussen,* 1890; Edgeworth, *Methods of Statistics,* Journal of the Statistical Society, Jubilee vol., 1885.

[The lectures were presented on six consecutive Wednesdays at 5:00 P.M., beginning 11 May 1892, admission free.]

Suggested Readings

The following works, cited in full in the Bibliography, have proved particularly valuable in writing this book and may be particularly useful to the reader wishing to pursue the subject further.

General References

There are few general histories of statistics; the best, despite their age, are those by Walker (1929) and Westergaard (1932). Walker's book emphasizes correlation; Westergaard's nonmathematical treatise emphasizes social statistics. Pearson (1978) is based on notes for lectures given in the 1920s, but it gives valuable insights into both the history of statistics and Karl Pearson himself. The theory of probability has received more systematic attention than has statistics, witness Hacking (1975), a study of the early philosophical evolution of probability, and Todhunter (1865), a still useful, nearly comprehensive outline of the literature on probability through the work of Laplace. Gouraud (1848) is out of date, but gives an interesting insight into how the subject was viewed at that time. Adams (1974) and David (1962) are two books on the history of probability that succeed in being accessible at a more popular level. The history of statistics has been best treated in journal articles; two notable collections are the volumes edited by Pearson and Kendall (1970) and by Kendall and Plackett (1977). In addition a long series of articles by Sheynin (those through 1983 are indexed in Sheynin, 1983) give a critical outline of much of the literature, and, like Todhunter (1865) can be a valuable entree to that literature. The major bibliographies of the history of statistics are those compiled by Merriman (1877), Harter (1974–1976, 1977), Kendall and Doig (1968), and Lancaster (1968). Lancaster's bibliography has been frequently supplemented in the *International Statistical Review.* The *Dictionary of Scientific Biography* (Gillispie, 1981b), the *International Encyclopedia of the Social Sciences* (Sills, 1968), and the *Encyclopedia of Statistical Sciences* (Johnson and Kotz, 1982–) are generally excellent sources of biographical information. The statistical articles in Sills (1968) were updated and republished as *The International Encyclopedia of Statistics* (Kruskal and Tanur, 1978); almost all of these articles have extensive bibliographies that include valuable historical references. Recently historians and philosophers of science have given increased attention to the history of statistics, with the studies of Cullen (1975), MacKenzie (1981), and Porter (1986) being the most relevant to the present book. Kuhn's (1961) essay on the role of quantification in science is particularly stimulating.

1. Least Squares and the Combination of Observations

The historical background of least squares has received at least passing notice in many accounts. Eisenhart (1964) and Plackett (1958) discuss general aspects of this background, Sheynin (1972) covers some of Euler's work, and Eisenhart (1961), Sheynin (1973b), and Stigler (1984) present accounts of different aspects of Boscovich's work. A portion of the discussion in this chapter of the early work of Laplace is expanded from material in Stigler (1973a, 1975a). Wilson (1980) gives a recent critical study that includes the history of investigations of the motions of Jupiter and Saturn. Gore's (1889, 1903) bibliographies of geodesy are narrow in focus but useful within that focus. The measurement of the meridian arc through Paris is discussed in Stigler (1981). In addition to the standard biographical accounts [including Gillispie's (1981a) extensive treatment of Laplace], see Gowing (1983) on Cotes, Hill (1961) on Boscovich, and Pearson (1929) and Crosland (1967) on Laplace.

2. Probabilists and the Measurement of Uncertainty

In addition to the works cited earlier (Hacking, 1975, in particular), the background to and context of Bernoulli's works are treated by Shafer (1978), Garber and Zabell (1979), Daston (1980), and Hald (1984a). The most thorough recent study of De Moivre is that by Schneider (1968); Walker (1934) remains the best purely biographical account. Hald (1984b) outlines (and presents a translation of) De Moivre (1711), and Archibald (1926) reprinted a facsimile of De Moivre (1733) (see also Pearson, 1926, and Daw and Pearson, 1972, in connection with that work). Seal (1949), Plackett (1958), and Sheynin (1973a) discuss Simpson's statistical work; biographical information can be found in Clarke (1929) and Lalande (1765). Barnard (1958) is the best biographical source on Bayes; it is accompanied by a reprinting of Bayes's *Essay*. Pearson (1925) also discussed Bernoulli's theorem.

3. Inverse Probability

Portions of that part of Laplace's work discussed in this chapter are also examined in Molina (1930) and Dale (1982), as well as in Todhunter (1865) and Sheynin (1976, 1977). Gillispie (1979) presents Laplace's previously unpublished 1777 memoir and two other pieces. Stigler (1978b) presents evidence on the dating of Laplace's early works; see also Baker (1975, pp. 436–437). Barnard (1958), Hacking (1965), Edwards (1978), Dale (1982), Shafer (1982), and Stigler (1982b, 1983) cover various aspects of the life and work of Bayes. For discussion of other related works of this period, see Daston (1979) on d'Alembert, Sheynin (1971) on Lambert, and Kendall (1961) on D. Bernoulli.

4. The Gauss–Laplace Synthesis

The dispute between Gauss and Legendre on the priority of the discovery of the method of least squares is documented in Plackett (1972) and Stigler (1977a, 1981). Plackett (1949), Seal (1967), Sprott (1978), and Sheynin (1979) review

Gauss's work on statistics. The discussion of Laplace's work on the tides of the atmosphere is expanded from Stigler (1975a). For the history of the subsequent development of the method of least squares in the nineteenth century, see Heyde and Seneta (1977), where the works of Bienaymé and Cauchy are emphasized. Seal (1967) also outlines some of this work. Stigler (1974a) discusses and translates some 1815 work of Gergonne on the use of least squares for polynomial interpolation. Farebrother (1985) discusses the history of the linear model before 1853.

5. Quetelet's Two Attempts

Laplace's analysis of the ratio estimator and the context of its use are discussed by Cochran (1978), Stephan (1948), and Pearson (1928). For Quetelet's life and work the best sources are Hankins (1908), Lottin (1912), Lazarsfeld (1961), Hilts (1973), and Porter (1985). Diamond and Stone (1981) have reprinted Florence Nightingale's notes on Quetelet. Poisson's work is reviewed by Sheynin (1978) and Bru (1981); Costabel (1981) gives a biographical account. Poisson's work on juries is studied by Gelfand and Solomon (1973); Daston (1981) examines this work in a broader intellectual context. Stigler (1982a) translates Poisson on the Poisson distribution. Ménard (1980) discusses Cournot's resistance to the application of probability to statistics, and Bernard Bru has appended a large number of useful notes to his 1984 edition of Cournot (1843). Baker (1975, chap. 6) discusses the general relationship between the philosophies of Comte and Condorcet. Wellens-de Donder (1966) gives an inventory of Quetelet's extensive correspondence.

6. Attempts to Revive the Binomial

Lexis's life and work are discussed in the article by Heiss (1978); and Heyde and Seneta (1977) explain the relationship between Lexian dispersion and earlier variations. Few modern textbooks treat this topic, which was once more widely covered, as in Coolidge (1925) and Uspensky (1937). Winsor (1947) discusses Bortkiewicz's most famous work, Bortkiewicz (1898). The articles Lexis (1879, 1886) and Bortkiewicz (1895, 1918) were translated into English during the early 1940s as part of a WPA project administered by the University of Minnesota. The translations were unfortunately never published, although a complete set of the typescripts is on file at the University of Minnesota's archives.

7. Psychophysics as a Counterpoint

The two books by E. G. Boring (1942, 1950) give a full account of the history of experimental psychology. Sanford (1888) is focused more narrowly upon the personal equation, and Adler (1966) and Jaynes (1981) present discussions of Fechner's life and work. Murray (1976) gives an historical account of works on memory; Stigler (1978d) calls attention to American work contemporary to Ebbinghaus.

8. The English Breakthrough: Galton

In addition to his own autobiography, *Memories of My Life* (1908), Galton has been the subject of two biographies. The first and most opulent was by Karl Pearson (1914–1930); it was a labor of love for Pearson and it similarly taxes the reader. The recent work by Forrest (1974) is much more accessible. The unsigned obituary in *Nature* is surely by Pearson (1911). Galton's statistical work has received increasing attention of late, in particular by Cowan (1972, 1977), Hilts (1973), and MacKenzie (1981). The work on correlation is discussed by Pearson (1920a), Lancaster (1972), and Plackett (1983). Hilts (1975) gives a useful guide to Galton's *English Men of Science.* Provine's (1971) study of the history of population genetics is relevant. Merrington and Golden (1976) is an invaluable guide to the Galton Archives.

9. The Next Generation: Edgeworth

Edgeworth's work on statistics is reviewed by Stigler (1978a). Bowley (1928) gives a bibliography of Edgeworth's statistical work, and Keynes (1933) and Kendall (1968) give excellent biographical accounts. The articles by E. S. Pearson (1965, 1967) and Stigler (1973b, 1980c) touch on Edgeworth, and Pratt (1976) discusses the relationship between Edgeworth's later work and R. A. Fisher's work on maximum likelihood. Edgeworth's work in economics is commented on by G. J. Stigler (1941, chap. 5; 1965). Contemporary work in America that is not discussed in this book is treated in Stigler (1978c) and reprinted in Stigler (1980a).

10. Pearson and Yule

E. S. Pearson (1938) and Eisenhart (1981) are full biographical accounts of Pearson; Morant (1939) is a nearly complete bibliography of his work. Various aspects of Pearson's work and its social background are discussed by Norton (1978), MacKenzie (1981), E. S. Pearson (1965, 1967), and Plackett (1983). Merrington et al. (1983) is a guide to the Pearson Archives. Särndal (1971) and Cramér (1972) treat the topic of frequency curves, the Edgeworth series in particular. Hald (1981) discusses contemporary work in Denmark by Thiele. Kendall (1952) is a good biographical account of Yule; Selvin (1976) discusses Yule's work on poor law statistics.

Bibliography

The works listed below have all been consulted in the edition or editions for which full bibliographical details are given. When a reprint has been accessible, that edition has been cited also, although the listing of reprints is far from complete. When a published translation is available, I have cited it in the bibliography and consulted it in preparing my own version. In general, translations presented in the text are my own; exceptions are indicated in the bibliography by notes stating that page references are to a translation.

Adams, William J. 1974. *The Life and Times of the Central Limit Theorem.* New York: Kaedmon.

Adler, Helmut E. 1966. Foreword to the 1966 translation of Fechner, 1860.

Adrain, Robert. 1808. Research concerning the probabilities of the errors which happen in making observations, etc. *The Analyst; or Mathematical Museum* 1 (4): 93–109. Probably published in 1809 (see Stigler, 1978c); reprinted in Stigler, 1980a, vol. 1.

Airy, George Biddell. 1861. *On the Algebraic and Numerical Theory of Errors of Observations and the Combination of Observations.* London: Macmillan (2nd ed., 1875; 3rd ed., 1879).

Aitken, George A. 1892. *The Life and Works of John Arbuthnot.* Oxford: Clarendon Press.

Arbuthnot, John. 1710. An argument for Divine Providence, taken from the constant regularity observ'd in the births of both sexes. *Philosophical Transactions of the Royal Society of London* 27: 186–190. Reprinted in Kendall and Plackett, 1977, pp. 30–34.

——— 1738. *Of the Laws of Chance,* 4th ed., revised by John Ham. London: B. Motte and C. Bathurst (1st ed., 1692).

Archibald, R. C. 1926. A rare pamphlet of Moivre and some of his discoveries. *Isis* 8: 671–683.

Baker, Keith M. 1975. *Condorcet—from Natural Philosophy to Social Mathematics.* Chicago: University of Chicago Press.

Barnard, George A. 1958. Thomas Bayes—a biographical note (together with a reprinting of Bayes, 1764). *Biometrika* 45: 293–315. Reprinted in Pearson and Kendall, 1970, pp. 131–153.

Baxter, J. H. 1875. *Statistics, Medical and Anthropological,* vol. 1. Washington: Government Printing Office.

Bayes, Thomas. 1764. An essay towards solving a problem in the doctrine of chances. *Philosophical Transactions of the Royal Society of London* for 1763, 53: 370–418. Reprinted with Barnard, 1958, in Pearson and Kendall, 1970, pp. 131–153.

Bernoulli, Daniel. 1778. Dijudicatio maxime probabilis plurium observationum discrepantium atque verisimillima inductio inde formanda. *Acta Academiae Scientiarum Imperialis Petropolitanae* for 1777, pars prior, 3–23. Reprinted in translation with Kendall, 1961.

Bernoulli, Jacob. 1713. *Ars Conjectandi.* Basil: Thurnisiorum.

Bernoulli, Jean III. 1785. Milieu à prendre entre les observations. Pp. 404–409 in *Encyclopédie méthodique: Mathématiques,* vol. 2. Paris: Panckoucke.

Berry, A. 1898. *A Short History of Astronomy.* London: John Murray. Reprinted, 1961; New York: Dover.

Bertillon, Louis-Adolphe. 1863. De la méthode dans l'anthropologie à propos de l'influence des milieux sur la coloration des téguments. *Bulletins de la société d'anthropologie de Paris* 4: 223–242.

——— 1876. Moyenne. *Dictionnaire encyclopédique des sciences médicales,* 2nd series, 10: 296–324. Paris: Masson & Asselin.

Bertrand, J. 1889. *Calcul des probabilités.* Paris: Gauthier-Villars (2nd ed. 1907). Reprinted, 1972; New York: Chelsea.

Bessel, Friedrich Wilhelm. 1818. *Fundamenta Astronomiae.* Regiomonti: Frid. Nicolovium.

——— 1876. Persönliche Gleichung bei Durchgangsbeobachtungen. Pp. 300–304 in *Abhandlungen von Friedrich Wilhelm Bessel,* vol. 3, ed. Rudolf Engelmann. Leipzig: Wilhelm Engelmann.

Bibliographie nationale: Dictionnaire des écrivains Belges et catalogue de leurs publications 1830–1880, vol. 3, N–U. 1897. Brussels: P. Weissenbruch.

Bienaymé, I. J. 1840. Probabilités. *Société philomatique de Paris, extrais des procès-verbaux des séances pendant l'année 1840,* pp. 37–43.

——— 1855. Sur un principe que M. Poisson avait cru découvrir et qu'il avait appelé loi des grands nombres. *Séances et travaux de l'Académie des sciences morales et politiques* 31: 379–389.

Black, R. D. C. 1977. *Papers and Correspondence of William Stanley Jevons,* vol. 4, *Correspondence 1873–1878.* London: Macmillan.

Booth, Charles. 1894. *The Aged Poor in England and Wales: Conditions.* London: Macmillan.

Boring, Edwin G. 1920. The logic of the normal law of error in mental measurement. *American Journal of Psychology* 31: 1–33.

——— 1942. *Sensation and Perception in the History of Experimental Psychology.* New York: Appleton-Century-Crofts.

——— 1950. *A History of Experimental Psychology,* 2nd ed. New York: Appleton-Century-Crofts.

Bortkiewicz, L. von. 1895. Kritische Betrachtungen zur theoretischen Statistik, II. *Jahrbücher für Nationalökonomie und Statistik,* 3rd series, 10: 321–360.

——— 1896. Kritische Betrachtungen zur theoretischen Statistik, III. *Jahrbücher für Nationalökonomie und Statistik,* 3rd series, 11: 671–705.

——— 1898. *Das Gesetz der Kleinen Zahlen.* Leipzig: Teubner.

——— 1909. Statistique. In *Encyclopédie des sciences mathématiques pures et appliquées* 1 (4): 453–480.

——— 1918. Homogeneität und Stabilität in der Statistik. *Särtryck ur Skandinavisk Aktuarietidskrift* 1–2: 1–81.

——— 1930. Lexis und Dormoy. *Nordic Statistical Journal* 2: 37–54.

Boscovich, Roger Joseph, and Christopher Maire. 1755. *De Litteraria Expeditione per Pontificiam ditionem ad dimetiendas duas Meridiani gradus.* Rome: Palladis.

——— 1770. *Voyage astronomique et geographique, dans l'état de l'église,* Paris: N. M. Tilliard. Includes a French translation of Boscovich and Maire, 1755.

Bouvard, A. 1827. Mémoire sur les observations météorologiques, faites à l'Observatoire Royal de Paris. *Mémoires de l'Académie royale des sciences de l'Institut de France* 7: 267–341.

Bowditch, Nathaniel. 1809. Observations of the comet of 1807. *Memoirs of the American Academy of Arts and Sciences* 3 (pt. 1): 1–17.

——— 1815. Elements of the orbit of the comet of 1811. *Memoirs of the American Academy of Arts and Sciences* 3 (pt. 2): 313–325.

——— 1832. *Mécanique Céleste,* vol. 2 of Laplace, 1829–1839; annotated translation of vol. 2 of Laplace, 1799–1805.

Bowley, A. L. 1901. *Elements of Statistics.* London: P. S. King.

——— 1928. *F. Y. Edgeworth's Contributions to Mathematical Statistics.* London: Royal Statistical Society. Reprinted, 1972; Clifton, N.J.: Augustus M. Kelley.

Bravais, A. 1846. Analyse mathématique sur les probabilités des erreurs de situation d'un point. *Mémoires presentés par divers savants à l'Académie royale des sciences de l'Institut de France* 9: 255–332. Presented to the Academy in 1838 and reported on by Poisson and Savary in *Comptes Rendus* 7: 77–78.

Bru, Bernard. 1981. Poisson, le calcul des probabilités et l'instruction publique. Pp. 51–94 in *Siméon-Denis Poisson et la science de son temps,* ed. M. Métivier, P. Costabel, and P. Dugac. Palaiseau: Ecole Polytechnique.

Buckle, Henry Thomas. 1857. *History of Civilization in England,* vol. 1. London: J. W. Parker.

Campbell, R. 1859. On a test for ascertaining whether an observed degree of uniformity, or the reverse, in tables of statistics is to be looked upon as remarkable. *Philosophical Magazine,* 4th series, 18: 359–368. Abstracted in the *Report of the British Association for the Advancement of Science* for 1859, pp. 3–4.

——— 1860. On a test for ascertaining whether an observed degree of uniformity, or the reverse, in tables of statistics, is to be looked upon as remarkable. *The Assurance Magazine, and Journal of the Institute of Actuaries* 8: 316–327.

Caneva, Kenneth L. 1981. Ohm, George Simon. In Gillispie, 1981b, vol. 10, pp. 186–194.

Cassini, Jacques. 1740. *Eléments d'astronomie.* Paris: Imprimerie Royale.

Chambers, E. 1779–1791. Expectation; Gaming; Probability. In *Cyclopaedia: or, an Universal Dictionary of Arts and Sciences,* new Ed., ed. A. Rees. London: Rivington et al.

Chapman, S. 1951. Atmospheric tides and oscillations. Pp. 510–530 in *Compendium of Meteorology,* ed. T. F. Malone. Boston: American Meteorological Society.

Clarke, A. R. 1858. *Ordnance Trigonometrical Survey of Great Britain and Ireland: Account of the Observations and Calculations, of the Principal Triangulation; and of the Figure, Dimensions and Mean Specific Gravity, of the Earth as Derived Therefrom.* London: Eyre and Spottiswoode.

——— 1880. *Geodesy.* Oxford: Clarendon Press.

Clarke, F. M. 1929. *Thomas Simpson and His Times.* New York: Waverly Press.

Clifford, William Kingdon. 1885. *The Common Sense of the Exact Sciences.* London: Kegan, Paul, Trench.

Cochran, W. G. 1978. Laplace's ratio estimator. Pp. 3–10 in *Contributions to Survey Sampling and Applied Statistics,* ed. H. A. David. New York: Academic Press. Reprinted, 1982; in Cochran's *Contributions to Statistics.* New York: Wiley.

Comte, Auguste. 1877. *Cours de philosophie positive,* vol. 4. Paris: Baillière (1st ed., 1839).

——— 1896. *The Positive Philosophy of Auguste Comte,* vol. 2, trans. H. Martineau. London: Bell.

Condorcet, Le Marquis de. 1785. *Essai sur l'application de l'analyse à la probabilité dés décisions rendues à la pluralité des voix.* Paris: Imprimerie Royale.

Coolidge, Julian Lowell. 1925. *An Introduction to Mathematical Probability.* Oxford: Clarendon Press.

Costabel, Pierre. 1981. Poisson, Siméon-Denis. In Gillispie, 1981b, vol. 15, pp. 480–490.

Cotes, Roger. 1722. Aestimatio Errorum in Mixta Mathesi, per Variationes Partium Trianguli Plani et Sphaerici. Part of Cotes's *Opera Miscellanea,* published with *Harmonia Mensurarum,* ed. Robert Smith. Cambridge.

Cournot, A. A. 1843. *Exposition de la théorie des chances et des probabilités.* Paris: Hachette. Reprinted, 1984; as vol. 1 of Cournot's *Oeuvres complètes,* ed. Bernard Bru. Paris: J. Vrin.

Cowan, R. S. 1972. Francis Galton's statistical ideas: the influence of eugenics. *Isis* 63: 509–528.

——— 1977. Nature and nurture: the interplay of biology and politics in the work of Francis Galton. *Studies in the History of Biology* 1: 133–208.

Cramér, Harald. 1972. On the history of certain expansions used in mathematical statistics. *Biometrika* 59: 205–207. Reprinted in Kendall and Plackett, 1977, pp. 437–439.

Crosland, M. 1967. *The Society of Arcueil.* Cambridge, Mass.: Harvard University Press.

Cullen, Michael. 1975. *The Statistical Movement in Early Victorian Britain: The Foundations of Empirical Social Research.* New York: Barnes & Noble.

Dale, A. I. 1982. Bayes or Laplace? an examination of the origin and early applications of Bayes' theorem. *Archive for History of Exact Sciences* 27: 23–47.

Daston, Lorraine J. 1979. D'Alembert's critique of probability theory. *Historia Mathematica* 6: 259–279.

——— 1980. Probabilistic expectation and rationality in classical probability theory. *Historia Mathematica* 7: 234–260.

——— 1981. Mathematics and the moral sciences: the rise and fall of the probability of judgments, 1785–1840. Pp. 287–309 in *Epistemological and Social Prob-*

lems of the Sciences in the Early Nineteenth Century, ed. A. N. Jahnke and M. Otte. Dordrecht: D. Reidel.

David, F. N. 1962. *Games, Gods & Gambling*. London: Charles Griffin.

Daw, R. H., and E. S. Pearson. 1972. Abraham De Moivre's 1733 derivation of the normal curve: a bibliographic note. *Biometrika* 59: 677–680. Reprinted in Kendall and Plackett, 1977, pp. 63–66.

De Forest, Erastus. 1882–83. On an unsymmetrical probability curve. *Analyst* 9: 135–142, 161–168; 10: 1–7, 67–74. Reprinted in Stigler, 1980a, vol. 1.

Delambre, J. B. J. 1799. *Méthodes analytiques pour la détermination d'un arc du méridien*. Paris: Duprat.

——— 1810. *Rapport historique sur les progrès des sciences mathématiques depuis 1789, et sur leur état actuel*. Paris: Imprimerie Impériale.

De Moivre, Abraham. 1711. De Mensura Sortis, seu, de Probabilitate Eventuum in Ludis a Casu Fortuito Pendentibus. *Philosophical Transactions of the Royal Society of London* (No. 329) 27: 213–264. Translated, with commentary, in Hald, 1984b.

——— 1718. *The Doctrine of Chances: or, A Method of Calculating the Probability of Events in Play*. London: W. Pearson.

——— 1725. *Annuities upon Lives*. London: W. Pearson.

——— 1730. *Miscellanea Analytica de Seriebus et Quadraturis*. London: J. Tonson and J. Watts.

——— 1733. *Approximatio ad Summam Terminorum Binomii* $\overline{a+b}|^n$ *in Seriem expansi*. Photographically reprinted in Archibald, 1926.

——— 1738. *The Doctrine of Chances*, 2nd ed. London: Woodfall.

——— 1743. *Annuities on Lives*, 2nd ed. London: Woodfall. A Dublin-printed "Second Edition" of 1731 is actually a reprinting of the 1st ed. with errata corrected and a table added.

——— 1756. *The Doctrine of Chances*, 3rd ed. London: Millar. Reprinted, 1967; New York: Chelsea.

De Morgan, Augustus. 1833–1843. Least Squares; Mean; Probability; Weight of Observations; and other articles. In *The Penny Cyclopaedia*, 27 vols. London: Charles Knight. De Morgan wrote a vast number of articles for this cyclopaedia. The already very long list in De Morgan, 1882, pp. 407–414, is incomplete.

——— 1837. Review of *Théorie Analytique des Probabilités*. *Dublin Review* 2: 338–354; 3: 237–248.

——— 1845. Theory of probabilities. Pp. 393–490 in *Encyclopaedia Metropolitana*, vol. 2, *Pure Mathematics*. London: B. Fellowes, et al. This article was written in 1836–37; see De Morgan, 1882, pp. 92–93.

De Morgan, Sophia. 1882. *Memoir of Augustus De Morgan*. London: Longman, Green.

Derham, W. 1754. *Physico-Theology: or, A Demonstration of the Being and Attributes of God, from his Works of Creation*, 12th ed. London: Innys and Richardson (1st ed., 1713).

Dewey, John. 1889. Galton's statistical methods. *Publications of the American Statistical Association* 7: 331–334.

Diamond, Marion, and Mervyn Stone. 1981. Nightingale on Quetelet. *Journal of the Royal Statistical Society* (A) 144: 66–79, 176–213, 332–351.

Dickson, J. D. Hamilton. 1886. Appendix [to Galton, 1886]. *Proceedings of the Royal Society of London* 40: 63–66.

Didion, Is. 1858. *Calcul des probabilités appliqué au tir des projectiles.* Paris: J. Dumaine.

Donkin, William F. 1844. *An Essay on the Theory of the Combination of Observations.* Oxford: Ashmolean Society.

Dormoy, Emile. 1874. Théorie mathématique des assurances sur la vie. *Journal des Actuaires Francais* 3: 283–299, 432–461. Reprinted in Dormoy, 1878, pp. 1–47.

——— 1878. *Théorie mathématique des assurances sur la vie,* vol. 1. Paris: Gauthier-Villars.

Ebbinghaus, Hermann. 1885. *Über das Gedächtnis.* Translated, 1913; as *Memory: A Contribution to Experimental Psychology,* trans. Henry A. Ruger and Clara E. Bussenius. New York: Teachers College, Columbia University. Reissued, 1964; New York: Dover Press. Page references are to the translation.

Edgeworth, Francis Ysidro. 1877. *New and Old Methods of Ethics, or "Physical Ethics" and "Methods of Ethics."* London: James Parker.

——— 1881. *Mathematical Psychics: An Essay on the Applications of Mathematics to the Moral Sciences.* London: C. Kegan Paul. Reprinted, 1967; New York: Augustus M. Kelley.

——— 1883a. The law of error. *Philosophical Magazine,* 5th series, 16: 300–309.

——— 1883b. The method of least squares. *Philosophical Magazine,* 5th series, 16: 360–375.

——— 1883c. On the method of ascertaining a change in the value of gold. *Journal of the Royal Statistical Society* 46: 714–718.

——— 1884. Review of *Investigations in Currency and Finance* by W. S. Jevons. *Academy* July 19, 1884: 38–39.

——— 1885a. Observations and statistics: an essay on the theory of errors of observation and the first principles of statistics. *Transactions of the Cambridge Philosophical Society* 14: 138–169. Abstracted in *Proceedings of the Cambridge Philosophical Society* 5: 310–312; corrigendum in *Proceedings of the Cambridge Philosophical Society* 6: 101–102.

——— 1885b. Methods of statistics. *Jubilee Volume of the Statistical Society,* pp. 181–217.

——— 1885c. On methods of ascertaining variations in the rate of births, deaths and marriages. *Journal of the Royal Statistical Society* 48: 628–649. Abstracted in the *Report of the British Association for the Advancement of Science* for 1885, pp. 1165–1166.

——— 1885d. Calculus of probabilities applied to psychical research, I. *Proceedings of the Society for Psychical Research* 3: 190–208.

——— 1886a. Progressive means. *Journal of the Royal Statistical Society* 49: 469–475.

——— 1886b. The law of error and the elimination of chance. *Philosophical Magazine,* 5th series, 21: 308–324.

——— 1886c. Problems in probabilities. *Philosophical Magazine,* 5th series, 22: 371–384.

——— 1887a. The empirical proof of the law of error. *Philosophical Magazine,* 5th series, 24: 330–342.

——— 1887b. On observations relating to several quantities. *Hermathena* 6: 279–285.

——— 1887c. *Metretike: or, The Method of Measuring Probability and Utility.* London: Temple.

——— 1887d. On discordant observations. *Philosophical Magazine,* 5th series, 23: 364–375.

——— 1888a. The mathematical theory of banking. *Journal of the Royal Statistical Society* 51: 113–127. Abstracted in the *Report of the British Association for the Advancement of Science* for 1886, pp. 776–778.

——— 1888b. The value of authority tested by experiment. *Mind* 13: 146–148.

——— 1888c. The statistics of examinations. *Journal of the Royal Statistical Society* 51: 346–368.

——— 1888d. Some new methods of measuring variation in general price. *Journal of the Royal Statistical Society* 51: 346–368.

——— 1889. Review of *Natural Inheritance* by F. Galton. *Nature* 39: 603–604.

——— 1890. The element of chance in competitive examinations. *Journal of the Royal Statistical Society* 53: 460–475, 644–663. Abstracted in the *Report of the British Association for the Advancement of Science* for 1890, p. 920.

——— 1892a. Correlated averages. *Philosophical Magazine,* 5th series, 34: 190–204.

——— 1892b. The law of error and correlated averages. *Philosophical Magazine,* 5th series, 34: 429–438, 518–526.

——— 1893a. A new method of treating correlated averages. *Philosophical Magazine,* 5th series, 35: 63–64.

——— 1893b. Exercises in the calculation of errors. *Philosophical Magazine,* 5th series, 36: 98–111.

——— 1893c. Note on the calculation of correlation between organs. *Philosophical Magazine,* 5th series, 36: 350–351.

——— 1893d. Statistical correlation between social phenomena. *Journal of the Royal Statistical Society* 56: 670–675. Abstracted in the *Report of the British Association for the Advancement of Science* for 1893, pp. 852–853.

——— 1894a. Asymmetrical correlation between social phenomena. *Journal of the Royal Statistical Society* 57: 563–568.

——— 1894b. The asymmetrical probability curve. *Proceedings of the Royal Society of London* 56: 271–272.

——— 1895. On some recent contributions to the theory of statistics. *Journal of the Royal Statistical Society* 58: 506–515.

——— 1896a. Bemerkungen über die Kritik meiner "Methoden der Statistik" von Dr. v. Bortkewitsch. *Jahrbücher für Nationalökonomie und Statistik,* 3rd series, 11: 274–277.

——— 1896b. Eine Erwiderung. *Jahrbücher für Nationalökonomie und Statistik,* 3rd series, 12: 838–845.
——— 1896c. Supplementary notes on statistics. *Journal of the Royal Statistical Society* 59: 529–539.
——— 1898. Miscellaneous applications of the calculus of probabilities, contd. *Journal of the Royal Statistical Society* 61: 119–131.
——— 1902. Error, law of. In *Encyclopaedia Britannica,* 10th ed., vol. 28 (supplement to 9th ed., vol. 4), pp. 280–291.
——— 1905. The law of error. *Transactions of the Cambridge Philosophical Society* 20: 36–65, 113–141.
——— 1908–1909. On the probable errors of frequency-constants. *Journal of the Royal Statistical Society* 71: 381–397, 499–512, 651–678; 72: 81–90.
——— 1911. Probability. In *Encyclopaedia Britannica,* 11th ed. New York: Encyclopaedia Britannica.
——— 1922. The philosophy of chance. *Mind* 31: 257–283.
——— 1925. *Papers Relating to Political Economy,* vols. 1–3. London: Macmillan.
Edinburgh Medical and Surgical Journal. 1817. Statement of the sizes of men in different counties of Scotland, taken from the local militia. 13: 260–264.
Edwards, A. W. F. 1978. Commentary on the arguments of Thomas Bayes. *Scandinavian Journal of Statistics* 5: 116–118.
Eisenhart, Churchill. 1961. Boscovich and the combination of observations. Chap. 7 in *R. J. Boscovich, Studies of His Life and Work,* ed. L. L. Whyte. London: Allen & Unwin. Reprinted in Kendall and Plackett, 1977, pp. 88–100.
——— 1964. The meaning of "least" in least squares. *Journal of the Washington Academy of Sciences* 54: 24–33.
——— 1981. Karl Pearson. In Gillispie, 1981b, vol. 10, pp. 447–473.
Elliott, Ezekiel B. 1863. *On the Military Statistics of the United States of America.* Berlin: International Statistical Congress.
Ellis, Robert Leslie. 1844. On the method of least squares. *Transactions of the Cambridge Philosophical Society* 8: 204–219. Reprinted in Ellis, 1863.
——— 1863. *The Mathematical and Other Writings of Robert Leslie Ellis M. A.,* ed. Walton, William. Cambridge: Deighton, Bell.
Encke, J. F. 1832. Über die Methode der Kleinsten Quadrate. *Berliner Astronomisches Jahrbuch für 1834,* pp. 249–312. Translated, 1841; in *Scientific Memoirs,* ed. Richard Taylor, vol. 2, pp. 317–369.
Encyclopaedia Britannica. 1771. Gaming. In vol. 2, pp. 643–644. Edinburgh: Bell and Macfarquhar.
Euler, Leonhard. 1749. *Recherches sur la question des inégalités du mouvement de Saturne et de Jupiter, sujet proposé pour le prix de l'anneé 1748, par l'Académie royale des sciences de Paris.* Reprinted, 1960, in *Leonhardi Euleri, Opera Omnia,* 2nd series, vol. 25, pp. 45–157. Basel: Turici. Page references are to the reprint.
——— 1768. *Lettres à une princesse d'Allemagne sur divers sujets de physique & de philosophie,* 3 vols. St. Petersburg: Académie imperiale des sciences.
——— 1778. Observationes in praecedentem dissertationem. *Acta Academiae Scientiarum Imperialis Petropolitanae* for 1777, pars prior, 24–33. Reprinted in translation, with Kendall, 1961.

Farebrother, R. W. 1985. The statistical estimation of the standard linear model, 1756–1853. *Proceedings of the First International Tampere Seminar on Linear Statistical Models and Their Applications*, pp. 77–99.

[Farr, William]. 1848. *Eighth Annual Report of the Registrar-General of Births, Deaths, and Marriages, in England*. London: W. Clowes and Sons.

Fechner, Gustav Theodor. 1831. *Massbestimmungen über die Galvanische Kette*. Leipzig: Brockhaus.

——— 1860. *Elemente der Psychophysik*, 2 vols. Vol. 1 translated, 1966; as *Elements of Psychophysics*, trans. Helmut E. Adler, ed. Davis H. Howes and Edwin G. Boring. New York: Holt, Rinehart & Winston. Page references are to this translation.

——— 1882. *Revision der Hauptpuncte der Psychophysik*. Leipzig: Breitkopf und Hartel.

Feller, William. 1968. *An Introduction to Probability Theory and Its Applications*, vol. 1, 3rd ed. New York: Wiley.

Fisher, Ronald A. 1922. On the mathematical foundations of theoretical statistics. *Philosophical Transactions of the Royal Society of London* (A), 222: 309–368.

——— 1928. On a distribution yielding the error functions of several well known statistics. *Proceedings of the International Mathematical Congress Held in Toronto, 1924*, 2: 805–813.

——— 1935. *The Design of Experiments*. Edinburgh: Oliver and Boyd.

——— 1959. *Statistical Methods and Scientific Inference*, 2nd ed. Edinburgh: Oliver and Boyd.

Fontenelle, Bernard. 1717. *The Lives of the French, Italian and German Philosophers, Late Members of the Royal Academy of Sciences in Paris*. London: W. Innys.

Forrest, D. W. 1974. *Francis Galton: The Life and Work of a Victorian Genius*. London: Elek.

[Fourier, Joseph]. 1829. Second mémoire sur les résultats moyens et sur les erreurs des mesures. *Recherches statistiques sur la ville de Paris et le département de la Seine* 4: ix–xlviii. Reprinted in Fourier, 1890, pp. 549–590.

Fourier, Joseph. 1890. *Oeuvres de Fourier*, vol. 2. Paris: Gauthier-Villars.

Francoeur, L.-B. 1840. *Astronomie pratique*, 2nd ed. Paris: Bachelier (1st ed., 1830).

Galloway, Thomas. 1839. *A Treatise on Probability*. Edinburgh: Adam and Charles Black.

Galton, Francis. 1853. *The Narrative of an Explorer in Tropical South Africa*. London: John Murray.

——— 1863. *Meteorographica, or Methods of Mapping the Weather*. London: Macmillan.

——— 1869. *Hereditary Genius: An Inquiry into its Laws and Consequences*. London: Macmillan (2nd ed., 1892).

——— 1874a. On a proposed statistical scale. *Nature* 9: 342–343.

——— 1874b. *English Men of Science: Their Nature and Nurture*. London: Macmillan.

——— 1875. Statistics by intercomparison, with remarks on the law of frequency of error. *Philosophical Magazine*, 4th series, 49: 33–46.

——— 1877. Typical laws of heredity. *Nature* 15: 492–495, 512–514, 532–533. Also published in *Proceedings of the Royal Institution of Great Britain* 8: 282–301.

——— 1879. The geometric mean, in vital and social statistics. *Proceedings of the Royal Society of London* 29: 365–367. (Accompanies McAlister, 1879.)

——— 1883. *Inquiries into Human Faculty and Its Development.* London: Macmillan.

——— 1885. Section H; Anthropology; Opening address. *Nature* 32: 507–510.

——— 1886a. Regression towards mediocrity in hereditary stature. *Journal of the Anthropological Institute* 15: 246–263.

——— 1886b. Family likeness in stature. *Proceedings of the Royal Society of London* 40: 42–73.

——— 1888. Co-relations and their measurement, chiefly from anthropometric data. *Proceedings of the Royal Society of London* 45: 135–145.

——— 1889. *Natural Inheritance.* London: Macmillan.

——— 1890. Kinship and correlation. *North American Review* 150: 419–431.

——— 1892. *Finger Prints.* London: Macmillan.

——— 1896. Application of the method of percentiles to Mr. Yule's data on the distribution of pauperism. *Journal of the Royal Statistical Society* 59: 392–396.

——— 1908. *Memories of My Life.* London: Methuen.

Garber, Daniel, and Sandy Zabell. 1979. On the emergence of probability. *Archive for History of Exact Sciences* 21: 33–53.

Gauss, Carl Friedrich. 1809. *Theoria motus corporum celestium.* Hamburg: Perthes et Besser. Translated, 1857, as *Theory of Motion of the Heavenly Bodies Moving about the Sun in Conic Sections,* trans. C. H. Davis. Boston: Little, Brown. Reprinted, 1963; New York: Dover. French translation of the portion on least squares, pp. 113–134 in Gauss, 1855.

——— 1816. Bestimmung der Genauigkeit der Beobachtungen. Pp. 109–117 in *Carl Friedrich Gauss Werke,* vol. 4. Göttingen: Königlichen Gesellschaft der Wissenschaften, 1880. French translation, pp. 141–153 in Gauss, 1855.

——— 1823. *Theoria Combinationis Observationum Erroribus Minimis Obnoxiae.* Göttingen: Dieterich. French translation, pp. 1–69 in Gauss, 1855.

——— 1855. *Méthode des moindres carrés. Mémoires sur la combination des observations* trans. J. Bertrand. Paris: Mallet-Bachelier.

Gelfand, Alan E., and Herbert Solomon. 1973. A study of Poisson's models for jury verdicts in criminal and civil trials. *Journal of the American Statistical Association* 68: 271–278.

Giere, Ronald N., and Richard S. Westfall, eds. 1973. *Foundations of Scientific Method: The Nineteenth Century.* Bloomington: Indiana University Press.

Gillispie, Charles Coulston. 1979. Mémoires inédits ou anonymes de Laplace sur la théorie des erreurs, les polynômes de Legendre, et la philosophie des probabilités. *Revue d'histoire des sciences* 32: 223–279.

——— 1981a. Laplace, Pierre-Simon, Marquis de. In Gillispie, 1981b, vol. 15, pp. 273–403.

Gillispie, Charles Coulston, ed. 1981b. *Dictionary of Scientific Biography,* 16 vols. New York: Charles Scribner's Sons.

Goldstein, Bernard R. 1983. The obliquity of the ecliptic in ancient Greek astronomy. *Archives Internationales d'Histoire des Sciences* 33: 3–14.

Goodman, Leo, and William Kruskal. 1959. Measures of association for cross classifications, II: further discussion and references. *Journal of the American Statistical Association* 54: 123–163. Reprinted, 1979; in their *Measures of Association for Cross Classifications*. New York: Springer-Verlag.

Gore, J. Howard. 1889. A bibliography of geodesy, 1st ed. Appendix no. 16 to the *Report of the Superintendent of the U.S. Coast and Geodetic Survey* for the year ending June 1887. Washington: Government Printing Office.

——— 1903. A bibliography of geodesy, 2nd ed. Appendix no. 8 to the *Report of the Superintendent of the U.S. Coast and Geodetic Survey* for the year 1902. Washington: Government Printing Office.

Gould, Benjamin Apthorp. 1869. *Sanitary Memoirs of the War of the Rebellion*, vol. 2, *Investigations in the Military and Anthropological Statistics of American Soldiers*. New York: U.S. Sanitary Commission.

Gouraud, C. 1848. *Histoire de calcul des probabilités depuis ses origines jusqu'à nos jours*. Paris: Auguste Durand.

Gowing, Ronald. 1983. *Roger Cotes—Natural Philosopher*. Cambridge: Cambridge University Press.

Graunt, John. 1665. *Natural and Political Observations Mentioned in a Following Index, and Made upon the Bills of Mortality*, 3rd ed. London: John Martyn and James Allestry (1st. ed. 1662).

Grove, W. R. 1865. *The Correlation of Physical Forces*, 4th ed. As reprinted in *The Correlation and Conservation of Forces*, ed. E. L. Youmans. New York: D. Appleton & Co.

Hacking, Ian. 1965. *Logic of Statistical Inference*. Cambridge: Cambridge University Press.

——— 1975. *The Emergence of Probability*. Cambridge: Cambridge University Press.

Hain, Joseph. 1852. *Handbuch der Statistik des Österreichischen Kaiserstaates*, vol. 1. Vienna: Tendler.

Hald, A. 1981. T. N. Thiele's contributions to statistics. *International Statistical Review* 49: 1–20.

——— 1984a. Nicholas Bernoulli's theorem. *International Statistical Review* 52: 93–99.

——— 1984b. A. de Moivre: 'De Mensura Sortis' or 'On the Measurement of Chance.' *International Statistical Review* 52: 229–262.

Halley, Edmund. 1693. An estimate of the degrees of the mortality of mankind, drawn from curious tables of the births and funerals at the city of Breslaw; with an attempt to ascertain the price of annuities upon lives. *Philosophical Transactions of the Royal Society of London* 17: 596–610.

Hankins, F. H. 1908. *Adolphe Quetelet as Statistician*. New York: Columbia University Press.

Harris, John. 1710. Play. In *Lexicon Technicum; or, An Universal English Dictionary of Arts and Sciences*, vol. 2. London: Brown, Goodwin, Walthoe, Nicholson, Tooke, Midwinter, Atkins, and Ward.

Harter, H. Leon. 1974–1976. The method of least squares and some alternatives. *International Statistical Review* 42: 147–174, 235–264, 282; 43: 1–44, 125–190, 269–278; 44: 113–159.

——— 1977. *A Chronological Annotated Bibliography on Order Statistics*, vol. 1, *Pre-1950*. Wright-Patterson Air Force Base, Ohio: Air Force Flight Dynamics Laboratory.

Hartley, David. 1749. *Observations on Man, His Frame, His Duty, and His Expectations*, 2 vols. London: Richardson.

Harvey, George. 1822. On the method of minimum squares, employed in the reduction of experiments, being a translation of the appendix to an essay of Legendre's, entitled, "Nouvelles Méthods pour la Détermination des Orbites des Comètes," with remarks. *The Edinburgh Philosophical Journal* 7: 292–301.

Hegelmaier, F. 1852. Über das Gedächtniss für Linearanschauungen. *Archiv für physiologische Heilkunde* 11: 844–853.

Heiss, Klaus-Peter. 1978. Lexis, Wilhelm. In Kruskal and Tanur, 1978, vol. 1, pp. 507–512.

Heyde, C. C., and E. Seneta. 1977. *I. J. Bienaymé: Statistical Theory Anticipated*. New York: Springer-Verlag.

Hill, E. 1961. Roger Boscovich, a biographical essay. Pp. 17–101 in *Roger Joseph Boscovich, S. J., F.R.S., 1711–1787; Studies of His Life and Work on the 250th Anniversary of His Birth*, ed. L. L. Whyte. London: Allen & Unwin.

Hilts, Victor L. 1973. Statistics and social science. In Giere and Westfall, 1973, pp. 206–233.

——— 1975. A guide to Francis Galton's *English Men of Science*. *Transactions of the American Philosophical Society* 65 (5): 1–85.

Hollerith, Herman. 1894. The electrical tabulating machine. *Journal of the Royal Statistical Society* 57: 678–689.

Hutton, C. 1795. *A Mathematical and Philosophical Dictionary*, 2 vols. London: J. Johnson.

Huygens, Christian. 1657. De Ratiociniis in Ludo Aleae. Pp. 517–534 in Frans van Schooten's *Exercitationum Mathematicarum*. Leiden: Johannis Elsevirii.

——— 1659. Van Rekeningh in Spelen van Geluck. Pp. 485–500 in Frans van Schooten's *Mathematische Oeffeningen*. Amsterdam: Gerrit van Goedesbergh. Dutch version of Huygens, 1657. Reprinted, together with a French translation, in 1920, in vol. 14, pp. 49–91, of *Oeuvres Complètes de Christiaan Huygens*. The Hague: Nijhoff.

Jaynes, Julian. 1981. Fechner, Gustav Theodor. In Gillispie, 1981b, vol. 4, pp. 556–559.

Jeffreys, Harold. 1939. *Theory of Probability*. Oxford: Clarendon Press.

Jevons, William Stanley. 1863. *A Serious Fall in the Value of Gold Ascertained, and Its Social Effects Set Forth*. Reprinted as chap. 2 of Jevons, 1884.

——— 1865. The variation of prices and the value of the currency since 1782. *Journal of the Royal Statistical Society* 28: 294–320. Reprinted in Jevons, 1884.

——— 1868. On the condition of the metallic currency of the United Kingdom, with reference to the question of international coinage. *Journal of the Royal Statistical Society* 31: 426–464.

——— 1869. The depreciation of gold. *Journal of the Royal Statistical Society* 32: 445–449. Reprinted in Jevons, 1884.

——— 1874. *The Principles of Science: A Treatise on Logic and Scientific Method,* 2 vols. London: Macmillan (2nd ed., 1877). Reprinted, 1958; New York: Dover.

——— 1881. Review of *Mathematical Psychics,* by F. Y. Edgeworth. *Mind* 6: 581–583.

——— 1884. *Investigations in Currency and Finance.* London: Macmillan (2nd ed., 1909; with different paging). Original version reprinted, 1964; New York: Augustus M. Kelley.

Johnson, Norman L., and Samuel Kotz. 1982–. *Encyclopedia of Statistical Sciences,* 6 vols. to date. New York: Wiley.

Kahle, Ludwig Martin. 1735. *Elemente Logicae Probabilium Methodo Mathematics in Usum Scientiarum et Vitae Adornata.* Halle: Officina Rengeriana. Reprints Bernoulli's (1713) proof.

Kendall, Maurice G. 1952. George Udny Yule, 1871–1951. *Journal of the Royal Statistical Society* (A), 115: 156–161. Reprinted in Stuart and Kendall, 1971, and in Pearson and Kendall, 1970, pp. 419–425.

——— 1961. Daniel Bernoulli on maximum likelihood [together with a reprinting of Bernoulli, 1778, and Euler, 1778]. *Biometrika* 48: 1–18. Reprinted in Pearson and Kendall, 1970, pp. 155–172.

——— 1968. Francis Ysidro Edgeworth, 1845–1926. *Biometrika* 55: 269–275. Reprinted in Pearson and Kendall, 1970, pp. 257–263.

Kendall, M. G., and A. Doig. 1968. *Bibliography of Statistical Literature Pre-1940.* Edinburgh: Oliver and Boyd.

Kendall, M. G., and R. L. Plackett. 1977. *Studies in the History of Statistics and Probability,* vol. 2. London: Griffin.

Keverberg, Baron de. 1827. Notes [appended to Quetelet, 1827]. *Nouveaux mémoires de l'Académie royale des sciences et belles-lettres de Bruxelles* 4: 175–192.

Keynes, John Maynard. 1921. *A Treatise on Probability.* London: Macmillan.

——— 1933. F. Y. Edgeworth. In *Essays in Biography.* New York: Harcourt, Brace & Co. Originally published in the *Economic Journal* (1926), 36: 140–153.

Kramp, Christian. 1799. *Analyse des réfractions astronomiques et terrestres.* Strasbourg: Dannbach.

Kruskal, William H., and Judith Tanur. 1978. *International Encyclopedia of Statistics,* 2 vols. New York: Free Press. Updated reprinting of the statistical articles from Sills, 1968.

Kuhn, Thomas S. 1961. The function of measurement in modern physical science. In Woolf, 1961, pp. 31–63.

Lacroix, Silvestre F. 1816. *Traité élémentaire du calcul des probabilités.* Paris: Courcier (2nd ed., 1822; Paris: Bachelier).

Lagneau, Gustave. 1870. Quelques remarques ethnologiques sur la répartition géographique de certaines infirmités en France. *Mémoires de l'Académie imperiale de médecine* 29: 293–317.

Lagrange, J. L. 1776. Mémoire sur l'utilité de la méthode de prendre le milieu entre les résultats de plusieurs observations; dans lequel on examine les avantages de cette méthode par le calcul des probabilités; & ou l'on resoud differens problèmes relatifs à cette matière. *Miscellanea Taurinensia* 5: 167–232. Reprinted in Lagrange, 1867–1892, vol. 2, pp. 173–236.

——— 1867–1892. *Oeuvres de Lagrange,* 14 vols. Paris: Gauthier-Villars. The correspondence with Laplace is in vol. 14.

Lalande, Joseph Jerome Le Francais. 1765. Remarques sur la vie et ouvrages de Mrs de la Caille, Bradley, Mayer, et Simpson. *Connoissance des mouvements célestes pour l'année commune* 1767, pp. 181–204.

——— 1771. *Astronomie,* 2nd ed., 3 vols. Paris: Desaint (3rd ed., 1792).

Lancaster, H. O. 1968. *Bibliography of Statistical Bibliographies.* Edinburgh: Oliver & Boyd.

——— 1972. Development of the notion of statistical dependence. *Mathematical Chronicle* (New Zealand) 2: 1–16. Reprinted in Kendall and Plackett, 1977, pp. 293–308.

Laplace, Pierre Simon. 1774. Mémoire sur la probabilité des causes par les évènemens. *Mémoires de l'Académie royale des sciences presentés par divers savans* 6: 621–56. Reprinted in Laplace, 1878–1912, vol. 8, pp. 27–65. Translated in Stigler, 1986.

——— 1777. Recherches sur le milieu qu'il faut choisir entre les résultats de plusieurs observations. In Gillispie, 1979, pp. 228–256.

——— 1781. Mémoire sur les probabilités. *Mémoires de l'Académie royale des sciences de Paris,* 1778, pp. 227–332. Reprinted in Laplace, 1878–1912, vol. 9, pp. 383–485.

——— 1782. Mémoire sur les suites. *Mémoires de l'Académie royale des sciences de Paris,* 1779, pp. 207–309. Reprinted in Laplace, 1878–1912, vol. 10, pp. 1–89.

——— 1785. Mémoire sur les approximations des formules qui sont fonctions de très grands nombres. *Mémoires de l'Académie royale des sciences de Paris,* 1782. Reprinted in Laplace, 1878–1912, vol. 10, pp. 209–291. Page references are to Laplace, 1878–1912.

——— 1786a. Mémoire sur la figure de la terre. *Mémoires de l'Académie royale des sciences de Paris,* 1783, pp. 17–46. Reprinted in Laplace, 1878–1912, vol. 11, pp. 3–32.

——— 1786b. Mémoire sur les approximations des formules qui sont fonctions de très grands nombres, suite. *Mémoires de l'Académie royale des sciences de Paris,* 1783. Reprinted in Laplace, 1878–1912, vol. 10, pp. 295–338. Pages references are to Laplace, 1878–1912.

——— 1786c. Sur les naissances, les mariages et les morts, à Paris, depuis 1771 jusqui'en 1784, et dans toute l'étendue de la France, pendant les années 1781 and 1782. *Mémoires de l'Académie royale des sciences de Paris,* 1783. Reprinted in Laplace, 1878–1912, vol. 11, pp. 35–46.

——— 1788. Théorie de Jupiter et de Saturne. *Mémoires de l'Académie royale des sciences de Paris,* 1785, pp. 33–160 (separate offprint issued 1787). Reprinted in Laplace, 1878–1912, vol. 11, pp. 95–239.

——— 1793. Sur quelques points du système du monde. *Mémoires de l'Académie royale des sciences de Paris,* 1789, pp. 1–87. Reprinted in Laplace, 1878–1912, vol. 11, pp. 477–558.

——— 1799–1805. *Traité de mécanique céleste,* vols. 1–4. Paris: Duprat (vols. 1–3) and Courcier (vol. 4). Reprinted in Laplace, 1878–1912, vols. 1–4. Translated in Laplace, 1829–1839.

——— 1810. Mémoire sur les approximations des formules qui sont fonctions de très grands nombres et sur leur application aux probabilités. *Mémoires de l'Académie des sciences de Paris,* 1809, pp. 353–415, 559–565. Reprinted in Laplace, 1878–1912, vol. 12, pp. 301–353.

——— 1811. Mémoire sur les integrales définies et leur application aux probabilités, et specialement à la recherche du milieu qu'il faut choisir entre les resultats des observations. *Mémoires de l'Académie des sciences de Paris,* 1810, pp. 279–347. Reprinted in Laplace, 1878–1912, vol. 12, pp. 357–412.

——— 1812. *Théorie analytique des probabilités.* Paris: Courcier (3rd ed., 1820, with supplements). Reprinted in Laplace, 1878–1912, vol. 7.

——— 1814. *Essai philosophique sur les probabilités.* Paris: Courcier. 6th ed., 1840, translated, 1902; as *A Philosophical Essay on Probabilities,* trans. F. W. Truscott and F. L. Emory. Reprinted 1951; New York: Dover.

——— 1823. De l'action de la lune sur l'atmosphere. *Annales de Chimie et de Physique* 24: 280–294. Reprinted with minor changes in Laplace, 1825, pp. 184–188, 262–268; and in Laplace, 1878–1912, vol. 5.

——— 1825. *Traité de mécanique céleste,* vol. 5. Paris: Bachelier. Photographically reprinted, 1969; as *Celestial Mechanics,* vol. 5. New York: Chelsea. Reprinted in Laplace, 1878–1912, vol. 5.

——— 1827. Mémoire sur le flux et reflux lunaire atmospherique. In *Connaissance des Tems pour l'an* 1830, pp. 3–18. Reprinted in Laplace, 1825, supplement, pp. 489–505; and in Laplace, 1878–1912, vol. 13, pp. 342–358.

——— 1829–1839. *Mécanique Céleste,* vols. 1–4, trans. N. Bowditch [extensively annotated translation of Laplace, 1799–1805]. Boston: Hilliard, Gray, Little, and Wilkins. Photographically reprinted, 1966; as *Celestial Mechanics,* vols. 1–4. New York: Chelsea.

——— 1878–1912. *Oeuvres complètes de Laplace,* 14 vols. Paris: Gauthier-Villars.

Lazarsfeld, Paul F. 1961. Notes on the history of quantification in sociology—trends, sources and problems. In Woolf, 1961, pp. 147–203. Also published in *Isis* 52: 277–333. Reprinted in Kendall and Plackett, 1977, pp. 213–269.

Legendre, Adrien Marie. 1798. Méthode pour déterminer la longueur exacte du quart du Méridien, d'après les observations faites pour la mesure de l'arc compris entre Dunkerque et Barcelonne. In Delambre, 1799, pp. 1–16.

——— 1805. *Nouvelles méthodes pour la détermination des orbites des comètes.* Paris: Courcier. Reissued with a supplement, 1806. Second supplement published 1820. A portion of the appendix was translated, 1929, pp. 576–579 in *A Source Book in Mathematics,* D. E. Smith, ed., trans. by H. A. Ruger and H. M. Walker, New York: McGraw-Hill; reprinted 1959 in 2 vols., New York: Dover.

Lexis, Wilhelm. 1876. Das Geschlechtsverhältnis der Geborenen und die Wahrscheinlichkeitsrechnung. *Jahrbücher für Nationalökonomie und Statistik* 27: 209–245.

——— 1877. *Zur Theorie der Massenerscheinungen in der menschlichen Gesellschaft.* Freiburg: Wagner.

——— 1879. Über die Theorie der Stabilität statistischer Reihen. *Jahrbücher für Nationalökonomie und Statistik* 32: 60–98.

——— 1880. Sur les moyennes normales appliquées aux mouvements de la population et sur la vie normale. *Annales de Demographie Internationale* 4: 481–497.

——— 1886. Über die Wahrscheinlichkeitsrechnung und deren Anwendung auf die Statistik. *Jahrbücher für Nationalökonomie und Statistik,* 2nd series, 13: 433–450.

——— 1903. *Abhandlungen zur Theorie der Bevölkerungs-und Moralstatistik.* Jena: Gustav Fischer.

Lindenau, Baron von. 1806. Über der Gebrauch der Gradmessungen zur Bestimmung der Gestalt der Erde. *Monatliche Correspondenz* 14: 113–158.

Livi, Ridolfo. 1896. Sulla Interpretazione delle Curve Seriali in Antropometria. *Atti della Societa Romana di Antropologia* 3: 21–52.

Lottin, J. 1912. *Quetelet, Statisticien et Sociologue.* Louvain: Institut Superieur de Philosophie.

MacKenzie, Donald A. 1981. *Statistics in Britain, 1865–1930: The Social Construction of Scientific Knowledge.* Edinburgh: Edinburgh University Press.

Marshall, Alfred. 1881. Review of *Mathematical Psychics,* by F. Y. Edgeworth, *Academy,* June 18, 1881, p. 457.

Maskelyne, Nevil. 1762. A letter from the Rev. Nevil Maskelyne, M.A.F.R.S. to the Rev. Thomas Birch, D.D. Secretary to the Royal Society: containing the results of observations of the distance of the moon from the sun and fixed stars, made in a voyage from England to the island of St. Helena, in order to determine the longitude of the ship, from time to time; together with the whole process of computation used on this occasion. *Philosophical Transactions of the Royal Society of London* 52: 558–577.

——— 1763. *The British Mariner's Guide.* London: J. Nourse.

Mayer, Tobias. 1750. Abhandlung über die Umwalzung des Monds um seine Axe und die scheinbare Bewegung der Mondsflecten. *Kosmographische Nachrichten und Sammlungen auf das Jahr 1748,* pp. 52–183.

McAlister, D. 1879. The law of the geometric mean. *Proceedings of the Royal Society of London* 29: 367–376. (Accompanies Galton, 1879).

Ménard, Claude, 1980. Three forms of resistance to statistics: Say, Cournot, Walras. *History of Political Economy* 12: 524–541.

Merriman, Mansfield. 1877. A list of writings relating to the method of least squares, with historical and critical notes. *Transactions of the Connecticut Academy of Arts and Sciences* 4: 151–232. Reprinted in Stigler, 1980a, vol. 1.

Merrington, M., and J. Golden. 1976. *A List of the Papers and Correspondence of Sir Francis Galton.* London: University College London.

Merrington, M., B. Blundell, S. Burrough, J. Golden, and J. Hogarth. 1983. *A List of the Papers and Correspondence of Karl Pearson (1857–1936) Held in the Manuscripts Room University College London Library.* London: University College London.

Merton, Robert K., David L. Sills, and Stephen M. Stigler. 1984. The Kelvin dictum and social science: an excursion into the history of an idea. *Journal of the History of the Behavioral Sciences* 20: 319–331.

Mill, John Stuart. 1843. *A System of Logic*, 2 vols. London: John W. Parker (2nd ed., 1846).

Molina, E. C. 1930. The theory of probability: some comments on Laplace's *Théorie analytique*. *Bulletin of the American Mathematical Society* 36: 369–392.

Montmort, Pierre Remond de. 1713. *Essay d'analyse sur les jeux de hazard*, 2nd ed. Paris: Jacque Quillau.

Morant, G. M. 1939. *A Bibliography of the Statistical and Other Writings of Karl Pearson*. London: Biometrika.

Müller, G. E. 1879. Über die Maassbestimmungen des Ortssinnes der Haut mittels der Methode der richtigen und falschen Fälle. *Archiv für die gesammte Physiologie des Menschen und der Thiere* 19: 191–235.

Murray, D. J. 1976. Research on human memory in the nineteenth century. *Canadian Journal of Psychology* 30: 201–220.

Nesselroade, J., S. M. Stigler, and P. Baltes. 1980. Regression toward the mean and the study of change. *Psychological Bulletin* 87: 622–637.

Newbold, Ethel M. 1927. Practical applications of the statistics of repeated events, particularly to industrial accidents. *Journal of the Royal Statistical Society* 90: 487–547.

Newcomb, Simon. 1903. *The Reminiscences of an Astronomer*. Boston: Houghton Mifflin. Reprinted in Stigler, 1980a, vol. 2.

Newman, D. 1939. The distribution of range in samples from a normal population, expressed in terms of an independent estimate of standard deviation. *Biometrika* 31: 20–30.

Newton, Isaac. 1726. *Philosophiae Naturalis Principia Mathematica*, 3rd ed. Translated, 1946; as *Sir Isaac Newton's Mathematical Principles of Natural Philosophy and his System of the World*, ed. Florian Cajori. Berkeley: University of California Press.

Norton, Bernard. 1978. Karl Pearson and statistics: the social origins of scientific innovation. *Social Studies of Science* 8: 3–34.

Ordnance Survey. 1967. *History of the Retriangulation of Great Britain 1935–1962*. London: Her Majesty's Stationery Office.

Pearson, Egon S. 1938. *Karl Pearson: An Appreciation of Some Aspects of His Life and Work*. Cambridge: Cambridge University Press.

——— 1965. Some incidents in the early history of biometry and statistics, 1890–94. *Biometrika* 52: 3–18. Reprinted in Pearson and Kendall, 1970, pp. 323–338.

——— 1967. Some reflexions on continuity in the development of mathematical statistics, 1885–1920. *Biometrika* 54: 341–355. Reprinted in Pearson and Kendall, 1970, pp. 339–353.

Pearson, Egon S., and M. G. Kendall, eds. 1970. *Studies in the History of Statistics and Probability*. London: Charles Griffin.

Pearson, Karl. 1888. *The Ethic of Freethought.* London: T. Fisher Unwin (2nd ed., 1901).

——— 1892. *The Grammar of Science.* London: Walter Scott (2nd ed., 1900; 3rd ed., 1911).

——— 1893a. Asymmetrical frequency curves. *Nature* 48: 615–616. Correction in *Nature* 49: 6.

——— 1893b. Contributions to the mathematical theory of evolution (abstract). *Proceedings of the Royal Society of London* 54: 329–333.

——— 1893c. Contributions to the mathematical theory of evolution. *Journal of the Royal Statistical Society* 56: 675–679.

——— 1894a. Contributions to the mathematical theory of evolution. *Philosophical Transactions of the Royal Society of London* (A), 185: 71–110. Page references are to the reprint in Pearson, 1956, pp. 1–40.

——— 1894b. Science and Monte Carlo. *Fortnightly Review* 50: 183–193.

——— 1895a. Contributions to the mathematical theory of evolution, II: skew variation. *Philosophical Transactions of the Royal Society of London* (A), 186: 343–414. Page references are to the reprint in Pearson, 1956, pp. 41–112.

——— 1895b. On skew probability curves. *Nature* 52: 317. Reprinted in Stigler, 1980a, vol. 2.

——— 1896. Mathematical contributions to the theory of evolution, III: regression, heredity and panmixia. *Philosophical Transactions of the Royal Society of London* (A), 187: 253–318. Page references are to the reprint in Pearson, 1956, pp. 113–178.

——— 1897. *The Chances of Death and Other Studies in Evolution,* vols. 1 and 2. London: Edward Arnold.

——— 1898. Mathematical contributions to the theory of evolution: on the law of ancestral heredity. *Proceedings of the Royal Society of London* 62: 386–412.

——— 1900. On the criterion that a given system of deviations from the probable in the case of a correlated system of variables is such that it can be reasonably supposed to have arisen from random sampling. *Philosophical Magazine,* 5th series, 50: 157–175. Reprinted in Pearson, 1956, pp. 339–357. Correction in *Philosophical Magazine,* 6th series, 1: 670–671.

——— 1905. Mathematical contributions to the theory of evolution, XIV: on the general theory of skew correlation and non-linear regression. *Draper's Company Research Memoirs,* biometric series 2. Page references are to the reprint in Pearson, 1956, pp. 477–528.

——— 1906. Walter Frank Raphael Weldon, 1860–1906: a memoir. *Biometrika* 5: 1–52. Reprinted in Pearson and Kendall, 1970, pp. 264–321.

——— 1911. Francis Galton (unsigned obituary). *Nature* 85: 440–445.

——— 1914–1930. *The Life, Letters and Labours of Francis Galton,* (3 vols. in 4 parts). Cambridge: Cambridge University Press.

——— 1920a. Notes on the history of correlation. *Biometrika* 13: 25–45. Reprinted in Pearson and Kendall, 1970, pp. 185–205.

——— 1920b. The fundamental problem of practical statistics. *Biometrika* 13: 1–16.

——— 1925. James Bernoulli's theorem. *Biometrika* 17: 201–210.

——— 1926. Abraham De Moivre. *Nature* 117: 551–552.

——— 1928. On a method of ascertaining limits to the actual number of marked members in a population of given size from a sample. *Biometrika* 20A: 165–174.

——— 1929. Laplace, being extracts from lectures delivered by Karl Pearson. *Biometrika* 21: 202–216.

——— 1941. The laws of chance, in relation to thought and conduct. *Biometrika* 32: 89–100.

——— 1956. *Karl Pearson's Early Statistical Papers.* Cambridge: Cambridge University Press (first issued, 1948).

——— 1978. *The History of Statistics in the 17th and 18th Centuries, against the Changing Background of Intellectual, Scientific and Religious Thought,* lectures from 1921–1933, ed. E. S. Pearson. London: Griffin.

Pearson, Karl, and L. N. G. Filon. 1898. Mathematical contributions to the theory of evolution, IV: on the probable errors of frequency constants and on the influence of random selection on variation and correlation. *Philosophical Transactions of the Royal Society of London* (A), 191: 229–311. Page references are to the reprint in Pearson, 1956, pp. 179–261.

Peirce, Benjamin. 1852. Criterion for the rejection of doubtful observations. *Astronomical Journal* 2: 161–163. Reprinted in Stigler, 1980a, vol. 2.

Peirce, Charles S. 1892. Review of *The Grammar of Science. Nation* 55: 15.

Peirce, Charles S., and Joseph Jastrow. 1885. On small differences of sensation. *Memoirs of the National Academy of Sciences for 1884* 3: 75–83. Reprinted in Stigler, 1980a, vol. 2.

Pigou, Arthur Cecil. 1910. Memorandum on some economic aspects and effects of Poor Law relief. Appendix 80, pp. 981–1000, in *Royal Commission on the Poor Laws and Relief of Distress,* appendix vol. 9. *Parliamentary Papers* for the Session 15 February 1910–28 November 1910, vol. 49. London: His Majesty's Stationary Office.

Plackett, Robin L. 1949. A historical note on the method of least squares. *Biometrika* 36: 458–460.

——— 1958. The principle of the arithmetic mean. *Biometrika* 45: 130–135. Reprinted in Pearson and Kendall, 1970, pp. 121–126.

——— 1972. The discovery of the method of least squares. *Biometrika* 59: 239–251. Reprinted in Kendall and Plackett, 1977, pp. 279–291.

——— 1983. Karl Pearson and the chi-squared test. *International Statistical Review* 51: 59–72.

Playfair, J. 1808. Review of *Traité de Méchanique Céleste [sic]* par P. S. La Place. *Edinburgh Review,* January 1808, 11 (22): 249–284.

Poincaré, Henri. 1896. *Calcul des probabilités.* Paris: Gauthier-Villars (2nd ed., 1912).

Poisson, Siméon Denis. 1824. Sur la probabilité des résultats moyens des observations. *Connaissance des temps pour l'an* 1827, pp. 273–302.

——— 1829. Suite du mémoire sur la probabilité du résultat moyen des observations, inséré dans la connaissance des temps de l'année 1827. *Connaissance des temps pour l'an* 1832, pp. 3–22.

——— 1836a. Note sur le calcul des probabilités. *Comptes rendus hebdomadaires des séances de l'Académie des sciences* 2: 395–400.

——— 1836b. Note sur la loi des grands nombres. *Comptes rendus hebdomadaires des séances de l'Académie des sciences* 2: 377–382.

——— 1837. *Recherches sur la probabilité des jugements en matière criminelle et en matière civile, précédés des règles générales du calcul des probabilités.* Paris: Bachelier.

Porter, Theodore M. 1985. The mathematics of society: variation and error in Quetelet's statistics. *British Journal of the History of Science* 18: 51–69.

——— 1986. *The Rise of Statistical Thinking, 1820–1900.* Princeton: Princeton University Press.

Pratt, John W. 1976. F. Y. Edgeworth and R. A. Fisher on the efficiency of maximum likelihood estimation. *Annals of Statistics* 4: 501–514.

Prony, R. 1804. *Recherches physico-mathématiques sur la théorie des eaux courantes.* Paris: Imprimerie Impériale.

Provine, William B. 1971. *The Origins of Theoretical Population Genetics.* Chicago: University of Chicago Press.

Puissant, L. 1805. *Traité de géodésie.* Paris: Courcier (2nd ed., 2 vols., Paris: Bachelier, 1819; 3rd ed., 2 vols., Paris: Bachelier, 1842).

——— 1807. *Traité de topographie, d'arpentage et de nivellement.* Paris: Courcier (2nd ed., 1820).

——— 1827. *Supplément au traité de géodésie.* Paris: Bachelier.

Quetelet, Adolphe. 1826. *Astronomie élémentaire.* Paris: Malher.

——— 1827. Recherches sur la population, les naissances, les décès, les prisons, les dépôts de mendicité, etc., dans le royaume des Pays-Bas. *Nouveaux mémoires de l'Académie royale des sciences et belles-lettres de Bruxelles* 4: 117–192.

——— 1828. *Instructions populaires sur le calcul des probabilités.* Brussels: Tarlier et Hayez.

——— 1832. Sur la possibilité de mesurer l'influence des causes qui modifient les élémens sociaux. Lettre à M. de Dr. Villermé. *Correspondance mathématique et physiques* 7: 321–346.

——— 1835. *Sur l'homme et le développement de ses facultés, ou Essai de physique sociale.* Paris: Bachelier.

——— 1837. Ouvrages nouveaux. [review of Poisson, 1837]. *Correspondance mathématique et physique* 9: 485–486.

——— 1842. *A Treatise on Man and the Development of His Faculties.* Edinburgh: Chambers. Translation of Quetelet, 1835.

——— 1846. *Lettres à S. A. R. le Duc Régnant de Saxe-Cobourg et Gotha, sur la théorie des probabilités, appliquée aux sciences morales et politiques.* Brussels: Hayez.

——— 1848. *Du système social et des lois qui le régissent.* Paris: Guillaumin.

——— 1849. *Letters Addressed to H. R. H. the Grand Duke of Saxe Coburg and Gotha, on the Theory of Probabilities as Applied to the Moral and Political Sciences,* trans. O. G. Downes. London: Layton. Translation of Quetelet, 1846.

——— 1852. Sur quelques propriétés curieuses que présentent les résultats d'une serie d'observations, faites dans la vue de déterminer une constante, lorsque les chances de rencontrer des écarts en plus et en moins sont égales et indépendantés les unes des autres. *Bulletins de l'Académie royale des sciences, des lettres et des beaux-arts de Belgique* 19 (pt. 2): 303–317.

——— 1853. *Théorie des probabilités.* Brussels: A. Jamar.
——— 1869. *Physique Sociale, ou, Essai sur le développement des facultés de l'homme,* vols. 1 and 2. Brussels: Muquardt.
——— 1870. *Anthropométrie, ou mesure des différentes facultés de l'homme.* Brussels: Muquardt.
Quetelet, Adolphe, and Ed. Smits. 1832. *Recherches sur la reproduction et la mortalité de l'homme aux différens ages, et sur la population de la Belgique.* Brussels: Chez Louis Hauman.

Reichesberg, Naúm. 1893. Der berühmte Statistiker, Adolphe Quetelet, sein Leben und sein Wirken. *Zeitschrift für schweizerische Statistik* 32: 418–460.
Rogers, William A. 1869. On the variability of personal equation in transit observations. *American Journal of Science and Arts,* 2nd series, 47: 297–307.

Sanford, Edmund C. 1888. Personal equation. *American Journal of Psychology* 2: 3–38, 271–298, 403–430.
Särndal, Carl-Erik. 1971. The hypothesis of elementary errors and the Scandinavian school in statistical theory. *Biometrika* 58: 375–391. Reprinted in Kendall and Plackett, 1977, pp. 419–435.
Schneider, Ivo. 1968. Der Mathematiker Abraham De Moivre. *Archive for History of Exact Sciences* 5: 177–317.
Schweber, Silvan S. 1977. The origin of the *Origin* revisited. *Journal of the History of Biology* 10: 229–316.
Seal, Hilary L. 1949. The historical development of the use of generating functions in probability theory. *Bulletin de l'Association des Actuaires Suisses* 49: 209–228. Reprinted in Kendall and Plackett, 1977, pp. 67–86.
——— 1967. The historical development of the Gauss linear model. *Biometrika* 54: 1–24. Reprinted in Pearson and Kendall, 1970, pp. 207–230.
Selvin, H. C. 1976. Durkheim, Booth and Yule: the non-diffusion of an intellectual innovation. *Archives Européennes de Sociologie* 17: 39–51.
'sGravesande, G. J. 1774. *Oeuvres philosophiques et mathématiques de Mr. G. J. 'sGravesande,* ed. J. N. S. Allamand, 2 vols. Amsterdam: Marc Michel Rey.
Shafer, Glenn. 1978. Non-additive probabilities in the work of Bernoulli and Lambert. *Archive for History of Exact Sciences* 19: 309–370.
——— 1982. Bayes's two arguments for the rule of conditioning. *Annals of Statistics* 10: 1075–1089.
Sheynin, Oscar B. 1968. On the early history of the law of large numbers. *Biometrika* 55: 459–467. Reprinted in Pearson and Kendall, 1970, pp. 231–239.
——— 1971. J. H. Lambert's work on probability. *Archive for History of Exact Sciences* 7: 244–256.
——— 1972. On the mathematical treatment of observations by L. Euler. *Archive for History of Exact Sciences* 9: 45–56.
——— 1973a. Finite random sums (a historical essay). *Archive for History of Exact Sciences* 9: 275–305.
——— 1973b. R. J. Boscovich's work on probability. *Archive for History of Exact Sciences* 9: 306–24.

——— 1976. P. S. Laplace's work on probability. *Archive for History of Exact Sciences* 16: 137–187.

——— 1977. Laplace's theory of errors. *Archive for History of Exact Sciences* 17: 1–61.

——— 1978. S. D. Poisson's work in probability. *Archive for History of Exact Sciences* 18: 245–300.

——— 1979. C. F. Gauss and the theory of errors. *Archive for History of Exact Sciences* 20: 21–72.

——— 1983. Corrections and short notes on my papers. *Archive for History of Exact Sciences* 28: 171–195.

Sills, David L., ed. 1968. *International Encyclopedia of the Social Sciences*, 18 vols. New York: Macmillan and Free Press.

Simpson, Thomas. 1740. *The Nature of Laws of Chance*. London: Edward Cave (reprint, 1792).

——— 1742. *The Doctrine of Annuities and Reversions*. London: J. Nourse.

——— 1743. *An Appendix, Containing Some Remarks on a Late Book on the Same Subject, with Answers to Some Personal and Malignant Misrepresentations, in the Preface thereof.* London: J. Nourse.

——— 1755. A letter to the Right Honorable George Earl of Macclesfield, President of the Royal Society, on the advantage of taking the mean of a number of observations, in practical astronomy. *Philosophical Transactions of the Royal Society of London* 49: 82–93.

——— 1757. *Miscellaneous Tracts on Some Curious, and Very Interesting Subjects in Mechanics, Physical-Astronomy, and Speculative Mathematics*. London: J. Nourse.

Sokal, M. M., ed. 1981. *An Education in Psychology: James McKeen Cattell's Journal and Letters from Germany and England, 1880–1888*. Cambridge, Mass.: MIT Press.

Somerville, Martha. 1873. *Personal Recollections, from Early Life to Old Age, of Mary Somerville*. London: John Murray.

Somerville, Mary. 1831. *Mechanism of the Heavens*. London: John Murray.

Spottiswoode, William. 1861. On typical mountain ranges: an application of the calculus of probabilities to physical geography. *Journal of the Royal Geographical Society* 31: 149–154.

Sprott, David A. 1978. Gauss's contributions to statistics. *Historia Mathematica* 5: 183–203.

Stephan, Frederick F. 1948. History of the uses of modern sampling procedures. *Journal of the American Statistical Association* 43: 12–39.

Stigler, George J. 1941. *Production and Distribution Theories: The Formative Period*. New York: Macmillan. Reprinted, 1968; New York: Agathon Press.

——— 1965. *Essays in the History of Economics*. Chicago: University of Chicago Press.

Stigler, Stephen M. 1973a. Laplace, Fisher, and the discovery of the concept of sufficiency. *Biometrika* 60: 439–445. Reprinted in Kendall and Plackett, 1977, pp. 271–277.

——— 1973b. Simon Newcomb, Percy Daniell, and the history of robust estimation, 1885–1920. *Journal of the American Statistical Association* 68: 872–879.

Reprinted in Kendall and Plackett, 1977, pp. 410–417, and in Stigler, 1980a, vol. 2.

——— 1974a. Gergonne's 1815 paper on the design and analysis of polynomial regression experiments. *Historia Mathematica* 1: 431–447.

——— 1974b. Cauchy and the witch of Agnesi: an historical note on the Cauchy distribution. *Biometrika* 61: 375–380.

——— 1975a. Napoleonic statistics: the work of Laplace. *Biometrika* 62: 503–517.

——— 1975b. The transition from point to distribution estimation. *Bulletin of the International Statistical Institute* 46 (2): 332–340.

——— 1977a. An attack on Gauss, published by Legendre in 1820. *Historia Mathematica* 4: 31–35.

——— 1977b. Eight centuries of sampling inspection: the trial of the Pyx. *Journal of the American Statistical Association* 72: 493–500.

——— 1978a. Francis Ysidro Edgeworth, statistician (with discussion). *Journal of the Royal Statistical Society* (A), 141: 287–322.

——— 1978b. Laplace's early work: chronology and citations. *Isis* 69: 234–254.

——— 1978c. Mathematical statistics in the early states. *Annals of Statistics* 6: 239–265. Reprinted in Stigler, 1980a, vol. 1.

——— 1978d. Some forgotten work on memory. *Journal of Experimental Psychology: Human Learning and Memory* 4: 1–4.

——— 1980a. *American Contributions to Mathematical Statistics in the Nineteenth Century.* (2 vols.) New York: Arno Press.

——— 1980b. Stigler's law of eponymy. *Transactions of the New York Academy of Sciences,* 2nd series, 39: 147–157.

——— 1980c. An Edgeworth curiosum. *Annals of Statistics* 8: 931–934.

——— 1980d. R. H. Smith, A Victorian interested in robustness. *Biometrika* 67: 217–221.

——— 1981. Gauss and the invention of least squares. *Annals of Statistics* 9: 465–474.

——— 1982a. Poisson on the Poisson distribution. *Statistics and Probability Letters* 1: 33–35.

——— 1982b. Thomas Bayes's Bayesian inference. *Journal of the Royal Statistical Society* (A), 145: 250–258.

——— 1982c. Jevons as statistician. *The Manchester School* 50: 354–365.

——— 1983. Who discovered Bayes's theorem? *The American Statistician* 37: 290–296.

——— 1984. Boscovich, Simpson, and a 1760 manuscript note on fitting a linear relation. *Biometrika* 71: 615–620.

——— 1986. Laplace's 1774 memoir on inverse probability. *Statistical Science* 1, in press. Includes a translation of Laplace, 1774.

Stirling, J. 1730. *Methodus Differentialis: Sive Tractatus de Summatione et Interpolatione Serierum Infinitarium.* London: Gul. Bowyer.

Stuart, A., and M. G. Kendall, eds. 1971. *Statistical Papers of George Udny Yule.* London: Griffin.

Süssmilch, Johann Peter. 1741. *Die göttliche Ordnung.* Portions of 3rd ed. reprinted, 1979; as *"L'ordre divin"—aux origines de la demographie,* trans. and ed., with commentary, J. Hecht, 2 vols. Paris: Institut National d'Etudes Demographiques.

Svanberg, Jöns. 1805. *Exposition des opérations faites en Lapponie, pour la détermination d'un arc du méridien en 1801, 1802, et 1803.* Stockholm: J. P. Lindh.

Swinburne, R. G. 1965. Galton's Law—formulation and development. *Annals of Science* 21: 15–31.

Titchener, Edward B. 1915. *Experimental Psychology: A Manual of Laboratory Practice,* vol. 2, pt. 2. New York: Macmillan.

Todhunter, Isaac. 1865. *A History of the Mathematical Theory of Probability.* London: Macmillan. Reprinted, 1949, 1965; New York: Chelsea.

——— 1873. *A History of the Mathematical Theories of Attraction and the Figure of the Earth,* 2 vols. London: Macmillan. Reprinted in one vol., 1962; New York: Dover.

——— 1879. On the arc of the meridian measured in Lapland. *Transactions of the Cambridge Philosophical Society* 12: 1–26.

Urban, F. M. 1910. The method of constant stimuli and its generalizations. *Psychological Review* 17: 229–259.

Uspensky, J. V. 1937. *Introduction to Mathematical Probability.* New York: McGraw-Hill.

Venn, John. 1866. *The Logic of Chance.* London: Macmillan (2nd ed., 1876; 3rd ed., 1888).

——— 1887. The law of error. *Nature* 42: 412–413.

Villermé, L. R. 1829. Mémoire sur la taille de l'homme en France. *Annales d'hygiène publique et de médecine légale* 1: 351–399.

Walker, Helen M. 1929. *Studies in the History of Statistical Method.* Baltimore: Williams & Wilkins. Reprinted, 1975; New York: Arno Press.

——— 1934. Abraham De Moivre. *Scripta Mathematica* 2: 316–333. Reprinted with De Moivre, 1756, 1967; New York: Chelsea.

Weldon, W. F. R. 1892. Certain correlated variations in Crangon vulgaris [correlation of shrimp measurements]. *Proceedings of the Royal Society of London* 51: 2–21.

Wellens-de Donder, Liliane. 1966. Inventaire de la correspondence d'Adolphe Quetelet déposée à l'Académie royale de Belgique. *Mémoires de l'Académie royale de Belgique* 37 (2): 1–299, and plates.

Werckmeister, Karl, ed. 1898–1899. *Das Neunzehnte Jahrhundert in Bildnissen,* 5 vols. Berlin: Photographishen Gesellschaft.

Westergaard, H. 1890. *Die Grundzüge der Theorie der Statistik.* Jena: Gustav Fischer.

——— 1932. *Contributions to the History of Statistics.* London: P. S. King. Reprinted 1969, New York: Augustus Kelley.

Whittaker, E. T., and G. Robinson. 1924. *The Calculus of Observations.* London: Blackie and Son.

Wilson, Curtis A. 1980. Perturbations and solar tables from Lacaille to Delambre:

the rapprochement of observation and theory, II. *Archive for History of Exact Sciences* 22: 189–304.

Winsor, Charles P. 1947. Das Gesetz der Kleinen Zahlen. *Human Biology* 19: 154–161.

Wolf, Rudolf. 1869–1872. *Handbuch der Mathematik, Physik, Geodäsie und Astronomie,* 2 vols. Zürich: Schulthess.

Woolf, Harry, ed. 1961. *Quantification—A History of the Meaning of Measurement in the Natural and Social Sciences.* Indianapolis: Bobbs-Merrill.

Yule, G. Udny. 1895. On the correlation of total pauperism with proportion of out-relief, I: all ages. *Economic Journal* 5: 603–611.

——— 1896a. Notes on the history of pauperism in England and Wales from 1850, treated by the method of frequency-curves; with an introduction on the method. *Journal of the Royal Statistical Society* 59: 318–357.

——— 1896b. On the correlation of total pauperism with proportion of out-relief, II: males over sixty-five. *Economic Journal* 6: 613–623.

——— 1896c. Remarks on Mr. Galton's note. *Journal of the Royal Statistical Society* 59:396–398.

——— 1897a. On the significance of Bravais' formulae for regression, &c., in the case of skew correlation. *Proceedings of the Royal Society of London* 60: 477–489.

——— 1897b. On the theory of correlation. *Journal of the Royal Statistical Society* 60: 812–854.

——— 1899. An investigation into the causes of changes in pauperism in England, chiefly during the last two intercensal decades, I. *Journal of the Royal Statistical Society* 62: 249–295.

——— 1900. On the association of attributes in statistics. *Philosophical Transactions of the Royal Society of London* (A), 194: 257–319.

——— 1907. On the theory of correlation for any number of variables, treated by a new system of notation. *Proceedings of the Royal Society of London* (A), 79: 182–193.

Index

Abel, Niels, 12
Academy of Sciences, Paris, 25–26, 31, 33
accidental causes, 175, 181
Adams, William J., 64n, 370, 374
additive effects model, 173, 249, 313–315
Adler, Helmut E., 243, 372, 374
Adrain, Robert, 117, 143n, 374
Airy, George Biddell, 284n, 294, 306, 374
Aitken, George A., 374
Albategnius, 6
Albert, Prince, 206
Allgemeine Geographische Ephemeridenz, 56n
analysis, mathematical, 43n, 51
analysis of variance, 154, 172–173, 179, 234, 309, 314, 349
ancestral heredity, law of, 273, 296, 304
ancillary statistics, 108n, 120
Andräe, C. C. G. von, 317
anticyclones, 267, 327
Arago, François, 162
Arbuthnot, John, 1, 63, 71, 221 (portrait), 225–226, 374
arc. *See* meridian arc
Archibald, Raymond Clare, 371, 374
arc length, 40–42, 43, 57–59
Argelander, F. W. A., 241
arithmetic mean, 14, 16, 112–113; principle of, 141. *See also* mean
Arzachel, 6
aspect, 275, 312
asymmetric distributions, 220, 330, 332. *See also* Edgeworth; Pearson; skewed distribution
Athenaeum, 170–171, 180
average man, 169–172, 201, 302; Cournot's criticism of, 171, 325; as ideal, 172, 214

B.A.A.S. *See* British Association for the Advancement of Science
Bacon, Francis, 1
Bacon, Roger, 1
Baker, Keith M., 195, 371–372, 374
ballistics, 317
Baltes, Paul, 296, 390
bankruptcy data, 308
Barnard, George A., 123, 371, 374
barometric data, 150, 156, 331
baseline, 52n
Baxter, J. H., 206, 219, 374
Bayes, Thomas, 88, 97–98, 100, 102–104, 108, 116, 122–133, 371, 374–375, 377, 381, 394, 396; critique of Simpson, 94–95, 100
Bayes's scholium, 127–128
Bayes's theorem, 102–103, 123, 126, 131–132, 141, 290
Bayesian inference, 6, 119, 190, 198, 310, 345. *See also* Bayes; Cournot; Edgeworth; Laplace; nonuniform prior distribution; Poisson; posterior distribution; uniform prior distribution
bees data, 311
Bernoulli, Daniel, 63, 88, 101–102, 110n, 371, 375, 386
Bernoulli, Jacob (or Jakob or James or Jacques), 62 (portrait), 63–70, 77–78; and *Ars Conjectandi*, 64, 70; and urn model, 64–67, 124–125; and binomial distribution, 64–67, 92; and inference, 65–67, 77–78, 122, 128, 134; proof of law of large numbers, 67–69, 386; and De Moivre, 70–72, 74, 77, 85; limitations to his result, 77–78, 80, 83; mentioned, 61, 87, 99–100, 103, 129, 158, 161, 193, 224–225; bibliographical references, 371, 375, 386, 391, 394, 396
Bernoulli, Jean I (or John or Johann), 63–64
Bernoulli, Jean II, 63
Bernoulli, Jean III, 50n, 63, 95, 101–102, 110n, 375

Bernoulli, Nicholas, 63–64, 71–72, 74, 77, 87, 226, 384
Bernoulli's law of large numbers, 64–70, 183–185
Berry, Arthur, 17n, 41, 375
Bertillon, Louis-Adolphe, 215, 217–218, 368, 375
Bertrand, Joseph, 185, 195, 375
Bessel, Friedrich Wilhelm, 202–205, 207, 213, 230n, 375; and personal equation, 241–242
beta distribution, 126, 132–133, 334. *See also* Dirichlet distribution; incomplete beta function
bias: due to measurement error, 296; integer, 286n; of estimate, 94–95, 106, 141, 147, 293–294
Bienaymé, I. J., 148, 185, 191, 197, 228n, 372, 375, 385
billiard table, Bayes's, 124–127
bills of mortality, 4, 89
bimodal distribution, 215, 217, 268, 332
binomial distribution, 66–87, 92, 122–134, 209–211, 222–226, 230, 246, 275–276, 333–334, 339–340. *See also* normal approximation to binomial
binomial theorem, 92
biology, statistics in, 351, 353
birth rate, 166–168, 232, 311
birth ratio, inference about, 134–135, 200, 222, 225–226, 365
bivariate normal distribution, 148–149, 206, 285, 287–288, 317, 324, 342, 344–345
Black, R. D. C., 301, 375
blind experiment, 253
Blundell, B., 389
blunder, 218, 321
Boole, George, 310
Booth, Charles, 345–347, 375, 394
Boring, Edwin G., 240, 243, 256, 261, 271, 372, 375
Bortkiewicz, L. von, 233, 236–238, 315, 372, 375–376, 380
Boscovich, Roger Joseph, 39, 42–55, 61, 95, 97, 140, 371, 376, 381, 385, 394, 396
Boscovich's method, 40, 51, 55
Bouguer, Pierre, 42
Bouvard, Alexis, 156–157, 162, 376
Bowditch, Nathaniel, 39, 52–55, 59–60, 376
Bowley, Arthur L., 341, 343, 373, 376
Box, George E. P., 357
Bradley, James, 202, 240–241, 387
Brahe, Tycho, 6
Bravais, Auguste, 83n, 353, 376, 398
Breslau tables, 86n, 89
British Association for the Advancement of Science, 162, 283, 313
brothers, data on, 292
Bru, Bernard, 372, 376–377
Buckle, Henry Thomas, 226–228, 376
Buffon, G.-L. L., Comte de, 178
Bull, John, 225
bungling, 218, 321
bureau des longitudes, 31
Burrough, S., 389
Butler, David E., 307n

Cambridge University, 43, 303, 304
Campbell, Robert, 226–228, 236, 376
Caneva, Kenneth L., 243, 376
Canton, John, 94–95
Cassini, Domenico, 6, 40, 42
Cassini, Jacques, 5, 6, 42, 376
Cassini de Thury, C.-F., 45
Cattell, James McKeen, 301, 395
Cauchy, Augustin-Louis, 148, 183n, 372, 396
Cauchy distribution, 183n, 261
Censorinus, 30, 38
census, 163, 165–166, 222, 360
central limit theorem, 134, 136–138, 144, 201–203, 205, 207, 209, 219, 274, 318
Chambers, Ephraim, 85, 376
Chandrasekhar, S., 8
Chapman, S., 155, 376
characteristic functions, 137
Chebychev's inequality, 69, 70n
chest measurements. *See* Scottish soldiers data
Cheysson, E., 368
chi-square, 237, 329, 335
chronograph, 242
Chuprov, A. A., 237
Clairaut, Alexis-Claude, 43n
Clarke, A. R., 158, 377
Clarke, F. M., 88n, 371, 377
Clifford, William Kingdon, 303, 377
Cochran, William G., 163n, 164, 372, 377
coefficient of correlation, 297, 319, 342. *See also* correlation; Edgeworth; Pearson
coefficient of divergence, 232
Columbia University, 301

combination of observations, 5, 28, 169
combinatorial method, Lexis's, 231
components of variance, 231, 293–294, 312–314, 360
computers, 158, 361
Comte, Auguste, 194–195, 197, 372, 377
conditional expectation, 290, 317–318, 321, 323
conditional inference, 108, 120
conditional probability, 115–116, 178, 196–197
conditional variance, 290, 294, 317
Condorcet, Marquis de, 103n, 123, 186, 193–195, 372, 374, 377
confidence interval, 187, 191, 361
consistent estimation, 65–66, 87, 93–94, 132, 183
constant causes, 175, 178–179, 181, 185
continuous density, 76, 96, 106–107, 126, 132
conviction rates, 174–182, 186, 188–189, 191
convolution. *See* generating functions
Coolidge, Julian Lowell, 237, 372, 377
Copernicus, Nicholas, 6
Cornell University, 301
correlated errors, 59–60, 151–152, 155. *See also* correlation
correlation, 8, 284, 290, 297–299, 304, 316–317, 342, 368; and causation, 348, 351; matrix, 324; spurious, 357; tables, 347
correlation coefficient, 297, 319–322, 342–343; multiple, 354; partial, 354; standard deviation of, 321–322, 325, 343, 345
Costabel, Pierre, 372, 377
Cotes, Roger, 16, 54, 371, 377, 384
Cournot, Antoine-Augustin, 171, 183, 191, 195–200, 207, 231, 306, 310, 325, 330, 359–360, 372, 377, 389
Cowan, Ruth S., 267, 281, 373, 377
crab data, 332, 335, 337
Cramér, Harald, 373, 377
crime rate, 170, 172, 232
Crosland, Maurice, 371, 377
cubit data, 319, 368
Cullen, Michael, 182n, 370, 377
Czuber, Emanuel, 317

Dale, A. I., 129–130, 371, 377
d'Alembert, Jean Le Rond, 31, 371, 377
Daniell, Percy, 395
Danti, Egnatio, 6
Darwin, Charles, 170n, 267, 304, 361
Darwin, Erasmus, 267
Darwin, George, 278–279, 301, 332, 338, 341, 344
Daston, Lorraine J., 371–372, 377
data tables: on obliquity of ecliptic, 6; on moon, 22; on Saturn, 34; on arc lengths, 43; on French arc, 59; on yearly conviction rates, 175; on conviction rates, subclassified, 176; on numbers convicted, 189; on barometric pressure, 156; on star positions, 204; on Scottish chests, 207; on Scottish stature, 208; on French conscripts, 216; on conscripts in Doubs, 218; on lifted weights, 250; on memory, 259; on family stature, 286; on brothers' stature, 292; on cubits, 319; on poor law unions, 347
David, F. N. 370, 378
da Vinci, Leonardo, 1
Davis, Charles Henry, 140
Daw, R. H., 71, 371, 378
death: normal, 223; premature, 223
death rate, 166–168, 311
De Forest, Erastus, 317, 333–335, 338, 378
degree of latitude, 40–41
degrees of freedom, 329n
de Keverberg dilemma, 163, 200, 357, 361. *See also* Keverberg, Baron de
Delambre, J. B. J., 51, 56, 140n, 378, 397
Delboeuf, J. L. R., 306
De Moivre, Abraham, 62 (portrait), 70–88; and annuities, 71, 83, 89–90; and binomial, 71–76, 78–84, 86–87; and Stirling, 72–73; and normal approximation, 73–76, 78–84, 136, 201; and square root rule, 77, 83; and Simpson, 88–92; dispute with Simpson, 89–90; and generating functions, 91–92, 118, 137; and Bayes, 98, 132; mentioned, 67, 100, 103, 117, 185, 205, 225, 281; bibliographical references, 371, 374, 378, 384, 391, 394, 397
De Morgan, Augustus, 16, 31, 157–158, 378
De Morgan, Sophia, 378
Derham, William, 226, 378
design of experiments, 22, 42, 244–245, 250, 254, 257, 396; randomized, 253. *See also* factorial design

Dewey John, 301, 304, 378
Diamond, Marion, 372, 379
Dickson, J. D. Hamilton, 288–289, 368, 379
Didion, Is., 317n, 379
Dionysus, 30, 38
Dirichlet distribution, 120. *See also* beta distribution
dispersion: normal, subnormal, and supernormal, 231
dispersion theory, 229. *See also* Dormoy; Lexis
distribution. *See* asymmetric d., beta d., bimodal d., binomial d., bivariate normal d., Cauchy d., Dirichlet d., double exponential d., error d., exponential d., gamma d., Gaussian d., hypergeometric d., Laplace d., lognormal d., mean, mixture, multinomial d., nonuniform prior d., normal d., Pearson's family of curves, Poisson d., posterior d., prior d., random d., rectangular d., skewed d., triangular d., uniform d.
Doig, A., 370, 386
Donkin, William F., 158, 379
Dormoy, Emile, 226, 229, 232–233, 376, 379
dot diagram, 268–269
double exponential distribution, 107, 111
Doubs, 215, 217–218, 364
doubtful cases, 244, 253–254
doubtful observations. *See* outliers
Duesing, Carl, 369
Dunkirk, 56, 59

earth: obliquity of ecliptic, 5, 6; figure of, 40–60
Ebbinghaus, Hermann, 239 (portrait), 254–261, 284, 379
Ecole Militaire, 12, 31
Ecole Normale, 12, 31
ecological correlation, 296
Eddington, Arthur S., 8
Edgeworth, Francis Ysidro, 266, 300 (portrait), 305–325; and modulus as scale parameter, 83, 310, 328, 364; as a Bayesian, 127, 309–310; and normal distribution, 203, 310; and Lexis, 237–238, 315; and Galton, 294, 306–307, 311–313, 315–316; and variance components, 294, 312–314; and significance tests, 308–311, 328; 1885 lectures, 309, 363; and additive effects model, 314–315; and Edgeworth's theorem on multivariate normal distribution, 316–318, 322–325, 329, 344; 1892 Newmarch lectures, 316, 327, 367; and correlation, 316, 348–349; and estimating correlation coefficients, 319–322, 342–343; errors by, 319–321; priority dispute with Pearson, 321–322, 343–347; and Pearson, 326–329, 333–335, 338–341; and degrees of freedom, 329; and test of fit, 329, 331, 334–335; and test of symmetry, 330; and skew curves, 330, 341; and Yule, 358; mentioned, 7, 219, 351, 354, 359, 361; bibliographical references, 373, 376, 379–381, 386, 389, 393, 396
Edgeworth series, 341
Edwards, A. W. F., 129, 371, 381
Eisenhart, Churchill, 94n, 303, 331, 371, 373, 381
elementary errors. *See* hypothesis of elementary errors
Elliott, Ezekiel B., 219, 258, 381
ellipses, Galton's, 285, 287–288
ellipticity of the earth, 44, 50, 59
Ellis, Robert Leslie, 157n, 381
Encke, Johann Franz, 241, 248, 381
Encyclopaedia Britannica, 25n, 85, 158, 381
Encyclopaedia Metropolitana, 158
entangled observations, 284n
equations of condition, 33, 52, 140
Eratosthenes, 6
Ernest, Duke of Saxe-Cobourg and Gotha, 206, 393
error distributions, 91, 96–97, 106–108, 110–111, 117, 120–122, 141–142, 148, 332–334
errors by statisticians. *See* bungling; Edgeworth; Galton; Laplace; Pearson; Quetelet
ethics, 306
Euler, Leonhard, 4, 17, 25–34, 43, 51, 55, 64, 98, 110, 133, 140, 166, 169, 206, 359, 371, 381, 394
evolutionary series, 229
experimental design. *See* blind experimentation; design of experiments; factorial design; randomization
exponential distribution, 107, 111

facility curve, 310
factorial design, 181–182, 245, 249
factor model, 335, 340
Farebrother, R. W., 372, 382
Farr, William, 181–182, 382
Favaro, A., 367
Fechner, Gustav Theodor, 239 (portrait), 240, 242–255, 261, 304, 306, 330, 372, 374, 382, 385
Fechner's law. *See* Weber–Fechner law
Feller, William, 70n, 382
Fermat, Pierre, 4, 63
fiducial, argument, 91, 101, 105, 125, 361
figure of the earth, 40–60
Filon, L. N. G., 342–343, 345, 392
fingerprints, 297–298
Fisher, Ronald A., 219, 341, 361, 373, 382, 393, 395; and the fiducial argument, 91, 101; and conditional inference, 108, 120; on Bayes, 124, 127, 129; on Lexis's coefficient, 237; and design of experiments, 244; criticizes Pearson, 345; and analysis of variance, 349
fixed setting, 184
Flamsteed, John, 17, 77, 240
Fleming, A., 241
fluctuation, 310
Fontenelle, Bernard, 63–64, 382
Forrest, D. W., 373, 382
Fourier, Joseph, 7, 162, 182, 306, 382
Fourier transforms, 137
Francoeur, L.-B., 39, 382
fraud, 215
Freeman, Peter, 338
French conscripts data, 215–216

Galilei, Galileo, 18
Galloway, Thomas, 158, 382
Galton, Francis, 265 (portrait), 266–299; and regression, 2, 7, 273, 282–283, 285–290, 294–297, 301, 359–360; quoted, 8; and Quetelet, 219, 267–268, 302; and normal distribution, 219, 268, 271; and Africa, 266; and graphical methods, 266–268, 293, 319–320; and fingerprints, 267, 298, 302; and heredity, 267; and statistical scale, 271, 298; and race, 272, and conditions for normality, 272, 274, 281; law of ancestral heredity, 273, 296, 304; and quincunx, 275–280, 285, 290, 313, 316, 360; and sweet peas data, 276, 281–283, 285, 296; and reversion, 282–283, 285, 301; and stature data, 283–293; and correlation, 284, 290, 297–299, 319–320, 324, 368–369; and bivariate normal, 285–289, 324; errors by, 291–293, 296; and variance components, 293–294, 360; reaction to, 300–305, 315–316; and Pearson, 302–305, 327, 332, 336–338, 342–345, 353; on social policy, 304; and Edgeworth, 306–307, 311–313, 315–316, 318, 320; and lognormal distribution, 330; and Yule, 346, 348, 351–352; bibliographical references, 373, 377, 378–380, 382–383, 385, 391, 397–398
Galton Laboratories, 276
gamma distribution, 333–334
gamma function, 333, 337
Garber, Daniel, 371, 383
Gassendi, Pierre, 6
Gauss, Carl Friedrich, 139 (portrait); priority dispute with Legendre, 2, 15, 145–146; and normal equations, 14, 350; and normal distribution, 117, 133, 141–142, 145, 202–203, 205; and discovery of least squares, 140–145; and principle of the arithmetic mean, 141, 157; and the elimination algorithm, 145, 157; and the Gauss-Markov theorem, 148; and estimating standard deviation, 230n; mentioned, 17, 24, 61, 138, 147, 158, 242, 246, 353n, 359; bibliographical references, 371–372, 383, 394, 395, 396
Gaussian distribution. *See* normal distribution
Gauss–Markov theorem, 148
Gelfand, Alan E., 188, 191, 372, 383
General Register Office, 181–182
generalized least squares, 154
generating functions, 91–92, 137–138, 201
geodesy, 15, 158. *See also* figure of the earth; meridian arc
geological analogy, 314–315
geometric approach, 43, 47, 98
Gergonne, J. D., 372, 396
Giere, Ronald N., 383
Giffen, Robert, 367
Gillispie, Charles Coulston, 118, 370, 371, 383
Glaisher, J. W. L., 338

Golden, J., 373, 389
Goldstein, Bernard R., 6n, 384
Goodman, Leo, 178, 384
goodness-of-fit, 329, 334–335
Gore, J. Howard, 371, 384
Gossen, H., 306
Gould, Benjamin Apthorp, 219, 258, 384
Gouraud, Charles, 64, 370, 384
Gowing, Ronald, 16, 371, 384
Graham, George, 182n
graphical methods, 166, 213, 267, 293, 308, 319–320, 327–328, 345, 367
Graunt, John, 4, 384
Greenwich Observatory, 12
Greenwich time, 18
Gresham professorship, 303, 326
grouped data, 213, 286n
Grove, William Robert, 297–298, 384
growth curves, 173

Hacking, Ian, 64, 127, 370–371, 384
Hain, Joseph, 312, 384
Hald, Anders, 72, 335, 371, 373, 384
Halley, Edmund, 17n, 25, 32, 34, 38, 86n, 89, 384
Ham, John, 374
Hankins, F. H., 162n, 171, 172n, 372, 384
Hargenvilliers, A. A., 216n
Harris, John, 225n, 384
Harter, H. Leon, 370, 385
Hartley, David, 98n, 132, 385
Harvey, George, 15n, 59n, 157, 385
Hegelmaier, F., 244–245, 385
Heiss, Klaus-Peter, 223n, 372, 385
Helmert, Friedrich Robert, 317
Helmholtz, Hermann von, 240, 306
Hertz, Heinrich Rudolf, 345
Heyde, C. C., 186, 197, 228n, 237, 372, 385
Hill, Elizabeth, 371, 385
Hilts, Victor L., 298, 372–373, 385
Hipparchus, 6, 17
histograms, 327
histories of statistics, 370
Hogarth, J., 389
Holland, John D., 94n
Hollerith, Herman, 361, 385
homme moyen. *See* average man
homogeneity, tests of, 222. *See also* Quetelet
homoscedastic data, 313
Horrocks, Jeremiah, 25
horse kick data, 236
Hotelling, Harold, 361
Humphreys, N. A., 369
Hutton, Charles, 88–89n, 385
Huygens, Christian, 4, 63, 71, 225, 385
hypergeometric distribution, 228, 334
hypothesis of elementary errors, 201–202, 274

importance, relative, 176, 178
incomplete beta function, 130
inhomogeneous trials, 183–184
in-relief, 347
insufficient reason, principle of, 111, 117, 127, 129. *See also* uniform prior distribution
insurance, 62
integer bias, 286n
International Statistical Congress, 162, 222
inverse probability, 101, 310. *See also* Bayes; Laplace
inversion formula for characteristic functions, 137–138

Jacobi, Carl Gustav, 12
Jastrow, Joseph, 253–254, 392
Jaynes, Julian, 243, 372, 385
Jeffreys, Harold, 127, 385
Jevons, William Stanley, 5, 298, 301, 306, 308, 330, 375, 385–386, 396
Johnson, Norman L., 370, 386
joke, 302
Jones, George W., 301n
Jouffret, E., 317n
judicial statistics, 174, 186, 188–189. *See also* conviction rates; jury models
Jupiter, 16–17, 25–27, 31–33, 39, 136; moons of, 25n
jury models, 186–187, 190

Kahle, Ludwig Martin, 386
Kant, Immanuel, 1
Kantian nominalism, 340
Kelvin, Lord. *See* Thomson
Kendall, M. G., 110n, 328, 345, 370–373, 386, 390, 396
Keverberg, Baron de, 163–166, 169, 200, 203, 222, 227, 245, 253, 357, 361, 386
Keynes, John Maynard, 237–238, 305, 373, 386
King's College, London, 306, 363
Kinnebrook, David, 240–241
Knorre, Karl, 241
Kotz, Samuel, 370, 386

Kramp, Christian, 202, 207, 248, 386
Kruskal, William H., 178, 370, 384, 386
Kuhn, Thomas S., 370, 386

Lacaille, N. L., 387, 397
Lacroix, Silvestre F., 157, 386
Lagneau, Gustave, 217, 386
Lagrange, Joseph Louis, 30, 32, 64, 103n, 182, 387; and the problem of choosing a mean, 85, 88, 97, 100–102, 117–119; and generating functions, 92, 137, 201; and multivariate normal distribution, 317
Lalande, Joseph Jérôme Le François de, 30–31, 89–90, 95n, 371, 387
Lambert, Johann, 30, 32, 88, 95, 100, 371, 394
Lancaster, H. O., 370, 373, 387
Laplace, Pierre Simon, 31, 99 (portrait); and least squares, 17, 61, 140–141, 146–154, 157–158, 317, 366; and the square root rule, 24, 84; and planetary motions, 30–39; generalizes Mayer's method, 31–39; and the figure of the earth, 44n, 45, 59; and the method of situation (Boscovich's method), 50–55, 330n; and inverse probability, 87–88, 94, 98, 100–105, 113–116, 310; and generating functions, 92, 137–138; and Bayes's theorem, 102–105, 122–124; and estimating location parameters, 105–109, 111–117; and uniform priors, 103, 106; and nonuniform priors, 103n, 135–136; and curves of error, 109–111, 117–122, 143; and the principle of insufficient reason, 111, 117, 127; and inference with a nuisance parameter, 112–116; errors by, 113–116, 151–157, 191; as a Bayesian, 119–120, 310; and inference about the binomial, 131–135; and the normal distribution, 133, 137, 143, 201; and inference about the birth ratio, 134–135, 226; and the central limit theorem, 134, 136–138, 144, 155, 201–203, 205, 207, 209; and ratio estimation, 136, 163–166; and significance tests, 151, 154–157, 311; and Quetelet, 162; Laplacian methods, 186–188, 190, 192, 225, 242; criticized by Comte, 194–195; later appeals to his conditions for normality, 222–223, 274, 330–331; as expositor, 314; mentioned, 12, 16, 97, 161, 173, 182, 193, 227, 246, 281, 306, 307, 353n, 359; bibliographical references, 370–372, 376, 377, 383, 387–388, 392, 395, 396
Laplace distribution, 107, 111
Laplace transforms, 137. *See also* generating functions
Laplace's method, 132–133
large data sets, 158
law of large numbers, 64–70, 183–185
law of small numbers, 236
Lazarsfeld, Paul F., 164n, 172n, 372, 388
least deviations regression, 36, 46, 50–51, 55
least squares, 36–37, 141, 260, 349–350, 358; estimates, calculation of, 158; invention of, 12–15, 17, 55–61, 141; translations, 15, 140, 157. *See also* Gauss; Laplace; Legendre; Yule
Lee, Alice, 340, 342n
Legendre, Adrien Marie, 11 (portrait), 12; priority dispute with Gauss, 2, 15, 145–146; and discovery of least squares, 12–15, 35–37, 55–61, 139–140; and figure of the earth, 55–60; mentioned, 17, 21, 30, 151, 349, 359; bibliographical references, 371, 383, 385, 388, 396
Leibniz, Gottfried, 63, 64, 145, 193
Levasseur, Emile, 367
Lexis, Wilhelm, 7, 221 (portrait), 222–226, 229–238, 315, 339, 372, 376, 385, 388–389
Lexis coefficient, 231, 237
libration of the moon, 16–18
lifted weights, 242, 244
Lindenau, Baron von, 15, 389
linear estimation, 35
linearity of regression, 289, 344, 349
Livi, Ridolfo, 218, 389
lognormal distribution, 330, 336
longitude at sea, 18, 25n
Longstaff, George B., 368
loss function, 46, 87, 109, 147
Lottin, Joseph, 372, 389
lunar tide, 151
lurking variables, 357
Luther, Martin, 303

McAlister, D., 330, 389
Macclesfield, Earl of, 90
MacKenzie, Donald A., 267, 304, 370, 373, 389

Maire, Christopher, 42–51, 376
Malthus, Thomas Robert, 170n
Manilius, 19–22, 30, 38
Marey, E., 367
Markov, A. A., 237
Markov process, 285
marriage statistics, 167, 230, 232, 314
Marshall, Alfred, 306, 356, 368, 389
Martin, W. R., 25n
Maskelyne, Nevil, 29, 240–241, 389
mathematical social science, criticism of, 165, 194, 197, 304, 356–357
Maupertuis, P. L. M. de, 42, 45
maximum likelihood estimate, 110, 141, 310, 342, 345
Maxwell, James Clerk, 306
Mayer, Tobias, 16–25, 27–31, 33–39, 46, 54–56, 61, 84, 130, 140, 359, 387, 389
Mayer's method, 29n, 31, 39, 55, 146–147
mean: choice of, 16, 90, 94, 101, 105, 112, 117, 141, 309, 363; distribution of, 97, 310. *See also* central limit theorem
mean error, 86, 109, 147
Méchain, P. F. A., 51, 56
median, 270–271, 293, 310; deviation, 293–294; generalized, 53
mediocrity, 271. *See also* regression
memory, 254
Men's and Women's Club, 303
Ménard, Claude, 372, 389
meridian arc: French, 12, 15, 51, 54, 57, 59–60; Lapland, 42, 45, 52–53; Paris, 42, 45; Peru (Quito), 42, 44–45, 60; Rome, 42, 50. *See also* arc length; baseline; data tables; module; toise
meridian quadrant, 55
Merriman, Mansfield, 158, 370, 389
Merrington, M., 373, 389
Merton, Robert K., 1n, 390
Metcalfe, H., 317n
meteorological data, 151, 220, 232, 331
method of average error, 242
method of averages, 29
method of moments, 331–334
method of right and wrong cases, 242, 244
method of situation, 50–51. *See also* Laplace
metric system, 12, 55
Michaelis, O. E., 317n
midparent, 284
Mill, John Stuart, 195, 369, 390
Mint, London, 3
missing values, 318
mixture: of normal distributions, 275, 280, 312, 332; of binomial distributions, 186, 191
Möbius, A. F., 247
module, 57, 59n, 60
modulus: in Bravais, 83; in De Moivre, 83; in Edgeworth, 83, 310, 328, 364
Moivre. *See* De Moivre
Molina, E. C., 371, 390
moments, method, 331–334
Moniteur Universel, 157
Monte Carlo roulette. *See* roulette
Montjouy, 56, 59
Montmort, Pierre Remond de, 71–72, 226, 390
moon, 16–20, 22, 150–151
moral certainty, 65, 67, 78
Morant, G. M., 303, 373, 390
most advantageous method, 151, 154
mountain range data, 219
Müller, G. E., 248, 253–254, 390
multinomial distribution, 118, 329
multiple comparisons, 155, 157, 199–200, 329
multiple correlation coefficient, 354
multiple range test, 157
multivariate normal. *See* normal distribution
murder rate, 170
Murray, D. J., 372, 390

Napoleon, 12, 31, 140n
natural selection, 282n
Naworth Castle, 298
Nesselroade, John, 296, 390
Newbold, Ethel M., 236, 390
Newcomb, Simon, 301n, 390, 395
Newmarch lectures. *See* Edgeworth
Newman, D., 157, 390
Newton, Isaac: and statistics, 3n, 6; Newtonian gravitational theory, 5, 17, 26, 30, 32, 39, 359; and the moon, 17n, 150; Newtonian science, 31, 110, 117; and the figure of the earth, 40, 42, 44, 50; geometric approach of, 43, 51, 98; and the binomial theorem, 92; mentioned, 63, 71, 145, 361; bibliographical reference, 390
Neyman, Jerzy, 87, 361
Nightingale, Florence, 372, 379

Nipher, Francis E., 256n, 260n
nonlinear design, 250n
nonsense correlation, 357
nonsense syllables, 256
nonuniform prior distributions, 103n, 135–136
normal approximation: to beta, 132–133; to binomial, 71, 73–76, 78–84, 136, 201
normal deaths, 223
normal dispersion, 231
normal distribution: introduction of, 73–76, 84, 133, 141–142; asymptotic series for, 80, 134; table of, 82, 202, 248; as error distribution, 117, 143–145, 201; multivariate, 118, 316–318, 323, 329; continued fraction expansion of, 134; Gauss's proof of, 141–142; names, 143, 201, 223, 310; bivariate, 148–149, 206, 285, 287–288, 317, 324, 342, 344–345; fit to data, 203–204, 206–207, 212, 216, 218, 268; as a quantal response model, 246–247; as a scaling device, 261, 271. *See also* central limit theorem
normal equations, 14, 36, 350
normal probability plot, 212
normality, conditions for, 272, 274, 281, 330–331
Norton, Bernard, 373, 390
nuisance parameters, 112

oblate spheroid, 40
obliquity of the ecliptic, 5
ogive, 268, 270
Ohm, G., 243, 376
Oliver, James E., 301n
optimality, 13, 46, 106, 109, 119, 141, 147–148, 310, 321, 342, 344, 349
Ordnance Office, 317n
Ordnance Survey, 158, 390
oscillatory series, 229
outliers, rejection of, 14, 54, 158, 337–338
out-relief, 346
Oxford University, 305

P-value, 151, 154–155, 187, 190, 198–200, 226
panmixia, 342, 353
Pappus, 5, 6
Paris Observatory, 6, 12, 150
Pascal, Blaise, 4, 63, 193
Pascal's triangle, 92
pauperism, data on, 335, 337, 340, 346–347, 355–357
Pearson, Egon S., 71, 110n, 303, 326–327, 329, 361, 370, 373, 378, 390
Pearson, Karl, 266, 302–303, 326 (portrait); writings on history, 76, 124, 127, 163n, 279, 296, 298, 301, 344n; and incomplete beta, 130; and mixtures of normal curves, 217, 332–333; and chi-square, 237, 329, 335; and correlation, 298, 342–345, 348–349, 351–354, 358; and Gresham professorship, 303, 326; and Galton, 303–305, 315–316; and Weldon, 305, 328, 332, 335–338; and Edgeworth, 315–316, 324–325, 327–329, 338–341; and priority dispute with Edgeworth, 321–322, 343–344; and estimating the correlation coefficient, 322, 342–343; Gresham lectures by, 326–329; and roulette, 327–329; and skew curves, 329, 331–338; family of curves, 333–335, 346; and Yule, 340, 344–346, 351–353, 358; lectures at University College, 342, 347; errors by, 343, 348; mentioned, 7, 219, 359, 361; bibliographical references, 370–373, 381, 390–392
Pearson's family of curves, 333–334, 346
Pearson's Type III curve, 333
Peirce, Benjamin, 140, 158, 392
Peirce, Charles S., 253–254, 327, 340, 392
Penny Cyclopaedia, 158, 217n
percentiles, 346
periodic inequalities, 32
periodic series, 229
personal equation, 240–242
physical method, Lexis's, 231n
physique sociale. *See* social physics
Picard, Jean, 45
Pidgin, Charles F., 367
Pigou, Arthur Cecil, 356–357, 392
pivotal quantity, 101, 108
Plackett, Robin L., 16, 143, 145, 317, 325, 329, 344n, 370–371, 373, 386, 392
Plato, 77
Playfair, John, 31, 392
Playfair, Lyon, 302
Poincaré, Henri, 157n, 392
Poinsot, Louis, 194–195, 197

Poisson, Siméon Denis, 182–194; and Laplace, 31, 182; and judicial statistics, 175n, 188–191; and Poisson distribution, 183; and Cauchy distribution, 183n; and law of large numbers, 183–185; and models for inhomogeneous trials, 183–185, 231; and jury models, 186–187; as a Bayesian, 187, 190; and Quetelet, 188, 190–194; error by, 191; and test of symmetry, 330; mentioned, 148, 225, 306; bibliographical references, 372, 375–377, 383, 392–393, 395, 396
Poisson distribution, 183, 236
polar excess, 43
poor law, 346, 356
population data, 166. *See also* census; ratio estimation
Portalis, Count, 188
Porter, Theodore M., 298, 370, 372, 393
posterior: distribution, 106–109, 119–120, 122, 126, 132, 141, 187–190, 198, 343; expected loss, 109, 147, 310; median, 109; consistency, 132–133; asymptotic normality, 132–133
power of a test, 151, 153
Pratt, John W., 373, 393
precision constant, 145, 246
Price, Richard, 98, 123, 127, 130–131
prior distributions. *See* nonuniform prior distribution; uniform prior distribution
priority disputes. *See* Edgeworth; Gauss; Legendre; Pearson
probability integral transformation, 213
probabiloid, 317
probable error, 180, 214, 230n, 242, 246, 257
probit analysis, 246
prolate spheroid, 40
Prony, G. F. C. M. R. de, 54, 393
Prophatius, 6
Provine, William B., 296, 373, 393
psychical research, 363
psychology, 239
psychophysics, 242–243
Ptolemy, 5, 6, 17
Puissant, Louis, 15, 39, 54, 157, 393
Pygmalion, 170, 214
pyramid of Cheops, 7
Pyx, Trial of the, 3

quadrature of the binomial, 79
quadrature of the normal density, 81
quadrature, of moon, 150–151
quantal response, 246
quartiles, 270
quasi-normal, 346
Queen: Anne, 225; Victoria, 206
Quetelet, Adolphe, 161 (portrait), 162–182; and test of homogeneity, 7, 222–223, 311; and Laplace, 162–165, 201; and Keverberg, 163–166; and ratio estimation, 163–165; and population statistics, 163–168; and graphical methods, 166–168, 213; and average man, 169–172, 214, 302; and social physics, 170–171, 193; and Cournot, 171, 200, 325; and statistical regularity, 172, 226, 228; and growth curves, 173; and types of causes, 174–175; and judicial statistics, 174–181, 188; errors by, 177, 180, 188; and analysis of variance, 179; and Poisson, 186, 188–194; and normal distribution, 201, 203, 205; method for fitting normal curves, 206–207, 209–216, 218–220; and Galton, 219, 267–268, 271, 281; and asymmetric distributions, 220, 330; and Lexis, 223, 231, 235; and Ebbinghaus, 255–258, 260–261; and Pearson, 335–336; mentioned, 158, 186, 238, 245, 272, 307, 312, 313, 357, 360, 369; bibliographical references, 372, 379, 384, 386, 389, 393–394, 397
Quetelismus, 203
quincunx, 275–280, 285, 290, 313, 316, 360

R^2 (multiple correlation coefficient), 354
random distribution, 120
randomization, 253
random sampling, 163–164
random setting, 184
ranks, 213
rare events, 236
ratio estimation, 136, 163–166
reaction times, 240
rectangular distribution. *See* uniform distributions
Regiomontanus, 6
regression, 2, 7–8, 272, 284, 295, 301, 349, 355, 360; lines, two, 288–289; via least squares, 349; filial, 289; fraternal, 290–291. *See also* correlation; Galton; least squares; reversion; Yule
Reichesberg, Naúm, 162n, 394

reparameterization, 44n, 59
residual, 24, 35; sum of squares, 153; variance, estimate of, 153
reversion, 273, 281–283, 301. *See also* regression
Ricardo, David, 359
Richer, Jean, 6, 40
right and wrong cases. *See* method of right and wrong cases
Robinson, G., 39, 397
robust estimation, 54, 267
robust regression, 36, 46, 50–51, 55
Rogers, William A., 241, 394
roulette, 327–329
Royal Institution, 276, 278, 281
Royal Society of London, 97–98, 123, 297, 305, 331–332, 336, 338, 341, 344, 348
Royal Statistical Society, 162n, 308, 314, 345–346, 348, 355

saddlepoint methods, 132–133
sample mean, distribution of, 97, 310. *See also* central limit theorem
sample survey, 163
Sanford, Edmund C., 240, 372, 394
Särndal, Carl-Erik, 373, 394
Saturn, 16–17, 25–27, 31–34, 38–39, 136
Savage, L. J., 361
scales, statistical, 271, 298
Schneider, Ivo, 371, 394
Schooten, Frans van, 385
Schweber, Silvan S., 170n, 394
Scottish soldiers data, 180n, 206–208, 212, 214
Seal, Hilary L., 317n, 325, 344n, 371–372, 394
secular inequalities, 32
Selvin, H. C., 373, 394
Seneta, Eugene, 186, 197, 228n, 237, 372, 385
sensation, measurement of, 243–244
serial correlation, 259, 324
sex ratio at birth. *See* birth ratio
sexual selection, 284, 304
'sGravesande, G. J., 226, 394
Shafer, Glenn, 64, 124, 371, 394
Sheynin, Oscar B., 72, 110n, 163n, 370–372, 394–395
Short, James, 54
data, 316
significance test, 151, 153–154, 226, 242, 260–261, 308–311, 328, 364
Sills, David L., 1n, 370, 390, 395
Simpson, Thomas, 88–98; and De Moivre, 85, 89–92; dispute with De Moivre, 89–90; and error distributions, 91–94, 95–97, 108, 110, 112; and generating functions, 91–92, 137; and Bayes, 94, 97–98, 123; and Lagrange, 118; mentioned, 87, 100, 359; bibliographical references, 371, 377, 387, 395
skew correlation, 344, 358
skewed distribution, 329–330. *See also* Edgeworth; Pearson
small numbers, law of, 236
Smith, Adam, 359
Smits, Ed., 166n, 394
social physics, 170–171, 193, 227
sociology, statistics in, 301
Sokal, M. M., 301, 395
Solomon, Herbert, 188, 191, 372, 383
Somerville, Martha, 395
Somerville, Mary, 38, 182, 395
spacings, ordered, 120
Spinoza, B., 303
spline, 97
Spottiswoode, William, 219, 395
Sprott, David A., 371, 395
spurious correlation, 357
square root rule, 3n, 7, 24, 77, 83–84, 180
stable laws, 310
stable series, 227, 235
stability, 230, 233
standard deviation, 77, 328
standardized variate, 321
stature, data on, 283
Stay, Benedict, 45
Stephan, Frederick F., 164n, 372, 395
Stigler, George J., 373, 395
Stigler, Stephen M., 1n, 3n, 31, 50n, 54, 83, 97, 98n, 100, 102, 103n, 129, 132, 143n, 145, 183, 230n, 238, 253, 256n, 260n, 294, 296, 304–305, 314, 330–331, 333, 338, 341n, 371–374, 390, 395–396
stillbirths, 224, 233
Stirling, James, 72–73, 396
Stirling's formula, 72, 133, 211
Stone, Mervyn, 372, 379
stratified sampling, 197
Struve, F. G. W., 241
Stuart, Alan, 396
subnormal dispersion, 231

suicide rates, 172, 232
supernormal dispersion, 231
Süssmilch, Johann Peter, 226, 396
Svanberg, Jöns, 53, 397
sweet peas data, 276, 281–283, 285, 296
Swift, Jonathan, 225
Swinburne, R. G., 296, 397
symmetry, test of, 256, 330
syzygy, of moon, 150–151

tabulating machines, 361
Tait, P. G., 306
Tanur, Judith, 370, 386
test of fit, 258. *See also* Edgeworth; goodness-of-fit; Pearson
test of significance. *See* significance test
Thiele, T. N., 335, 373, 384
Thomson, William (Lord Kelvin), 1, 306, 390
tides of atmosphere, 148, 151
time series, 309, 312, 315, 324
Titchener, Edward B., 244, 253–254, 397
Todhunter, Isaac, 43n, 45, 53, 72, 225n, 303, 370–371, 397
toise, 43, 59–60
trend, 233, 309, 312, 315
triangular continuous distribution, 96
triangular discrete distribution, 92
triangulation, 52n
triangulation of British Isles, 158
Trinity College, Dublin, 305
Tunbridge Wells, 97
two-way classification, 238, 249, 313–315
typical series, 229

unbiased estimation. *See* bias
undulatory series, 229
uniform continuous distribution, 96
uniform discrete distribution, 92
uniformity, test of, 228
uniform prior distribution, 103, 106, 127–129, 310. *See also* insufficient reason; nonuniform prior distribution
union, poor law, 346
University College London, 303, 306, 316, 345
University of Minnesota, 372
University of Pennsylvania, 301
Urban, F. M., 261, 397
urn models, 65–70, 124–125, 131, 225, 238
Uspensky, J. V., 64n, 66, 237, 372, 397

variable causes, 175, 179, 181
variance, estimate of, 153, 310
variance components, 293–294, 314, 360
Venn, John, 291, 302, 304, 307, 310, 327–328, 330–331, 335, 368, 397
Venus, transit of, 122
verdict, probability of correct. *See* jury models
versed sine, 42, 44, 46
Vierordt, Karl von, 244
Villermé, L. R., 174n, 216n, 393, 397
Virgil, 311, 314
vital statistics. *See* birth rate; birth ratio; death

Walbeck, H. J., 241
Walker, Helen M., 71, 242, 370–371, 397
Walras, Léon, 306, 389
Waltherus, 6
wasps data, 311
Weber, Ernst Heinrich, 243–244
Weber–Fechner law, 243, 248, 250
weighted analysis, 52, 54, 59, 147
weighted mean, 16
Weismann, A., 304
Weldon, W. F. R., 217, 305, 316, 328, 332, 335–338, 342, 344, 369, 391, 397
welfare, 347
Wellens-de Donder, Liliane, 372, 397
Werckmeister, Karl, 397
Westergaard, H., 327–328, 369–370, 397
Westfall, Richard S., 383
Whiston, William, 25n
Whittaker, E. T., 39, 397
Wilson, Curtis A., 371, 397
Winsor, Charles P., 236, 372, 398
Wishart, David, 341n
Wolf, Rudolf, 39, 398
Woolf, Harry, 398
Wundt, Wilhelm, 240, 306

Yule, George Udny, 326 (portrait), 345–358; and measures of association, 178, 361; and Pearson, 333n, 334–335, 340, 342, 345–349, 351–353, 358; and correlation, 344, 346, 348, 354; and linearity of regression, 344–345, 349; and pauperism, 345–348; and least squares, 349–353, 359; and regression, 353–358; mentioned, 7, 266; bibliographical references, 373, 386, 394, 396, 398

Zabell, Sandy, 371, 383
Zach, Franz Xaver von, 15